Quenching the Prairie Thirst

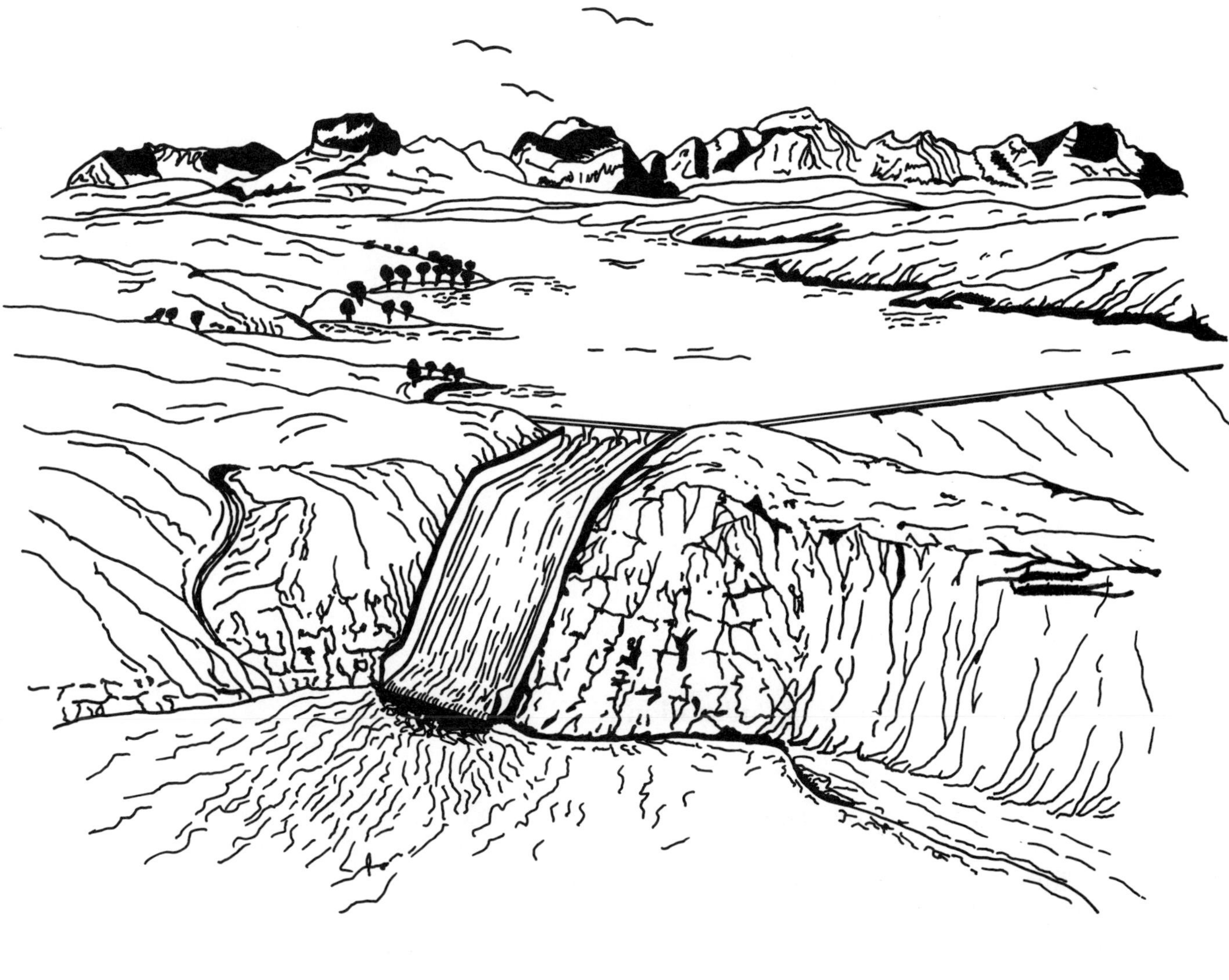

A History of the
Magrath Irrigation District
Raymond Irrigation District
Taber Irrigation District
St. Mary River Irrigation District

First printing, 2000

Canadian Cataloguing in Publication Data

Gilpin, John F., 1947-
Quenching the Prairie Thirst

Includes bibliographical references and index.
ISBN 1-55056-780-2
1. St. Mary Irrigation Project—History. 2. Irrigation—Alberta, Southern—History. I. St. Mary River Irrigation Project. II. Title.
HD1741.C22G55 2000 333.91′3′0971234 C00-0910823-8

Published by St. Mary River Irrigation Project
1201-36 Street North, Box 278
Lethbridge, Alberta
T1J 3J7

Printed and bound in Canada by Friesens Corporation, Altona, Manitoba, R0G 0B0

Table of Contents

Dedication

This book is dedicated to the men and women who have
contributed to realizing the vision
of the St. Mary Project

Acknowledgements

This book was written with a research grant from the Alberta Historical Resources Foundation. Assistance was also provided by the Irrigation Branch of the Department of Agriculture, Food and Rural Development which provided four of the maps in the colour section of the book. The Prairie Farm Rehabilitation Administration provided valuable historical records and a number of photographs for this publication. Assistance was also provided by many library and archives including the Sir Alexander Galt Museum and Archives in Lethbridge, the Glenbow Archives in Calgary, the Provincial Archives of Alberta in Edmonton and the Experimental Farm Library in Lethbridge. The assistance of these institutions is gratefully acknowledged

Abbreviations

AICo	Alberta Irrigation Company
AR&CCo	Alberta Railway and Coal Company
AR&ICo.	Alberta Railway and Irrigation Company
CNWICo.	Canadian North-West Irrigation Company
IJC	International Joint Commission
LDS	Church of Jesus Christ of Latter-Day Saints
LSEP	Lethbridge South Eastern Project
MID	Magrath Irrigation District
NWC&NCo.	North Western Coal & Navigation Company
PFRA	Prairie Farm Rehabilitation Administration
RID	Raymond Irrigation District
SMRD	St. Mary and Milk Rivers Development
SMRID	St. Mary River Irrigation District
TID	Taber Irrigation District

Preface

The history of the St. Mary Project is the realization of a vision, conceived in the 1890s, to make hundreds of thousands of acres of land southeast of the St. Mary River more productive through irrigation. Farmers, urban businessmen, corporate executives, diplomats, bureaucrats, engineers, politicians, foreign investors, farm labourers and scientists from Canada and around the world have contributed to this achievement. Today that vision has been realized in a system with the capacity to irrigate over 518,000 acres of land.

Canada's oldest and largest irrigation system is made up of four irrigation districts along with Alberta Environment and the Blood Tribe. Alberta Environment owns and operates the headworks which extend from Waterton Dam to Ridge Reservoir. The main canal runs east for an additional 483 miles to Medicine Hat. The system also includes 12 storage reservoirs ranging in size from the St. Mary Reservoir, which stores 299,000 acre-feet, to Fincastle Reservoir, which stores 4,000 acre-feet of water.

The United Irrigation District and the 19,000 acre Blood Tribe Agricultural Project, which are located west of the St. Mary River, are the first distribution systems which take their water from these headworks. The Magrath Irrigation District operates a distribution system with 82 miles of open ditches and 29 miles of pipelines, serving 18,300 acres of land northwest of the Milk River Ridge around Magrath. The Raymond Irrigation District serves 46,000 acres which extend north of the Milk River Ridge from Welling in the west to Stirling and Crow Indian Lake in the east. The principal crops of the Magrath and Raymond Irrigation Districts are alfalfa and barley for export and feed in the livestock industry. The St. Mary River Irrigation District covers an area between Lethbridge and Medicine Hat and irrigates 370,000 acres of land by means of 1,280 miles of canals and pipelines. Crop production in the SMRID reflects the increasing heat units available as you move from west to east. The Lethbridge-Coaldale area is a major producer of sugar beets. The Bow Island area is well known for the production of beans, corn and sunflowers. Alfalfa and oilseed crops such as canola are the principal crops grown in the Medicine Hat area. The Taber Irrigation District has 216 miles of canals and pipelines which serve 82,000 acres. It produces a wide variety of speciality crops with corn being the most famous. Magrath, Raymond, Coaldale, Taber, Burdett, Grassy Lake, Foremost, Bow Island and Seven Persons depend on the St. Mary Project for their water supply. The St. Mary Project also shares the water of the Waterton and Belly Rivers with the United, Mountain View, Leavitt and Aetna Irrigation Districts.

This history of the St. Mary Project deals with the development of its various components chronologically as the entire development moved from conception to initial construction and then through expansion to its present size. The four mountain districts are included because their development changed the federal government's plan to concentrate irrigation development east of the St. Mary River. These districts also supported the development of the sugar beet industry and the community efforts made to ensure the overall development of irrigation in southern Alberta. Government also played a significant role, starting in 1895, in planning and expanding the system to ensure that Canada made use of its portion of water from the St. Mary River. The scientific research in support of irrigation farming begun by William Fairfield in Lethbridge in 1901, and continued by the Dominion Experimental Station, also forms part of this book. To make this history of the St.Mary Project complete, the technological revolution in irrigation after World War II has been included. The managers and water users in the St. Mary Project have made significant contributions to its development. The economic growth generated by the system and its impact on the social history of southern Alberta are illustrated through an account of the sugar beet industry. This industry has been closely identified with the St. Mary Project over its entire history. Irrigation has made an important contribution to the creation of an ethnically diverse society in southern Alberta. The final theme developed in this book is the connection between irrigation and the environment. The most obvious example of this relationship is the slope of the land which allows the water to run in a northwesterly direction.

The history of the St. Mary Project begins with the geological events which created its physical setting. This setting is diverse, beginning in the ice fields of Glacier National Park and ending on the arid plains west of Medicine Hat. The natural link between the mountains and the plains is a network of rivers which originate in northern Montana and flow east through deep valleys eventually emptying into Hudson Bay via the South Saskatchewan or the Gulf of Mexico via the Mississippi. The geography of the St. Mary Project is further diversified by variations in soil, precipitation and temperature. These factors have combined to provide the means and the necessity for irrigation to support intensive agriculture in southern Alberta.

The Blackfoot Nation was the first to occupy the plains in the vicinity of Lethbridge in historical times. From the First Nations, the focus of the book shifts to exploration and settlement as a result of the transcontinental expansion of Canada and the United States. The 1818 decision to use the 49th parallel as the boundary between British North America and the United Sates divided the St. Mary and Milk River basins and ensured international competition for their waters. The American threat was an important catalyst for federal government involvement in efforts to develop irrigation based on the St. Mary River from the 1890s through to the 1950s. Canadian success in laying claim to its share of this water, ironically, has come about with the help of settlers and engineers from the United States. The establishment of coal mining and the construction of a railway network began in the 1880s. These activities established the city of Lethbridge and made the Alberta Railway and Coal Company the largest private owner of land in the area southeast of Lethbridge.

The introductory chapter describes the development of irrigation in the West. Irrigation began in Alberta in the 1870s but the projects built up to the1890s were generally small and had no place in the federal government's plan for western settlement. William Pearce was the first to advocate that irrigation could and should be developed on a grand scale as part of an overall approach to the settlement of the West. This idea led to the passage of three pieces of legislation in 1894, the most important of which was the North-west Irrigation Act.

The St. Mary Project was the first irrigation development in Canada to achieve the objectives of Pearce and the federal government. The planning, construction, operation and expansion of the project under the auspices of the Alberta Irrigation Company and its successors, between 1894 and 1907, are the subjects of separate chapters. The chapter on planning notes that after various false starts, the St. Mary Project was successfully launched by Elliott Galt as a corporate venture with the cooperation of the Church of Jesus Christ of Latter-Day Saints, a group of British investors, Colorado irrigation engineer George Anderson and the federal government. Because of its size, its attraction of foreign investment and immigrants and its influence on other projects, the St.Mary Project transformed irrigation from a marginal aspect of Alberta agriculture into an industry with an impact on the economic, social and political history of Alberta and Canada.

The CPR's acquisition of a controlling interest in the Alberta Railway and Irrigation Company in 1908 began the second era of development for the St. Mary Project. The CPR chose to emphasize the colonization of the land through its various settlement and agricultural programs and leave extension of the canal and lateral systems to the farmers and the federal government. Mennonites and other eastern Europeans arrived in the late 1920s attracted by CPR efforts to obtain workers for beet fields. The CPR continued to depend on the water users for the operation and maintenance of the distribution system in the Lethbridge-Coaldale area. This approach led to conflict between neighbours and haphazard lateral maintenance.

Limited expansion of the project was achieved in 1920s through the creation of farmer- operated irrigation districts provided for in the Irrigation Districts Act of 1915. The Act was passed as a result of the efforts of farmers in the Barnwell and Taber area who started the development of an irrigation system in 1908. The CPR also lent its support to their efforts since they facilitated their objective of turning over the operation of the distribution system to the water users. The TID was created in 1915 but actual construction did not start until 1919 because of delays caused by World War I.

Interest in the creation of additional districts increased dramatically at the end of World War I as a result of drought and a fall in commodity prices. Six districts as far east as Bow Island and as far south as Warner were planned for the area southeast of the existing St. Mary system, but only the Magrath and Raymond areas were successful in actually creating irrigation districts. Unlike Taber, these two districts took over the administration of water contracts formerly administered by the CPR. Both districts added new water users through the extension of the lateral systems. Negotiating changes in contracts with the CPR was a major preoccupation for both the Raymond and Magrath Irrigation Districts through to 1946.

The lack of any financial incentive for the CPR to undertake the planning and construction of a system that would fully utilize Canada's portion of the water from the St. Mary River forced the federal government to assume responsibility for expansion of the St. Mary Project after the ratification of the International Waterways Treaty in 1909. The Department of the Interior prepared a plan to permit Canada to achieve this objective and called it the Lethbridge South Eastern Project. This plan, however, remained only a blueprint until the 1940s because of the lack of private or public funds for its construction.

The expansion of the St. Mary Project came to a halt by the end of the 1920s as the system reached the limits of its ability to provide water. The water shortage problem became particularly acute during the 1930s, with 1936 being the most difficult year because of record low flows in the St. Mary River. Added to the water shortages was a debt problem in many districts caused by the inability of water users to pay the rates because of falling commodity prices.

The conditions which threatened the existence of the St. Mary Project during the 1930s, however, proved to be the means of its salvation. The creation of the Prairie Farm Rehabilitation Administration was the first of a number of positive steps taken to solve the water crisis in the West. Originally started in 1935 as a temporary measure, it had become, by 1938, a permanent organization within the Canada Department of Agriculture. This agency resurrected the dormant LSEP and used it as the basis for its plan to provide water storage facilities for the St. Mary Project. The Ewing Commission of 1936, the South Alberta Water Conservation Council, established in 1939, and the 1942 report of the St. Mary and Milk Rivers Water Development Committee focussed further attention on the need to build reservoirs for the St. Mary Project. The construction of the Fresno Dam, near Havre, Montana, provided a major incentive for the building of water storage on the St. Mary Project.

The Jensen and St. Mary River Dams were the first projects undertaken as a result of the new willingness of government to pay for the construction of irrigation infrastructure. These projects set the stage for the total transformation of the St. Mary Project after 1946. This change took place in two phases with 1968 being the division point between the two. From 1949 to 1968 the federal and provincial governments spent approximately 50 million dollars to extend the St. Mary Project east to Medicine Hat and west to the Waterton River. The federal dam construction program permitted spring run-off to be stored and used through the summer. In 1968, the provincial government took the initiative with respect to irrigation development, passing new legislation and implementing an ambitious funding program aimed at the rehabilitation of the existing system. The new legislation created the St. Mary River Irrigation District and introduced a number of administrative changes to the way existing irrigation districts operated. This second round of government expenditures on the St. Mary Project permitted the rebuilding of the main canal, construction of Forty Mile Coulee Reservoir and the rehabilitation of the irrigation distribution system.

ONE
IRRIGATION IN WHOOP-UP COUNTRY

The St. Mary Irrigation Project begins at the continental divide in northern Montana and ends in the Alberta portion of the Palliser Triangle. The water supply for the project comes from the St. Mary, Belly and Waterton Rivers which start as lakes fed by melting glaciers. These rivers flow into the Oldman River and eventually on to the Hudson Bay via the South Sakatchewan River. The Milk River system which is not part of the St. Mary Irrigation Project, is the other river which plays an important role in its history. The south, middle and north branches of the Milk River system derive their water supply from springs scattered through the American portion of the Milk River Ridge rather than from glacier fed lakes. The volume of water in these streams is thus subject to sudden fluctuations governed by precipitation over its widely extended drainage basin. The north and south branch flow into Canada as separate streams before joining at a location west of the Town of Milk River via the Mississippi.

The natural vegetation of the Palliser Triangle is short grass prairie and precipitation insufficient to support the type of agriculture of eastern North America.[1] This is the result of the rain shadow effect which describes the role the mountains play in preventing the moist air from the Pacific from reaching the prairies. The average annual precipitation varies from 117 inches at Lethbridge to 12 inches at Medicine Hat. On average, about 60 percent of the total precipitation falls between May and August but evapo-transpiration is up to three times higher than rainfall. The famous chinook winds carry away snow as water vapour drying the land further. Chinook winds begin as flows of moist air from the Pacific Ocean forced upward by the mountains. Cold at increased altitudes condenses water which falls as snow at the highest elevations of the Rockies thus creating the snow pack which sustains the River systems of the West Variations in precipitation and temperatures combine to create cycles of dry and wet years. Droughts occurred in 1889 to 1894, 1917 to 1926, 1929 to 1937 and 1983 to 1988.

The Blackfoot Nation (Blackfeet in American usage) originally occupied the region from the Saskatchewan River to the southernmost headwaters of the Missouri River and eastward along the Missouri from the Rockies to the mouth of the Yellowstone River. They maintained no permanent habitations but lived in skin tepees which, with their other belongings, were easily packed on travois pulled by dogs, and after the 17th century, by horses, for transport from place to place. The Blackfoot Nation is made up of four tribes, three of which are now located in Canada while the

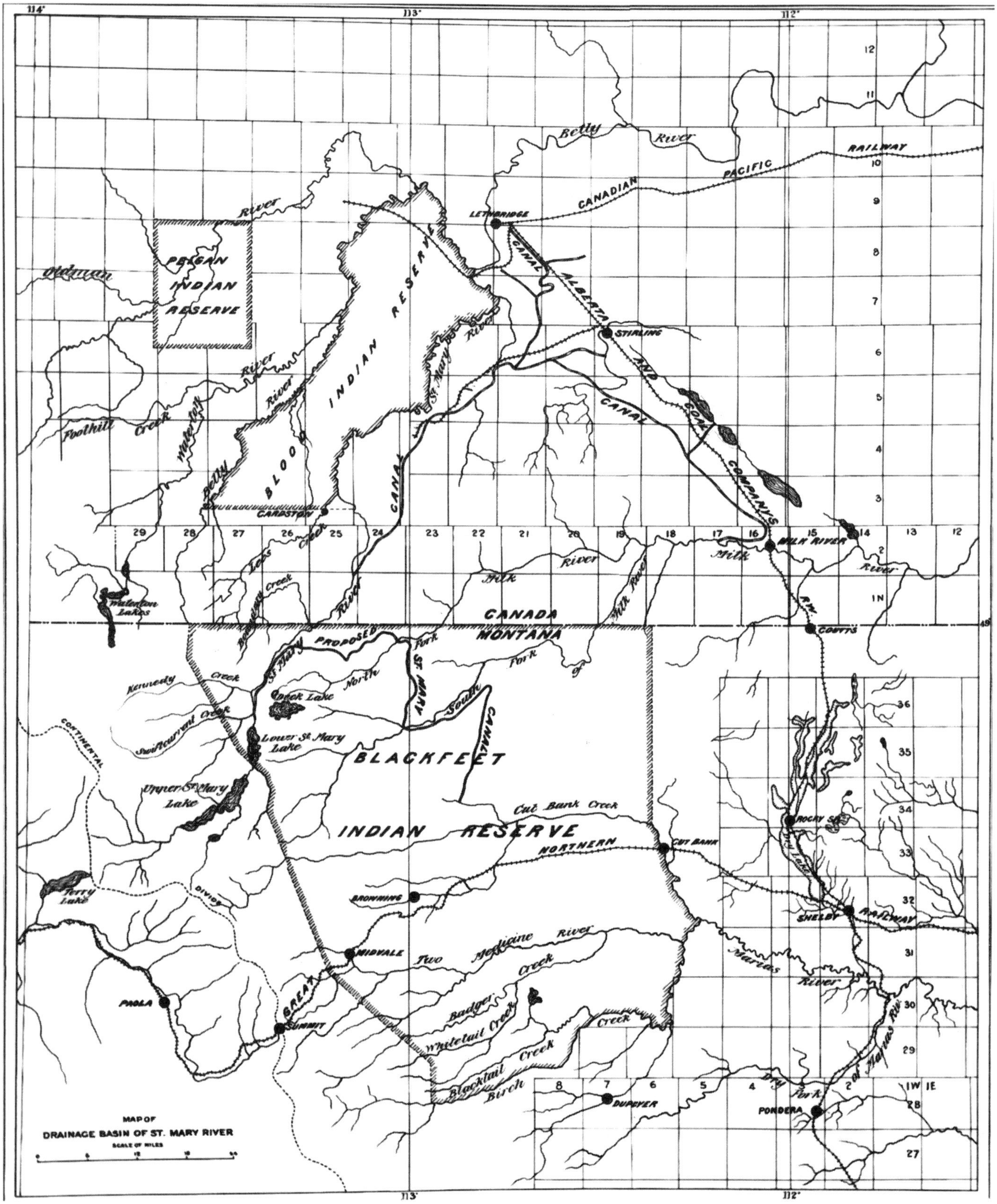

Map One This map indicates the location of the St. Mary, Belly, Waterton and Milk Rivers and the border which divided their watersheds between Canada and the United States. Also indicated are the irrigation systems constructed and planned by the two countries. (Sir Alexander Galt Museum & Archives, Lethbridge, Alberta)

fourth is in northern Montana.

The most important tribe, from the point of view of the history of the St. Mary Project, is the Kainai or Blood who derived their name from the way they painted their faces and arms with bright red paint made from clay or possibly berries. Following the signing of Treaty 7 in 1877, the Bloods lived with the other Blackfoot tribes at Blackfoot Crossing east of Calgary. When this arrangement proved unsatisfactory, they surrendered their lands in return for a separate reserve on the Waterton River. The arrangement for the new reserve was made between Chief Red Crow and the local Indian agent, Norman Macleod. Red Crow's initial request for the land between the Waterton and Belly Rivers was rejected on the grounds that there was no wood for cabins and the soil was mostly gravel. The land between the St. Mary and Belly Rivers was judged more suitable in terms of the supply of wood and the quality of the soil. In the fall of 1880, about 800 members of the tribe moved to the reserve. In 1881, they were joined by an additional 2,500 members of the tribe who had been in Montana.

European exploration of the Great Plains region was begun by Spanish explorer Coronado in the 16th century. The British became involved in 1690 when Henry Kelsey, an employee of the Hudson's Bay Company, was sent from York Factory on Hudson Bay to the interior with a party of Stone Indians to explore the country and safeguard the supply of furs by pacifying the Indians. He established a base camp at Deering Point (near the present town of The Pas, Manitoba) and in the summer of 1690 and 1691 he made two trips into what is now northern Saskatchewan. He may have been the first European to see the Northern Great Plains. Following his trip, the British made no further intrusions into the region until 1754 when Anthony Henday, also an employee of the Hudson's Bay Company travelled from York Factory to the interior. He reached a Blackfoot encampment near Red Deer on 14 October 1754. He spent the winter of 1754-1755 trading for furs before his return to York Factory.

Between the trips of Kelsey and Henday, the French undertook a number of expeditions into the interior from their base in Quebec. Pierre Caultier de La Verendrye established posts as far west as the Mandan villages on the Missouri in modern day Montana by 1738 in his search for the Pacific Ocean. From this base he explored the land to the north, discovering the Saskatchewan River in 1741. In 1742 two of his sons set out from one of these posts to become the first Europeans to reach the Rocky Mountains.

The North West, XY and the Hudson's Bay Companies continued the British effort to explore and control the northwest following the conquest of Quebec in 1759. These companies established posts along the North Saskatchewan River and further north because of the wealth of furs in this region. Expansion of the fur trade onto the plains region to the south was limited to the establishment of a small number of posts and brief visits from company officials. Chesterfield House on the South Saskatchewan River near present-day Empress, Alberta, and Peigan Post or Old Bow Fort, near modern day Calgary, were established by the North West Company. David Thompson, while an employee of the Hudson's Bay Company, wintered with the Blackfoot in 1787-88 and Peter Fidler, an employee of the Hudson's Bay Company, lived among the Blackfoot along the 49th parallel near the foothills of the Rockies in 1792.

Following the amalgamation of the North West and Hudson's Bay Companies in 1821, the southern posts were abandoned, leaving Rocky Mountain House and Fort Edmonton as the most southerly fur trade outposts on the plains north of the 49th parallel. After 1821, the Hudson's Bay Company's objective was to extend the Canadian fur trade into the upper Missouri and the Columbia River basins.

American interest in the northwest was evident shortly after the Revolution. In 1783 Thomas Jefferson proposed a cross-continental expedition to George Rogers Clark and, in 1786, advanced a plan to send John Ledyard across Europe to North America. In 1801, when Jefferson became President, he immediately pursued the plan to send an expedition up the Missouri. The Lewis and Clark Expedition of 1804-1806 was the result. This expedition had the objective of investigating commercial routes to the Pacific and establishing an American fur trade in the upper Missouri. Other American explorations of the Great Plains were made by Zebulon Pike in 1806-1807 and Stephen H. Long in 1819-1820. As originally planned, Long's expedition was to ascend the Missouri River to establish military posts to protect the fur trade and to counteract the influence of the British in the northern region. Later, the plan was changed and Long was instructed to follow the Platte west toward the

Rocky Mountains and return to the Mississippi by way of the Arkansas and Red Rivers.

The Lewis and Clark Expedition was followed up the Missouri by American fur traders. One of these, Manuel Lisa, went as far as the junction of the Yellowstone and Bighorn Rivers in 1807 where he established a post. He later helped to organize the Missouri River Fur Company which, along with the American Fur Company, dominated the fur trade in the upper Missouri until the 1840s. The American Fur Company established posts on the Missouri as far north as modern Fort Benton. The southwest corner of the Canadian prairies remained more or less untouched as fur trading spurred development in the Parkland and northern forests. This changed when the industrial revolution in Europe created a demand for buffalo hide to make drive belts for machinery. American buffalo hunters and traders made fortunes trading guns, blankets and whisky for the valuable hides. The effects of the liquor and the distance from the forces of law and order led to the name Whoop-Up Country. The whisky traders' fort on the Oldman River near present day Lethbridge was called Fort Whoop-Up. Lawlessness and the degradation of the native people by American whisky traders led to the formation of the North-West Mounted Police in 1873.

The dominant image of the Great Plains which emerged out of early exploration activities of the Spanish and the Americans was that of the "Great American Desert."[2] The concept reached the height of its acceptance in the 1850s, but its origins can be traced as far back as Coronado who noted in a letter to the King of Spain that he had travelled across a desert for 72 days. Stephen Long, in his report of his 1819-1820 trip, noted that a desert 500-600 miles long extended along the base of the Rocky Mountains from north to south and that:

> it is almost wholly unfit for cultivation, and of course uninhabitable by people depending upon agriculture for their subsistence. Although tracts of fertile land considerably extensive are occasionally to be met with, yet the scarcity of wood and water, almost uniformly prevalent, will prove an insuperable obstacle in the way of settling the country.[3]

Long, as well as other observers, did not see this impenetrable region as a disadvantage since it might "serve as a barrier to prevent too great an extension of our population westward, and secure us against the machinations or incursions of an enemy that might otherwise be disposed to annoy us in the part of our frontier."[4]

The use of the term "Great American Desert" was not extended into the Canadian portion of the Northern Great Plains because of a lack of knowledge about the region. The area between the Saskatchewan and the 49th parallel was blank on British and Canadian maps of this period. What little was known tended to support the idea of a bountiful prairie because of the immense herds of buffalo.[5]

SURVEYS AND DEVELOPMENT

The view that the "Great American Desert" would be a barrier to the settlement of the West was challenged during the decade when this idea was most popular. The Americans began the process by sending surveyors to locate a route for a Pacific railway in 1850. The surveys in northern Montana were directed by Isaac Stevens, Governor of the Washington Territory. The surveyors made the first documented trip into the region which is now Glacier National Park.

The settlement of what would become Montana began in 1859 when construction commenced on the Mullan Road which connected Fort Benton with Walla Walla, Washington. It was completed just in time for the Montana gold rush of 1862. In 1865, Montana's first legislative assembly was held in Bannack and by 1865, the state's population was estimated at 150,000. The population dwindled with the end of the gold rush, but mining and agriculture were firmly established in the region by then. Ranching began in the 1870s when the Montana range land was stocked via the Chisholm Trail from Texas. Railroad development began in 1880 with the arrival of the Utah Northern from Salt Lake City. The Northern Pacific and the Great Northern Railway through the state were completed by 1893.

British and Canadian interest in investigating and possibly developing the West developed in the 1850s for a number of reasons, but particularly the fear that the activities of the Americans were jeopardizing British, and ultimately Canadian, control of this region. Also, the license giving the Hudson's Bay Company exclusive trading rights in the region was about to expire. The British government had to decide whether to allow the company to continue to have exclusive control of the area or open up the region to settlement to accommodate the western expansion ambitions of the new

Dominion of Canada. These factors combined in the 1850s to produce a parliamentary investigation and expeditions into the region between 1857 and 1860 organized both by the British and Canadian governments.

The British parliamentary committee on the future of the Hudson's Bay Company held hearings in 1857. The witnesses who gave evidence to the committee about the nature of the country included Sir George Simpson, Governor of the Hudson's Bay Company. He considered the area unsuitable for settlement because of the scarcity of fuel, the inferiority of the soil and the uncertainty of crop production. The advocates of settlement, who had less direct knowledge of the area, made counter claims for its settlement potential based on the American experience in Minnesota. The committee made no attempt to reconcile these opposing views but it did recommend the annexation of the territory to Canada and the end of the fur trade.

The first systematic collection of information on the Canadian prairies was made by the Palliser Expedition. Captain John Palliser was an Irish aristocrat, who had spent 1847-1848 at Fort Union and other posts on the upper Missouri River. Palliser originally intended to make the trip at his own expense but his trip was transformed from an individual adventure into a scientific enterprise by the Royal Geographical Society, which asked the British government for a grant to cover the costs of sending additional members on the expedition. The added staff would permit the expedition to expand its investigation and make better use of the materials gathered. Two of these people were to be scientific assistants while the other two were to be members of the Royal Engineers. The British government was persuaded of the value of this plan and provided the requested financing. The individuals added to the expedition were Eugene Bourgeau, Thomas Wright Blakiston, James Hector and John William Sullivan.

The Palliser expedition confirmed the accuracy of existing maps of western Canada as well as adding to the knowledge of the region by mapping the mountains in southern Alberta for the first time. It collected astronomical, meteorological, and magnetic observations and made detailed studies of geographic features and geologic formations. Specimens and records of fauna and flora were collected and sent back for study and classification by experts in England, Canada and the eastern United States.

One of the Palliser expedition's most important observations was that the prairies were made up of a sequence of clearly defined belts of vegetation. The northern zone was heavily wooded. South of this boreal forest the original forest had been destroyed by fires. These fires created a parkland region with fertile soil partially covered with poplar, willows and a few pines. A treeless area, estimated at 80,000 square miles and triangular in shape, was located south of this transitional zone. Palliser equated the lack of trees with aridity and considered the prairie an extension of the "Great American Desert". The effects of grasshoppers and the vast herds of buffalo on the region's meagre vegetation reinforced Palliser's unfavourable view of the region. Palliser's view of the potential of the region for settlement was strongly influenced by its lack of trees for construction materials and fuel. This was considered a more significant problem than the lack of precipitation. To Palliser, the prairie west was a contrast between the "least valuable portion of the prairie country . . . lying along the southern branch of the Saskatchewan, and southward from thence to the boundary line" and the "partially wooded country, abounding in lakes and rich natural pasturage, in some parts rivalling the finest park scenery of our own country."[6]

Palliser came to the conclusion that the potential for settlement on the western plains was high as long as the settlement took place in the transitional parkland zone where the natural environment supplied food for both the settlers and their cattle, along with building supplies and fuel in the form of wood. Palliser was vividly conscious of the difficulties settlers would face during the bitter winter on the dry, treeless plains. He did not know the length of the growing season, but had observed damaging hailstorms and lack of precipitation. The difficulties of settlement had been particularly evident because the expedition itself had experienced a lack of wood, water and grass during their travels through the country.

In 1857 and 1858, while Palliser was pursuing his explorations, the Canadian government sent its own expeditions to investigate the same questions of transportation and settlement potential. These expeditions included geologist and naturalist, Professor Henry Youle Hind. His conclusions about the nature of the land in the West were similar to those of Palliser. The fertile parkland belt had good soil, adequate precipitation and the fuel supplies necessary to support settlement. The short grass prairie had inadequate precipitation, inferior soils and fuel was non-existent. In his view, the region was "a treeless plain, with a light and sometimes drifting soil, . . . not in its present condition fitted for the permanent habitation of civilized man."[7]

George M. Dawson was the next person to examine the region and evaluate its potential for settlement. In 1873, he was appointed Geologist and Botanist to the International Boundary Commission travelling through southern Alberta. During the following two years, Dawson described landscape features, including the Milk River Ridge, and identified three deterrents to settlement: distance from markets, grasshoppers and the lack of trees. He attributed the lack of trees to their destruction by periodic grass fires which he considered part of a cycle with loss of forest leading to a decrease in precipitation which in turn contributed to the further destruction of the forests. Dawson was optimistic that the cycle could be reversed and the forest restored, thus increasing precipitation:

> Considering that the amelioration of the conditions in these regions must be a work of time, it cannot be too soon begun, and must proceed either as a public work, or one encouraged and protected by the Government. Woods and belts of timber, already existing, should be protected and maintained; areas formerly wooded, replanted; and the planting of woodland de novo pushed forward in every suitable locality. It cannot be doubted that the results of such a policy will be, not only the redemption of great areas of wild, open prairie, but a very sensible improvement of climate.[8]

THE CANADIAN PACIFIC RAILWAY AND THE SOUTHERN ROUTE

The westward expansion of Canada, like that of the USA, overwhelmed any concerns about the unsuitability of this region for settlement. The most important step in the Canadian occupation of its portion of the Great Plains was the decision of the Canadian Pacific Railway to abandon the northern route through the fertile belt and build across the plains. This decision was made for a number of reasons, including the CPR's desire to avoid rewarding land speculators. It was also supported by Professor John Macoun, Botanist to the Engineer-in-Chief of the CPR. Macoun concluded that the apparent aridity of this area was associated with the heavy cretaceous clays which baked in the sun and became impervious to rain. Once the soil was broken and cultivated, the precipitation was ample, he believed. Like Dawson, Macoun felt that it was within the capacity of the settlers and the government to improve their prospects for success. Based on this view, he increased the size of the area suitable for agriculture and pasture and reduced the size of Palliser's Triangle to only 20,000 square miles. Macoun was also one of the first people to draw attention to the northward sweep of the summer isotherms and the large proportion of annual precipitation falling in the warm season. His explanation for this phenomenon was that the warm air rising from the Great American Desert allowed winds from the Gulf of Mexico to carry their moisture north for the benefit of the Canadian West.

The construction of the CPR across the prairies opened up the possibilities for the development of a wide range of resources, one of which was the coal deposits along the Oldman River in what would become the Lethbridge area. Coal mining was begun on a modest scale in the Lethbridge area in 1870 by Nicholas Sheran. Assistant Indian Commissioner, Elliott T. Galt, noticed Sheran's mine during a visit to the area in 1879. Galt's family played an important role in the political and economic history of Canada, starting with the arrival in 1824 of his grandfather, John Galt, from Scotland to serve as the Secretary and Manager of the Canada Land Company, which had extensive land holdings in Upper Canada. Alexander Tilloch Galt, John Galt's son, and Elliott Galt's father, was a prominent businessman in Upper Canada, closely associated with the development of the Grand Trunk Railway. He was a Father of Confederation and Canada's first High Commissioner to Britain, which provided him with the opportunity to meet potential investors for any development in western Canada.

Elliott and Alexander Galt realized the potential value of the coal in southern Alberta but also realized that improvements in the transportation system in the West were required before full scale development could take place. The solution to this problem came with the CPR's decision to take the southern route across the Canadian prairies, which provided the Galts with both a market for the coal and a means of transporting it to consumers further east and west. In the spring of 1882, Alexander Galt incorporated the North Western Coal & Navigation Company (NWC&NCo.)

Elliott Torrence Galt
(Glenbow Archives, Calgary, NA-1611-1).

in England. Its objectives were to explore, survey and examine any lands in the North West Territory for the purpose of developing the resources located thereon.[9] The company was also authorized to cultivate these lands and promote their colonization. Development of transportation systems and acquisition of water rights were also objectives of the company. The largest shareholders were William H. Smith, William Lethbridge and William Ashmead Bartlett Burdett-Coutts. William H. Smith founded the news agency and book selling firm of W. H. Smith and Sons. William Lethbridge was his close friend and senior employee. Burdett-Coutts was a wealthy American residing in England whose wife, Baroness Angela Burdett-Coutts, was the richest heiress in England.

The coal mining operations of the NWC&NCo. began in October 1882 when a drift mine was opened on the flats of the Oldman River under the supervision of mine superintendent William Stafford. A small town called Coalbanks developed on the flats close to the mine site. The company intended to transport the coal using the Oldman and South Saskatchewan Rivers. The company built two stern-wheel river steamers, the *Baroness* and the *Alberta,* and brought in a small tugboat called the *Minnow* and a fleet of barges from Winnipeg. Water levels in the Oldman and South Saskatchewan Rivers were too low to allow coal to be carried as planned to Medicine Hat in 1883 and 1884.

Transportation of coal by river barge was declared impractical and a railway was planned to solve the transportation problem. In 1884, an act was passed by the Parliament of Canada to permit the NWC&NCo. to build a railway from a point on the CPR in the North-West Territories near Medicine Hat to the company's mines on the Oldman River with an extension further west to Fort Macleod.[10] The company was given six years to build the railway to its mines and seven years to complete the extension to Fort Macleod.

At the same time that the NWC&NCo. was authorised to build a railway, the Galts also incorporated a second company called the Alberta Railway and Coal Company.[11] This company was authorised to build the same railway in the event that the NWC&NCo. failed to do so. The creation of this second company turned out to be unnecessary as the NWC&NCo. completed the railway in September 1885, although it was built using a narrow gauge because sufficient funds could not be raised for a standard gauge track. In addition to solving their transportation problems, the completion of the railway gave the Galts their first opportunity to acquire land. Under the regulations set up to encourage western settlement, the NWC&NCo. received 700,800 acres for the 109.5 miles of track it built between Dunmore and Lethbridge.

In 1886, Alexander Galt continued the expansion of the railway network in southern Alberta with the incorporation for the second time of a company called the Alberta Railway and Coal Company (AR&CCo.).[12] The second incorporation involved a different group of investors who were either Canadian or American rather than British and included Alexander and Elliott Galt, Sir Roderick Cameron of New York, William Miller Ramsey of Montreal, William G. Conrad of Fort Benton, Montana, Samuel T. Hauser of Helena, Montana, Walter Shanly of Montreal and Donald Watson Davis of Fort Macleod. The company was authorised to lay out, construct and operate a railway from a point on the railway of the NWC&NCo. at or near Lethbridge to the boundary line between Canada and the United States. The Great Falls and Canada Railway was created to construct the line to Great Falls.[13]

On completion the AR& CCo. received 413,568 acres. In 1891, the assets of the NWC&NCo were transferred to the new AR&CCo and it ceased to exist. The AR&CCo. thus controlled a total of 1,114,368 acres in land grants received by the two companies. This land was initially obtained in alternate townships in order to facilitate its use by ranchers. In 1896 the holdings were further consolidated to facilitate irrigation.

The development of AR&CCo. land began with the creation of a new townsite on the plain above the mine entrance and the community of Coalbanks. The plan incorporated wide streets and a site for a park. The town was named after William Lethbridge, one of Galt's partners in the creation of the NWC&NCo. By October 1885, 60 buildings were under construction including six stores, five saloons, four billiard rooms, two barbershops, one hotel and a livery stable. By the fall of 1885, the new townsite had a voice, as a newspaper, the *Lethbridge News*, was established by E.T. Saunders, a former member of the North West-Mounted Police, who had helped found the *Macleod Gazette*. Ontario-born Charles A. Magrath, the Land Commissioner of the NWC&NCo., was very active in the community and became its first mayor.

Charles A. Magrath
(Glenbow Archives, Calgary, NC-2-657)

The sale of the NWC&NCo's. vast holdings of land began immediately after the arrival of the railway. The original intention of Alexander Galt was to use this land for ranching. The 4 June 1886 issue of the *Lethbridge News* advertised for sale or lease 400,000 acres of "choice grazing lands in blocks of townships all of which [were] well watered and within easy reach of Lethbridge." The same advertisement also noted the availability of a small quantity of land that had been prepared for farming purposes in the immediate vicinity of Lethbridge and Lethbridge town lots. Further particulars were available from C.A. Magrath who was also issuing permits to cut hay on the company's lands. The initial land sales were made to Sir R.W. Cameron, a director of the AR&CCo., who purchased two parcels of ranch land containing over 500,000 acres at $1.25 an acre. Overall, however, sales were disappointing, only an eighth of the company's land had been sold by 1889.

THE IRRIGATION OPTION

Irrigation was not a part of the federal government's policy for the development of the West. This is evident in the total lack of references to not only irrigation but even to water in the *Dominion Lands Act* of 1872 and in the lack of surveys to determine the extent of the water resources. Irrigation, therefore, developed on a limited scale as community based projects or as projects undertaken by ranching companies to improve pasture land and garden plots. The first documented irrigation project in Alberta was undertaken in 1878 when American ranchers built diversion ditches on land which later became part of the Peigan Reserve. Other projects were undertaken by the members of the Mormon community. Mormon immigration to Canada was begun by Charles Ora Card who came first to British Columbia and then, in 1886, to southwestern Alberta in search of possible settlement sites. Having selected land on Lee Creek, Card and his associates organized the relocation of a group of settlers from Utah to the Cardston area in 1887.

The Mormon community at Cardston built two projects under the direction of Card and Josiah Hammond. These irrigation projects were not distinguished by their size or by their date of construction but rather by their influence in popularizing the idea of irrigation, which was unfamiliar to the majority of immigrants to the Canadian West. During the 1890s Mormon irrigation efforts were limited because their community was close to the mountains where rainfall is relatively plentiful.

The two projects used Lee Creek which flows south of the town. The first project was constructed by 15 July 1890 but did not go into operation until June1894. It was designed as a multi-purpose system to drive a grist mill as well as to provide irrigation for adjacent farmers and owners of town lots. The second project, which was referred to as the Pioneer Irrigation Canal, irrigated approximately 800 acres on the south side of Lee Creek. It required the construction of a ditch as well as a flume to carry water across Lee Creek from the headgate on the north side of the stream. Both systems operated until 1902 when they were damaged by floods and not rebuilt.

Irrigation development on the Bow River was begun by John Glenn, a squatter who took advantage of the annual floods to irrigate his land at the confluence of Fish Creek and the Bow River south of Calgary. In 1879, he relocated to a spot further up the Elbow River where he constructed a ditch to irrigate his land as well as that of a neighbour.

WILLIAM PEARCE AND THE BIRTH OF AN INDUSTRY

The federal and territorial governments adopted a policy on water development as part of the strategy for western settlement due primarily to the influence of William Pearce, the Mormon community and a drought between the late 1880s and the mid-1890s. Pearce had joined the Department of the Interior as a surveyor in 1874 and was appointed to the newly created Dominion Lands Board in 1882. This board was responsible for making regulations, recommending legislation, formulating resource development policies and supervising the exploitation of all land, timber, mineral and water resources throughout the northwest. As Inspector of Land Agencies, William Pearce's duties included the organization and supervision of the various land offices being established in the West in anticipation of the wave of settlers who would arrive on completion of the CPR. He also had a special relationship with Prime Minister, Sir John A. Macdonald, who asked Pearce to report directly to him on matters relating to western growth. Pearce took his duties as an advisor on western development policy seriously. He travelled extensively in the region and obtained information from settlers through personal interviews and questionnaires.

William Pearce
(Glenbow Archives, Calgary, NA-339-1)

Pearce's interest in obtaining information to help him formulate policies that would best suit the settlement of western Canada took him to the American West. In 1881, he went to Colorado and Utah where he studied Mormon irrigation and settlement methods. He was impressed with their ability to combine irrigation and dryland farming techniques to maximize the productivity of the land. It was hoped sub-soil transference of water would make the dryland adjacent to irrigated land more productive.

Pearce also became very interested in the hamlet system of land settlement which had been used extensively by the Mormons in developing their communities on the American frontier. The system involved concentrating the rural population at one location rather than scattering them out, as under the traditional survey system used in Canada and the United States. The residents of each community were provided with plots of land within the hamlet and in the surrounding agricultural area.

By 1883, Pearce was convinced that the similarities in geography and climate between Utah and southern Alberta were sufficient to warrant the use of Mormon irrigation and land settlement practices in the Canadian West. He detailed the system in a report entitled "Hamlet System Combined with Irrigation," describing the creation of irrigation-based communities devoted to the production of dairy products, pork, poultry, eggs, vegetables and fruit. Each hamlet would consist of 30 to 50 families located in the centre of a circular area of 3,000 acres of irrigable land surrounded by 3,000 to 12,000 acres of dryland for pasture.[14] Within this system of hamlets, one would serve as the location for a creamery and other facilities which would process the milk and other produce.

Pearce's effort to promote irrigation within the Department of the Interior began in 1885 when he included a section on irrigation in his annual report. He also took the opportunity to give speeches on the topic to the annual meetings of the Association of Dominion Land Surveyors. In these speeches he emphasized the absolute necessity of a new approach to settling the arid regions of the West and the suitability of the hamlet system of settlement to meet the social and economic needs of this region.

In his first speech to the Dominion Land Surveyors Association in November of 1889, Pearce argued that settlement policies were not appropriate for the semi-arid region of the Canadian West and minimized the economic benefits to be derived from the settlement of the area. The policies failed to recognize the nature of the water resources in the region and the absolute necessity of irrigation if the area was to be productive. The problems were made even worse by the arrival of settlers from the eastern provinces of Canada and Great Britain where water was abundant. They had no knowledge of or interest in irrigation. However:

> had our settlers come in any material numbers from irrigated districts bringing with them for application here their knowledge and experience of the practices in vogue in Montana, Arizona or any districts requiring an artificial water supply, it is probable that the number of our settlers would be greater than at present, and their advancement in production and wealth would be enormously increased.[15]

Pearce further warned that time was running out for the federal government to implement a settlement policy which would maximize the use of the semi-arid region's water resources. He recommended development of a comprehensive system of settlement and irrigation. Unless this was done, Pearce warned:

> the progress of settlement will be slow, and the country necessarily so sparsely settled that our people will not for a long time to come enjoy the privileges of schools, churches and social intercourse which are so essential to the well-being of all communities, and which might be made available to them if the right plans were designed.[16]

Pearce devoted the rest of the speech to outlining his vision for the development of the semi-arid region. He felt that the future of agriculture in the area must continue to be chiefly pastoral, with the raising of cattle and horses, sheep, hogs, poultry and the manufacture of dairy products the staple industries for all time.

Pearce's ideas were not favourably received by the government, which feared that any discussion of irrigation would discourage immigration to western Canada. After his speech in January 1891 to the Association of Dominion Land Surveyors, Pearce was discouraged from making further public statements on irrigation. As part of this effort to suppress his views, he was refused permission to deliver a speech on the progress of irrigation in Canada to the International Irrigation Congress in the United States.

Irrigation was promoted during the 1880s by members of the Mormon community whose arrival in the country was welcomed by Pearce. As he noted in his 1887 report "the Mormons, who settled this year south of Fort Macleod, have had such experience, and should their location prove to be one requiring irrigation, the example they will show of what can be done in that way will prove invaluable."[17] The Mormons and Pearce did not coordinate their efforts to influence federal policy, but acted independently to achieve a mutually held goal. The Mormon effort in the 1880s was much more limited than those of Pearce and was directed towards the immediate area of Mormon settlement. Their first opportunity to directly influence federal policy came in 1888 when a delegation consisting of Francis M. Lyman, John W. Taylor and Charles Ora Card went to Ottawa to obtain several rights including the privilege of moving plural families to Alberta. During their meeting with Prime Minister, Sir John A. MacDonald, they discussed their practical experience with irrigation techniques in Utah.

The efforts of Pearce and the Mormons to encourage federal government support for irrigation

Charles Ora Card
(Glenbow Archives, Calgary, NA-1759-1)

John W. Taylor
(Glenbow Archives, Calgary, NC-7-534)

were unsuccessful in the 1880s, but changing circumstances made the idea more attractive. A drought, which began in the late 1880s and continued into the mid-1890s, was the new ally in Pearce's efforts to encourage irrigation. The drought encouraged out-migration by settlers and increased the difficulties of persuading new settlers to come to southern Alberta, potentially sabotaging the federal government's plans to settle the West. As a result of this drought, interest in irrigation rapidly gained momentum in the early 1890s, particularly with the urban business communities in Medicine Hat, Lethbridge, Fort Macleod and Calgary. Reflecting this, newspapers in these communities took an increasing interest in the work of Pearce. The paper Pearce presented to the Dominion Land Surveyors Association in November of 1889 was reproduced and favourably commented upon by the *Lethbridge News, Macleod Gazette* and the *Medicine Hat Times*. The *Medicine Hat Times* commented that:

> The subject is one which must prove interesting, alike to settlers, railroad companies and the Dominion Government. Irrigation on a small scale had been tried for market gardening purposes at Calgary, Lethbridge, Medicine Hat, Maple Creek and other places in the Territories, and in every case the verdict is that it pays. How much then would it pay to have a system of irrigation such as that proposed by Mr. Pearce, by which water would be supplied to each settler at a much lower rate than he could procure it by his own individual efforts. We quite agree with the *Lethbridge News* and the *Macleod Gazette* in calling for a convention of delegates from the different centres of the territories to discuss the question and to urge upon the government the necessity of immediate action.[18]

E. T. Saunders, editor of the *Lethbridge News*, became the most enthusiastic supporter of irrigation, which he considered, along with railway construction, the basis of the future prosperity of Lethbridge. This enthusiasm for irrigation prompted the *Lethbridge News* to suggest that a congress be organized to discuss the question of irrigation and to urge the government to take action. The *Lethbridge News* not only supported Pearce's ideas but also began to make its own comparisons between the potential for irrigation in the Lethbridge area and the successful development of irrigation in states such as Utah.

The government of the North-West Territories brought further pressure to bear on the federal government to take action. On 18 December 1891, Charles Magrath, who was by then a member of the Territorial Assembly, introduced a motion to appoint a special Irrigation Committee to investigate the issue and make recommendations. The Committee was duly created and Magrath became its chairman. The committee's 12 January 1892 report to the Territorial Assembly was influenced by the 1889-1890 *Annual Report of the United States Irrigation Survey* and the views of its director, Major John Wesley Powell. Powell's report stressed the need for a comprehensive survey to determine the location of the land to be irrigated and the locations of the headworks and the canal systems needed for the irrigation. Surveys, according to this report, were needed before vested interests appropriated the best sites for irrigation works. The Territorial Irrigation Committee suggested that similar surveys be undertaken in western Canada for the same reasons commencing in the summer of 1892 and covering an area between the Cypress and Porcupine Hills and north from the international border to the Bow River.

The committee also reminded the Territorial Assembly of the American threat on the St. Mary River which was increasing. A joint resolution of the Senate and House of Representatives of the State of Montana had requested that the US federal government survey and estimate the cost of diverting the St. Mary River to the Milk River. This project, concluded the Territorial Irrigation Committee, "simply means the total annihilation of the St. Mary River, as far as Alberta is concerned, and consequently causes serious interference with the vested rights of the settlers of that section of the district."[19] The committee also recommended that the Canadian government contact the American government "with a view to an understanding on the matter."[20]

The combined effects of the drought and pressure from within and from without the Department of the Interior had the desired effect by the spring of 1892. A.M. Burgess, Deputy Minister of the Department of the Interior, asked Pearce to prepare a detailed report on irrigation and irrigation law in order to familiarize the other officials of the department with the issues involved in the administration of water rights. Pearce's report was intended as a discussion paper in preparation for the drafting of legislation which was to be placed before Parliament in the spring of 1893. Pearce brought all of his previous experience and research to bear in the writing of this report and recommended the suppression of private riparian rights and their acquisition by the Crown. Reserving these rights for the Crown would give the government greater control over the development of irrigation. The report added conviction to the impression the Minister of the Interior, T.M. Daly gained of the need for irrigation on a recent trip to the West which had included a stop in Calgary. He called Pearce to Ottawa to incorporate his ideas into draft legislation. The bill was debated during the 1893 session of Parliament but not passed that year because of the number of other bills requiring action. During the delay, the federal government had copies of the draft bill widely distributed and sent J. S. Dennis Jr., a senior member of the Department of the Interior, on a fact finding mission to California, Utah, Colorado and Montana.

The delay in the passage of the draft irrigation bill in 1893 prompted the convening of the irrigation convention originally suggested in 1891. It was organized by the Calgary Agricultural Society and held in March of 1894. The two day convention heard papers on various aspects of irrigation, established committees to review the federal draft irrigation legislation, considered the formation of a permanent organization to promote irrigation and debated various resolutions concerning irrigation development. The convention had two immediate results: the creation of the North West Irrigation League and the organization of a delegation to Ottawa to meet with Prime Minister, John Thompson. The North West Irrigation League was organized on the basis of local branches with a central council to provide overall direction. The delegation included Charles Magrath and Richard Pilling. According to William Pearce, Pilling was the most effective member of the delegation since "he could talk from personal experience on irrigation, and that was what took the ear of Sir John Thompson."[21]

The continued pressure on the federal government from the West and its own growing interest in irrigation brought about the successful passage of the North-west Irrigation *Act* in 1894. This legislation introduced a number of important concepts derived from Australia and the United States. The idea that the state should control the riparian rights was taken from Australia while the idea that the date of application should be used by the state in the distribution of those rights was American. The concept of the "duty of water" was also obtained from the United States.

A second act relating to irrigation passed in 1894 by the federal government was designed to assist the CPR in its efforts to develop a large project east of Calgary. The "Bill Respecting the Land Subsidy for the CPR Company" reflected the federal government's view that irrigation could be used to resolve its problems in meeting its obligations under the CPR contract to provide more land which met the criteria of being fairly fit for settlement. This legislation provided for the creation of a consolidated block of CPR land between Medicine Hat and Calgary which was not subject to the land survey method used elsewhere in the West. During the second reading of the bill, T. M. Daly, indicated irrigation was now of greater interest to the federal government. In his view the land was:

> arid and requires irrigation; and the terms of the original grant specified that the land that the company shall get be land fairly fit for settlement. There is no question as to this land being fairly fit for settlement if it were irrigated; but as it is, it is not fairly fit for settlement. The Company has experimented at different stations along the line of railway in the tract I speak of within the last eight or nine years by having gardens made. About every third or fourth year there is a rainfall, and then they have a good crop; but in other years they have not. That demonstrates that it is impossible to do anything with the land except by irrigation and this bill will enable the company to irrigate the land.[22]

Following approval of the bill by the House of Commons and the Senate it was given Royal Assent on 23 July 1894. The terms of this legislation reflected the influence of William Pearce who had been actively promoting irrigation to the CPR.

A third piece of legislation was passed in 1894 when the territorial government enacted legislation providing for the creation of farmer-operated irrigation districts. This legislation, which was based on the Wright Act of California, was requested at the 1893 Irrigation Convention on behalf of the farmers in the Springbank area west of Calgary.

Despite the widespread interest in irrigation and the flurry of legislation it produced, irrigation remained small scale until 1898. This lack of action was the result of concerns on the part of the CPR about the terms of the North-west Irrigation Act and the end of the drought. The projects undertaken at this time included one by John Quick. In 1895, he built a ditch on the north fork of Sheep Creek to irrigate 160 acres. In the spring of 1897, Messrs. Findlay, McDougall and Riley in association with the High River Horse Ranch Company completed a ditch to take water from the Highwood River to irrigate 5,400 acres. A second ditch in the High River area, originally constructed by Isaac Potter, was subsequently enlarged and extended by R. A. Wallace in 1897 to irrigate 1,000 acres.

The creation of companies which successfully implemented irrigation schemes was a feature of irrigation development in the Calgary area during the 1890s. The Calgary Hydraulic Company and the Calgary Irrigation Company were both incorporated by Acts of the Parliament of Canada in 1893. By 1896, the Calgary Hydraulic Company had built a system capable of irrigating 1,900 acres. The Calgary Irrigation Company, which was established by William Pearce as a further attempt to promote irrigation, was the most successful of the irrigation companies up to 1898 in terms of the size of the undertaking and the sophistication of its works. The 25,000 acres of land irrigated, along with the fact that it attempted to raise capital in England for irrigation development in Canada, made it the precursor of development after 1898.

Development of small scale irrigation projects in the Oldman River basin also continued in the 1890s. These projects were undertaken by Richard Pilling, another member of the Mormon community, who was actively engaged in irrigation development and promotion. His most ambitious project involved the damming of Snake Creek and the diversion of the water into a natural reservoir for storage. The water was then to be released back into Snake Creek for transport to a location south west of the townsite of Aetna where it was once again diverted to irrigate lots in the townsite.

Pilling was one of three people who proposed small projects based on the St. Mary River itself. Pilling's St. Mary project involved diverting water from the St. Mary River by use of a water wheel he designed to irrigate 57 acres of his own land. James

Kearl and Jesse L. Smith also proposed diversions from the St. Mary using water wheels to irrigate 98.5 acres and 87 acres respectively. Three projects of a similar scale were proposed for Boundary Creek.

The Waterton River was used by the Cochrane Ranche Company as the source of water for a canal 2 miles long which irrigated 200 acres. The largest project using water from the Waterton River was built by the North-West Mounted Police at its post at Standoff. The work of building the 3,200 yards-long canal was performed by prisoners so the only expenditure required was $4.50 for lumber to build a sluice gate. Based on the need for 6 men for 40 days at $1.25 per day and two teams at $5.00 per day for 40 days, North-West Mounted Police Superintendent Sam Steele estimated that the use of prisoners had saved $700.00. The system irrigated 300 or 400 acres of pasture and made Steele an enthusiastic supporter of irrigation which he felt would be particularly useful to the ranchers to maintain pasture land.

Indian Agent, James Wilson, also proposed in 1894 that a portion of the Blood Reserve be irrigated to solve the hay shortage caused by the drought on the reserve. An investigation by the Department of the Interior indicated that some 300 to 400 acres of pasture could be irrigated with an expenditure of $2,000.00. No further action was taken when Wilson changed his mind about feasibility of the project.

TWO
PLANNING THE ST. MARY PROJECT 1891-1898

The shift to large scale irrigation development in western Canada took place in 1898 when construction began on the St. Mary Project. A large scale irrigation project using the St. Mary River was first suggested by William Pearce in 1891. Two prominent members of the Mormon community, John W. Taylor and Charles Ora Card, made the first of at least three attempts to begin construction when they acquired land from the AR&CCo. in 1891. Their failure to interest the Church of Jesus Christ of Latter-Saints (LDS) in the Project and the lack of alternative investors resulted in failure. In 1896 the fortunes of the St. Mary Project were reversed, largely through the negotiating skills of Elliott Galt, who was able to bring together the federal government, the LDS Church, engineer George Anderson and a group of British investors to initiate construction. The improved economic conditions of the latter part of the 1890s was one of the reasons Galt achieved success.

DEVELOPMENT PROPOSALS

The idea of a Canadian version of the St. Mary Project originated in 1891 with William Pearce, John W. Taylor and Charles Ora Card. Pearce's role, at this time, was to simply suggest that the St. Mary River should be used to irrigate the land southeast of Lethbridge and to raise the issue of the American threat to Canadian use of this river. In his 1891 report as Superintendent of Mines, Pearce noted that if "the St. Mary River could be cheaply brought on to irrigate this tract . . . it would be second to none in Canada." [23] In this same report, Pearce pointed out that the Americans had their own version of the St. Mary Project. He described the plans of the Great Northern Railway Company to "dam the St. Mary River south of the international boundary, cut a canal from above the dam to the Milk River, carry the water down the Milk River and irrigate along the Great Northern line in Montana."[24] Pearce took no further action on this issue at this time, because he was aware of the intention of members of the Mormon community in Cardston to implement a plan for a large irrigation project in the area.

In December 1891, John W. Taylor and Charles Ora Card purchased a total of 720,000 acres of land from the AR&CCo. Taylor was a senior official of the LDS Church who arrived after the Cardston settlement had been established in 1887. In addition to his church duties, Taylor was a businessman and a visionary when it came to the development of Canada in general and Alberta in particular. His son, Samuel Woolley Taylor, recounted in his *Family Kingdom* that his father was dedicated to the implementation of a dream of the development of Canada. John W. Taylor:

> set an example by his own spectacular faith in its future. Ranches, irrigation projects, oil wells, cattle, real estate - he was in everything.

> He couldn't keep out of anything that looked good, and everything looked wonderful to him. From the moment he awoke from the dream he lived, slept, and ate Canada. He preached Canada in church and out of church.[25]

The AR&CCo. was anxious to cooperate because it had been able to dispose of only a small portion of the 1,592,368 acres it had received as a land grant for its railway construction program. The AR&CCo. not only sold land to Taylor and Card but also secured an amendment to its Act of Incorporation to permit the lands to be used for irrigation purposes. This change permitted the company to excavate, construct, maintain, and operate irrigation ditches or canals in that portion of the district of Alberta lying south of the 50th parallel of latitude and to collect such rates or charges for water supplied for irrigation or other purposes as fixed by a by-law of the company. The decision of Galt to undertake this irrigation project, however, proved to be unacceptable to some of the shareholders in the AR&CCo. thus resulting in the creation of the Alberta Irrigation Company (AICo) in 1893 to undertake the irrigation project in cooperation with Taylor and Card. The first shareholders and provisional directors were Alexander Galt of Montreal, Elliot Galt of Lethbridge, Isaac David Haines of Logan, Utah, Charles Magrath of Lethbridge, Alexander Ferguson of Ontario and Donald Davis of Fort Macleod. The company's head office was to be in Lethbridge and its initial capital stock was $400,000. The company was given the same powers with respect to irrigation development as the AR&CCo. plus additional powers with respect to the purchase of land and the acquisition of water. The act of incorporation also required the company to construct any dams with an apron or slide to permit the passage of saw logs and timber which might be floated down the river. The company was incorporated for the purpose of supplying water for irrigation, manufacturing and other purposes to municipalities, corporations and persons and it was given three years to start construction on the project and a total of 10 years to complete the entire project.

Card and Taylor had originally planned to interest the leaders of the LDS Church at Salt Lake City in their plan and to use their influence to promote further Mormon colonization in southern Alberta. In a letter to the First Presidency of the Church dated 21 July 1892, John W. Taylor explained the deal with the AR&CCo. and the advantages of settlement in Alberta:

> I would recommend that the lands be handled by the Latter-Day Saints, that they accept the offer of the franchise from the railroad company in connection with the subsidy lands; that they build the canals and get possession of that fertile belt of country while it is offered to them . . . There are more tillable lands in the purchase that was made on the 9th day of last December, for our people, than there is in the territory of Utah. I believe that every acre of land in the purchase and in the subsidy, will increase at the rate of $1.00 per acre per year for ten years and that the State of Alberta will become one of the greatest grain districts in North America.[26]

Despite Taylor's enthusiasm, he was unable to interest the LDS Church leaders at that time. One reason for this lack of interest was that the persecution of the Mormons in the United States had lost its impetus, thus making the necessity of emigration to Canada less necessary. Despite this lack of interest, both Card and Taylor continued to promote their idea and make the necessary plans for the implementation of the scheme which included the tentative planning of a town which was to be located near Pothole Creek, close to present day Magrath.

The AR&CCo. received at least two other proposals from individuals who wanted to develop large scale irrigation projects. While these proposals were all unsuccessful, they did have the effect of solving one of the potential problems in the development of an irrigation project given the land distribution policy of the federal government. The AR&CCo. had received its land in alternative townships in an effort to facilitate its use for ranching purposes. To facilitate its use in an irrigation project, it had to be further consolidated into a solid block. Galt made his request for the further consolidation of his land holdings southeast of Lethbridge in the fall of 1895 when two American land promoters purchased 700,000 acres contingent upon their consolidation into a compact block of 500,000 acres located adjacent to the irrigation canal.

DEFENDING CANADIAN INTERESTS

The failure of the AICo. to initiate construction of the St. Mary Project was not a major concern of the federal government until October 1894 when William Pearce returned from the International

Irrigation Congress in Denver, Colorado with information about the American interest in developing its own St. Mary Project. In a letter to the Minister of the Interior, dated 23 October 1894, he recounted a conversation he had had with W. W. Follet. He had been a member of Colonel E.S. Nettleton's expedition which had been sent in 1891 to look into and report upon the feasibility of diverting the waters of the St. Mary River into the bed of the Milk River as requested by Montana. Follet advised Pearce that, while the project was expensive, it was "within the limits of financial practicability and that he understood Senator Power of Montana was promoting it in the interest of that state."[27] Pearce felt that there would be "three fold more advantages" if water from the St. Mary River was used to irrigate the territory covered by the AICo.'s scheme than if it was diverted in Montana. He thus concluded that if the AICo. was not going to take steps to utilise the said waters that:

> early next season the Government make the survey in detail of a canal large enough to carry all the water of the St. Marys River from its bed on to what lands it could be most profitably and beneficially applied. With that data, probably capitalists could be induced to take it up on terms advantageous to the Public but what the writer has more particularly in view, is, that steps should be taken to secure immediately and without doubt the waters of the St. Mary River to Southern Alberta.[28]

The efforts of Pearce to initiate irrigation on the St. Mary River came just after the federal government had decided to support irrigation in the West with the passage of the North-west Irrigation Act in 1894. The federal government had also decided to undertake surveys to determine the extent of the water supply available for irrigation purposes, the nature of the land through which irrigation canals could be built and potential reservoir sites. The federal government was thus able to incorporate Pearce's suggestions into the survey program already underway.

The irrigation survey work was divided into divisions A and B. Division B, under the direction of Arthur O. Wheeler, was responsible for the area now covered by the St. Mary Project. Wheeler began the work of his division at a location northwest of Calgary. From this spot, he proceeded south and east to Fort Macleod. His 1895 report described the land east of the St. Mary River including the Milk River Ridge. The report also identified the location for a potential reservoir which could be filled using water from the St. Mary River. This site, which is now referred to as Ridge Reservoir, was developed in the early 1950s. Wheeler reported on the rivers in the area including the St. Mary and discussed the location for a canal diversion which could be used to irrigate the Lethbridge area.

In 1895, the work of the irrigation survey was extended when J.S. Dennis, Chief Inspector of Surveys, was directed to survey a canal "to divert the waters of the St. Mary River in southern Alberta, with the object of creating a vested right on our side of the International Boundary, before the Americans divert the waters of this stream to their side of the line." [29] Based on these surveys, the federal government designed a canal with a capacity of 500 cfs to take water from the St. Mary River at Kimball, six miles north of the border, and transport it northeast around the Milk River Ridge to a location near present day Stirling. The total amount of land which could be irrigated was about 71,500 acres. The 1895 *Annual Report of the Department of the Interior* noted that "this tract is situated on the northwest slopes of the Milk River Ridge, and lies at a considerably higher elevation than the plains in the vicinity of Lethbridge, which it is hoped may be ultimately served by an enlargement of the canal."[30] The anticipated cost of construction was $213,000.00.

In addition to designing a canal, the federal government also ensured that it would have an adequate supply of water. On 21 September 1897 an order-in-council was passed reserving the water of the St. Mary River for future irrigation development. The order-in-council increased the amount of land that could be irrigated from the St. Mary canal to 250,000 acres and reserved a quantity of water equal to 500 cfs during the low water stage and 1,000 cfs during the high water stage of the stream. The federal government had no intention of building the system itself but wanted to ensure that the water from the St. Mary River would not be used for a number of small projects but for one large undertaking. The federal government anticipated that the AICo. or its successor would be the recipient of this water right.

The Canadian government also used diplomatic means to protect the water supply of the St. Mary Project from the American threat. Pearce and J.S. Dennis were sent to the 1895 International

Irrigation Congress at Albuquerque, New Mexico. At this meeting the Canadian delegates, with some difficulty, arranged to have a motion passed requesting "the appointment of an International Commission to act in conjunction with the authorities of Mexico and Canada in adjudicating the conflicting rights which have arisen, or may hereafter arise, on streams of an international character." [31] Following this meeting, the Canadian government sent a diplomatic note to the American government, via the British Ambassador, expressing its willingness to act upon the suggestion of the Irrigation Congress. The response of the American Secretary of State was that, while he was interested in this important subject, he was unable "to give expression to the views of his government upon the subject at the present time." [32] The American threat in 1895, however, temporarily subsided because the American federal government did not have a comprehensive program for irrigation development in the West.

REVIVAL OF THE ALBERTA IRRIGATION COMPANY

The hopes of the federal government that the AICo. would eventually resume its involvement in the St. Mary Project came true in 1896. In that year, a series of successful initiatives were taken by Elliott Galt and Charles Magrath which would bring success. Galt and Magrath were prompted into action by the fact that the time limit for the initiation of construction by the AICo., as specified in its original Act of Incorporation, was about to expire. In 1896 steps were taken to ensure that the AICo. would have an ongoing role in developing irrigation included revising the original act incorporating the company, negotiations with the LDS Church to construct the system, an engineering evaluation of the project, negotiations with the federal government for the rebate of survey fees and negotiations with potential investors in England. These actions brought the LDS Church more directly into the project and dealt with the

L-R W.D. Barclay, local manager of the Bank of Montreal; Charles Magrath, Land Commissioner of the Alberta Railway and Coal Company; George G. Anderson, Denver consulting engineer; Elliott Galt, President of the Alberta Railway and Coal Company, and Jeremiah Head, consultant from England during visit to Lethbridge in 1898.
(Glenbow Archives, Calgary, PD-301-19)

This is the site on the St. Mary River where water was diverted for use in the St. Mary Project as it appeared before construction began in 1898. The X marks the spot where the first intake was constructed. It was hoped that the use of a secondary channel of the river would provide some protection for the diversion structure from floods and ice. (Glenbow Archives, Calgary, PD-301-1)

problem of lack of confidence on the part of the British investors which had plagued the project during the early 1890s.

The AICo.'s amended act of incorporation gave it until 1899 to begin construction and until 1906 to complete construction. Alexander Galt died in September of 1893 and was replaced on the Board of Directors by Donald W. Davis, William Barclay and Charles Ora Card. The addition of Card's name to the revised act was an important indication of the new era of cooperation between the AICo. and the LDS Church after 1896.

In 1897 Elliott Galt and Charles Magrath, with the encouragement of the Mormon community in Cardston, contacted the First Presidency of the LDS Church with the objective of negotiating a contract for the construction of the first section of the canal by settlers brought into the country. The Galts were anxious to interest the Mormons in the project because they had faced a number of problems as irrigation entrepreneurs on the western Canadian frontier in the 1890s. They would have to organize a settlement program to recruit settlers who knew something about irrigation farming and who would remain on the land during the initial phase of development. It was necessary for the irrigation company to generate revenue from the sale of land and water as soon as possible to cover the costs of construction and operation as well as giving a profit to the investors. They thus had to recruit settlers and establish viable communities on the frontier in a very short time.

The Galts came to appreciate the contributions which the Mormons could make to achieving this objective. The Mormons moved onto the frontier as a single social unit rather than as individuals thus creating an instant community in a new region which reduced the rigours of the settlement process. Contented settlers were more likely to remain in the area and on the land thus contributing to the viability of the project. The Mormon settlers were also prepared to take land in return for building the system which would help reduce construction costs.

The opinion which the AR&CCo. had of the Mormons was expressed by Charles Magrath who was the Land Commissioner for the AICo. In a pamphlet entitled *The Galts: Father and Son*, written in 1936, Magrath noted that "the early Cardston settlers, accustomed to irrigation in Utah, were constantly drawing our attention to the importance of irrigating the territory adjacent to the foothills of the Rockies."[33]

In this same document, Magrath also explained the basis of the cooperation between the Galts and the Mormons. Like Pearce, Magrath appreciated the Mormons' knowledge of irrigation and their ability to settle the frontier in a way that ensured success:

> In my opinion, the movement of the Mormons to southern Alberta was of inestimable value in opening up that section of our West. They understood irrigation and, having made Lethbridge their market town, we were continually told of the wealth that could be created by the diversion of some of our waters that were wasting down our rivers, to the lands in the immediate neighbourhood of our little town. It was a fortunate coincidence for that section of the country that on one side we had the Galts who understood that fundamental of colonization - care of the newcomers until well settled, and on the other, the Mormons so familiar with irrigation and by actual experience, with the difficulties of getting established in a new country. It was only a question of bringing about the co-operation between the two interests in order to get results, and the results are to be seen in southern Alberta today.[34]

The previous involvement of the LDS Church in major construction projects in the American West including the Denver and Rio Grande Railway also made them desirable partners to the AICo.

The negotiations were successfully concluded between 29 September, 1897 and 14 April, 1898. The arguments of John W. Taylor earlier in the decade that Alberta was a land of opportunity now had more influence in Utah because there were economic problems for many Mormons living on marginal agricultural land.

GEORGE G. ANDERSON

In 1897, the AICo. took another step towards building the system when it retained the services of Colorado irrigation engineer, George G. Anderson, to undertake a review of the project as designed by the federal government. Anderson, was a partner in a firm well respected for its technical knowledge and integrity. Amongst the group that respected his abilities but not impressed with his personality was Pearce who noted in a letter to CPR President Sir William Van Horne that the firm of Campbell & Anderson "has perhaps the best reputation of any engineering firm on this continent regarding irrigation not only from a technical stand point, but they have kept themselves free from any wild cat enterprises or swindles on the . . . investing public."[35]

Anderson's November 1897 report was a complete review of the project. It covered water supply, character of the lands and climate, the engineering required to deal with the topographical conditions of the area, construction of the canal, estimated cost of construction, duty of water, estimate of operating expenses, source of revenue and the prospects for settlement. Anderson's discussion of the water supply for the St. Mary Project acknowledged that insufficient data existed to permit any conclusions about the minimum flow of the river during the period when the discharge would be utilized for irrigation purposes. He, therefore, relied on the conclusion already drawn by the federal government that a very considerable quantity of water could be diverted from the stream for irrigation purposes. Conspicuously absent from his comments on water supply was any reference to its control by the United States.

With respect to the suitability of the lands and climate for irrigation agriculture, Anderson came concluded that "with the addition of sufficient moisture, the soil [was] well suited to the cultivation of the various crops of the temperate zone, especially to all cereals and grasses. The climate [was] a favourable one being free from late frosts and marked by an equitable temperature, devoid of rapid and extreme variations."[36]

The engineering required to build an irrigation system, given the topographical features of the area was the largest portion of the report. Anderson found that the St. Mary Project presented him with a unique challenge because the topographical conditions in this locality were rarely encountered by the irrigation engineer. The main engineering challenge was to dissipate elevation cheaply and economically rather than the usual problem for irrigation engineers, which is to conserve elevation by building canals on gradients as low as possible.

At the point of the irrigation intake, the St. Mary River has an elevation of about 3,850 feet while Lethbridge is at 2,958 feet and Stirling is about 3,100 feet. Anderson's solution was to use the natural water courses. Pinepound Creek and the west branch of Pothole Creek were to be the main "arteries" of the system to conduct the waters of the St. Mary River to the farmland. The use of these natural water courses reduced the amount of construction required and thus "presented the best means for the solution of what might otherwise prove to be a certainly expensive and perhaps impossible proposition."[37] This was his most important change to the federal government's 1895 plan which used canals throughout its entire length.

The only possible objection to using natural channels, which Anderson dealt with at some length in his report, was erosion of the channels in these natural water courses. Based on his examination of these channels, Anderson concluded that erosion would only occur for a short time before the flowing water encountered a layer of earth resistant to erosion. The erosion which did occur before the channel stabilized "would be deposited at the various bends of the tortuous channels of both Creeks."[38] If erosion was greater than expected, the various headgates in the system could be modified to pass the silt through. The speed of the water could be kept at a sufficient velocity to carry the silt right through the system resulting in its "discharge upon the irrigated lands, where it would be uniformly distributed and not injurious."[39]

Based on his idea of using a combination of ditches and natural water courses, Anderson designed a system which started near Kimball, the site identified by the federal government's surveys as a suitable location for an intake. The cost of the canal to Pinepound Creek, including the diversion structure on the St. Mary River was $62,000.The first area of potential irrigable land was north of this canal, east of the St. Mary River, south of the present village of Spring Coulee and west of Pinepound Creek. The main lateral to serve this area could be built at an estimated cost of $10,500. The second stretch of canal began south of the future site of the village of Spring Coulee and went around the village to the east in order to reach the head of the west arm of Pothole Creek. The estimated construction cost of $12,800 made it the cheapest of the three sections of canal. Anderson estimated the cost to build the lateral to serve the area referred to as division 2, which was bounded on the west by Spring Coulee, on the north by the St. Mary River, on the east by the Pothole River and on the south by the Milk River Ridge, was $13,200. This area includes much of the land which is now in the Magrath Irrigation District. Division 3 included the land east of Pothole Coulee from what is now the Welling area to Stirling. This division, along with its distribution laterals cost an estimated $18,000. This area includes most of the land in the Raymond Irrigation District. With the addition of $5,500 to build a telephone line 70 miles long, the total construction cost was $122,000. Engineering expenses plus incidentals added a further $12,200 bringing the total to $134,000. Once the system was built, Anderson estimated the annual operating cost to be $14,580. This amount included renewal of canal structures, the telephone line and main canal.

Anderson estimated that the rates which would be charged to pay for the building and operation of the system would be in line with similar projects in the United States where rates averaged $1.00 per acre. Anderson optimistically predicted that "as settlement increases and the development of the irrigable area progresses, a gradual reduction of this charge can be made without injury to either the Canal Company or Consumer."[40] The area irrigated by the system depended on the size of the canals and the value chosen for the duty of water. Based on an original capacity of 500 cfs and a duty of water of 150 acres, as defined in the North-west Irrigation Act, Anderson estimated that theoretically 75,000 acres could be irrigated.

Lastly Anderson discussed the prospects for settlement. The most important problem facing the company in starting the project was "the difficulty of securing pioneer settlers who have some means, considerable knowledge of the special features of agriculture under irrigation, and a large community to follow them upon success following the original efforts."[41] The Mormons, in Anderson's opinion, would solve this problem because:

> these people are peculiarly well adapted for the pioneer work of an enterprise of this kind. They have had an extended experience in irrigation farming and that experience is at the command of all their people. Their efforts are always combined, and not scattering, as communities they are under able leadership, and these features command success in new countries where single, scattering effort, usually without experience,

> though with perhaps the best intention, meets only with dismal failure. In my observation, these people are as individuals, frugal, thrifty and industrious; as an organization, well disciplined and harmonious. In new fields they have the special advantage of great numbers to draw upon if their pioneers are satisfied with the conditions and prospects. As a rule they make good, law-abiding citizens and from my observations of what they have been able to do in other irrigated sections I am confident that a business arrangement with them will result to the best interests of this enterprise and this section of the country.[42]

Anderson indicated further that once the preliminary work of settlement was well started other settlers could be recruited in such countries as Scotland. To start with them originally, Anderson cautioned:

> would be unwise, largely on account of their lack of familiarity with irrigation, and the business of inducing immigration from such countries must be carried on with much prevision and forethought, which cannot ordinarily be devoted to it, but could be in this case with the assurance that the pioneer work was in the hands of an organization as well equipped as the Mormon Church, and with the feeling that the enterprise was being tided over its earlier history on a self-supporting basis.[43]

FEDERAL GOVERNMENT COOPERATION

Armed with the terms of a tentative agreement with the LDS Church and the favourable report from Anderson, Elliott Galt next contacted Clifford Sifton, the Minister of the Interior. In a letter dated 15 December 1897, Galt explained how the improved commercial conditions of the country had prompted the AR&CCo. to renew its efforts to develop an irrigation project. Galt then described his and Charles Magrath's trip to Salt Lake City in November of 1897 where they:

> placed [themselves] in direct communication with the Heads of the Mormon faith there, feeling that if we could arrive at some understanding with them as to the colonization of our lands, we might perhaps be enabled to do the necessary financing, the result being, I am glad to say, the evolution of a proposal by the Heads of the said faith to take a contract from this company . . . to construct the Canal System upon the basis of receiving payment therefor, as to one half in cash, and the remaining half in irrigated lands, upon which for their workmen and their families to settle.[44]

Galt also drew Sifton's attention to Anderson's report noting two points. The first was the reports:

> stress upon the special advantages to be secured by making an arrangement with the parties indicated, for the reason that they are accustomed to the use of irrigated lands, and that in point of industry, thrift and other qualifications enjoyed by them, they, as pioneer settlers cannot be excelled.[45]

The other point was Anderson's cost estimate for the project which Galt considered low because "we shall have to advance monies not only to bring in the settlers, but also to help them in various ways to make a good start upon their farms, in order to assure their success."[46]

Galt then proceeded to link the confidence which the British investors had in the new AICo. to the support which the project would receive from the government. "I have represented to our principal people in England" Galt continued "that you think so well of the project that you are disposed to recommend the government to grant us some material assistance to aid us in carrying through the program as outlined . . . ."[47] The type of assistance requested by Galt was a rebate of the fee charged by the government to survey the lands which amounted to 10 cents per acre. Given this concession, Galt anticipated the company would be prepared to start the project immediately.

The Minister being "favourably impressed by the practicability of the scheme" recommended that the request be granted. In an order-in-council passed on 6 January 1898, the government agreed to remit these fees if the company had spent at least $200,000 within three years and had implemented the settlement scheme as outlined in the agreement with the LDS Church.

This decision had the desired result. The investors, however, required one final assurance that the project was viable and sent Jeremiah Head to Canada to investigate. He came to the same conclusions about the project and the Mormons as had Anderson.

THE CONSTRUCTION AND LAND SETTLEMENT CONTRACT

The contract covered the construction of an irrigation canal, its management thereafter and the settlement of two towns. The original 1895 plan of the federal government for the St. Mary Canal as modified by Anderson was the basis of the construction program. The Mormon settlers received their wages half in cash and the balance in land at the rate of $3 per acre. The land was to be provided with a perpetual water right. Construction was to start no later than September 1, 1898 and by December 31, 1899 the contractors had to have earned at least $100,000. Each of the two towns had to have a minium population of 250 people. The payment for earthwork on the proposed canal was to be 12.5 cents per cubic yard for ordinary soil or dirt from those sections on the level. A higher rate was to be paid for rock work, cement, gumbo and side-hill work. The LDS Church agreed to earn $75,000 in cash and 25,000 acres in land within two years from 1 April 1898. The land was to be in two equal blocks as nearly square as possible and on the canal. The centre of one block was to be within 20 miles and the other within 35 miles of Lethbridge. Hamlets to be established at the centre point of each block were to contain at least 50 families, averaging five souls each prior to 1 April 1898. If the hamlets were not established on schedule the settlers would have to pay $5 per acre rather than $3 per acre for the land. Upon the establishment of 50 families in each hamlet, settlers were to be permitted to take water from the canal for a fee of $2 per acre per year. The settlers were permitted to work on irrigation works or railway construction to pay their water rates. The future management of the system was covered in the provision that the main canal was to be the responsibility of the parties using the water in proportion to their interests. The company thus envisaged a system which would to a large extent, be run by the water users themselves. To aid in the travel of settlers to the area, the company agreed to advance railway fares for families from Utah as long as the distance travelled did not exceed the distance from Salt Lake City to Lethbridge. The advance was to be deducted from the payments made for canal construction.

THREE
CONSTRUCTION BEGINS
1898 -1900

After seven years of planning and many false starts, the first period of construction of the St. Mary Project began in the late summer of 1898. Over the next two years, the AICo., whose name was changed to the Canadian North-West Irrigation Company (CNWICo.) in 1899, quickly accomplished a number of important tasks which compensated for earlier delays. This first period of construction was undertaken in two phases, the first of which was the completion of the canals and related structures through to the new village of Stirling. The second phase involved the completion of the Lethbridge branch of the main canal. The construction of this section was subject to negotiations between the CNWICo. and the Town of Lethbridge because it was an addition to the plan originally prepared by the federal government and George Anderson and was not included in the original negotiations with the LDS Church.

Original intake of the St. Mary Project shortly after construction in 1899. (Glenbow Archives, Calgary, NA-922-10)

KIMBALL TO STIRLING

Construction of the St. Mary Project began on 28 August 1898 when Charles Ora Card plowed the first three furrows of a section of the main canal near the future location of the diversion structure at Kimball. The construction program was managed by Charles Ora Card and George Anderson until November, 1898 when Milton D. Hammond, a construction engineer from Providence,

Main canal approximately a mile from the intake at Kimball in 1900. The St. Mary River is in the background. (Glenbow Archives, Calgary, NA-922-11)

This structure south of Spring Coulee diverted water from Pinepound Creek into a canal which took it around the east side of the village to Pothole Creek. (Glenbow Archives, Calgary, NC-2-161)

This box flume which took water across Willow Creek (also referred to as Rolph Creek) was approximately 5 miles from the intake at Kimball. It was made wide enough to allow the passage of dredges used in the cleaning and widening of the main canal.
(Sir Alexander Galt Museum & Archives, Lethbridge, Alberta)

This spillway transferred the water from the canal east of Spring Coulee into the west branch of Pothole Creek.
(Sir Alexander Galt Museum & Archives, Lethbridge, Alberta)

Utah arrived to take over the task. In addition to his supervisory duties, Hammond had contracts to work on specific sections of the canal. Hammond's arrival permitted Card to attend to other duties, the most important of which was recruiting settlers to work on the project. Anderson continued as the consulting engineer and was responsible for location of the canal in advance of the Mormon workers. He also supervised the building of various structures along the canal that were not part of the contract with the LDS Church.

The construction program concentrated on the completion of divisions 2 and 3 of the plan proposed by George Anderson since they were located where the two townsites were to be developed as part of the Mormon settlement program. Division 1 south of the village of Spring Coulee was dropped from any future development plans.

The first workers on the canal came from the Alberta Mormon community since Mormons in Utah did not immediately respond to the opportunity to come to Alberta. In an effort to encourage interest, the LDS church undertook an

Diversion structure at Magrath looking northeast in 1912. Town of Magrath is on the horizon. (Glenbow Archives, Calgary, NC-2-159)

extensive advertising campaign which described the opportunities for advancement in Alberta. The *Deseret News* of Salt Lake City played an important role in this effort by publishing pamphlets instructing people on how to prepare for colonization and printing a constant flow of encouraging articles. Alberta was a good place to make comfortable homes and work on the canal would provide "good opportunity to accumulate the means to pay for them with out incurring the bondage of debt."[48] Card contributed articles which emphasized the opportunities for advancement including the availability of land through the homestead system which he said was "a bonanza for our boys." John W. Taylor and Charles Ora Card also travelled throughout Utah promoting Alberta as a place to settle and work.

Diversion structure at Magrath looking from the east in 1912. (Glenbow Archives, Calgary, NC-2-158)

When the campaign to encourage people to come to Alberta failed to provide a sufficient number of settlers to complete the project, the LDS Church called men and women to missions in Alberta. This approach resulted in the rapid growth of the two agricultural villages which had to be established as part of the contract with the AR&CCo. The communities of Stirling and Magrath developed rapidly beginning in the spring of 1899. The family of Theodore Brandley was the first group to arrive at Stirling, named by the AR&CCo. after one of the investors in

Main canal between Magrath and Welling under construction in 1899. (City of Lethbridge Library)

Main canal between Magrath and Welling. (Glenbow Archives, Calgary, NA-922-8)

the company. Brandley came from Richfield, Utah along with 24 members of his family as well as Joe Peters, who planned to work on the MacIntyre ranch. Development of Magrath also started in the spring of 1899 with the arrival of Charles Heber Dudley. Magrath was named after Charles Magrath, the Land Commissioner of the AR&CCo.

Main Canal near Welling.(Glenbow Archives, Calgary, NC-2-157)

Junction of the main canal and the Raymond Lateral. Water went through the headgates to the right to Raymond and Stirling and over the structure on the left into Nine Mile Coulee. (Glenbow Archives, Calgary, NA-922-17)

Box flume on the Raymond Lateral across Kipp Coulee. (Glenbow Archives, Calgary, NA-922-6)

The development of these communities represented the implementation of the hamlet system of settlement which William Pearce had advocated in the 1880s. The design of Stirling and Magrath was based on the Plat of Zion which incorporated Joseph Smith's ideas on town planning. Each community was to be one mile square and oriented north-south. The land within was subdivided into 10 acre blocks which were further subdivided into 1/2 acre lots. The land around the village was extensively subdivided to provide farm land for the residents of the village. The water supply for Stirling was initially provided by wells but in 1905 a system of ditches was built and connected to the irrigation canal.

The first section of the system consisted of a canal 10.5 miles long running from the intake at Kimball to Pinepound Creek (also called Spring Coulee). This canal was built 20 feet wide on the bottom to carry 5 feet of water on a gradient of 2.112 feet per mile. The sides of the canal were sloped one and one half to one. The banks were built one and one half feet above the high water line and were 8 feet thick. The natural channel of Pinepound Creek was used for a further 10 miles during which time the water dropped 280 feet in elevation. A second canal, one and three quarter miles long, was constructed east of the future site of the village of Spring Coulee to take water from Pinepound Creek to the west branch of Pothole Creek. The natural channel of Pothole Creek was used for 12 miles with a further loss in elevation of 386 feet. A third canal, 21 miles long, took the water out of Pothole Creek and transported it east to Stirling. The system included two diversion structures in addition to the main diversion structure at Kimball.

The system was completed as far as Stirling during the fall of 1899. Water flowed via Magrath to Stirling on 14 November of that year. The completion of this section was marked by a visit to the project by Clifford Sifton, the Minister of the Interior. He arrived in Lethbridge on 18 November 1899 and was given the grand tour by George Anderson along with 60 businessmen from Lethbridge.

With construction under way, by September 1899, Elliott Galt and Charles Magrath again asked the federal government for assistance building the project. Galt's request to Sifton noted that construction work had actually commenced by a reorganized AICo. operating on $200,000 worth of stock subscribed by various London investors. Land had been transferred from the AR&CCo. to the AICo. Given the progress, Galt requested that the survey fee rebate, which was to be given to the company after three years, be given to the AICo. immediately. This money, Galt pointed out, would allow the AICo. to pay for a portion of the land it had acquired from the AR&CCo. Such a payment, Galt argued, would provide extra funds to cover the costs of bringing in settlers to the benefit of the entire region:

> I may say that we are finding it necessary to bring in the first settlers at our own expense and we shall have to advance to them further amounts for dwellings, barns, fences, etc. so

> as to give the pioneers every chance to make a rapid success of their efforts, which we hope will have the effect of stimulating further immigration, in consideration whereof and also as to the demonstrating of the feasibility of reclaiming arid lands in this western country by means of irrigation, I trust you may think it proper to favourably consider my request.[49]

The government responded with a partial rebate of expenditures by the AICo. of $50,000.

THE LETHBRIDGE BRANCH

The CNWICo. began construction of the St. Mary Project with the long term goal of building the system through to the area east of Lethbridge. The higher cost of building the system into Lethbridge, however, resulted in the decision of the company to end the canal at Stirling during this first period of construction. The extension of the system north to Lethbridge and beyond was first discussed publicly on 1 December 1898 when the *Lethbridge News* carried an interview with Elliott Galt. The interview was contained in an article with the heading "Irrigation - $500,000 to be Expended and 500,000 Acres Irrigated - Main Canal to be Finished Next Year and Laterals Year After." Galt's comments, however, were in sharp contrast to the massive development suggested in the headline. Galt stressed the cautious approach and limited goals of the company for the immediate future. He recalled that the AR&CCo. had been approached by a number of promoters with irrigation schemes for the reclamation of their lands between Cardston and Lethbridge but that negotiations had come to naught over the issue of finding markets for the products of the land grown with the aid of irrigation. The decision to build the Crow's Nest Pass Railway and the company's success in finding "responsible" people to undertake the work had made commencement of the project possible. The company only intended to build the system to Stirling and to use Mormons to settle the land along the canal. The company did not "propose to expend more money upon capital account until a large portion of such land was occupied by settlers."[50] The inclusion of Lethbridge in the CNWICo.'s development plans, therefore, would "depend upon the extent to which the Municipality would co-operate."[51]

Main Canal coming into Lethbridge.(Glenbow Archives, Calgary, NA-506-1)

Flume bringing water into Lethbridge looking northwest from 7A Avenue and 13 Street South (Glenbow Archives, Calgary, NA-922-4)

The *Lethbridge News* immediately started a vigourous campaign to ensure that the "municipal cooperation" sought by Elliott Galt would be forthcoming. Its first step was an editorial which repeated what it had said many times before about the value of irrigation to Lethbridge. Irrigation, it argued, would replace the "illimitable expanse of unproductive prairie" which was occupied by "an occasional ranch and infrequent herder with but the scattering flocks of sheep and herds of cattle and the sparse population which these industries create" with the "continued succession of tilled fields rich with their burden of grains and grasses of all varieties, and a thickly-settled population of contented and prosperous farmers."[52] Lethbridge could become the service centre for the new towns and villages which would develop in this area as well as serving the needs of the farmers located immediately adjacent to their own community. Irrigation could make Lethbridge the commercial and industrial centre of southern Alberta. The citizens, however, had to "seize the opportunity at once, and to allow nothing to interfere with their set purpose to make such arrangements with the Irrigation Company as will best secure that end."[53]

The *Lethbridge News* found an immediate ally in the Lethbridge Board of Trade which passed a motion at their meeting the following day urging support for the negotiation of an agreement with the AICo. The preamble to this motion also noted that Lethbridge could not take for granted its inclusion in the irrigation development plans of the CNWICo. since the main canal of the AICo. would only come within 20 miles of Lethbridge unless a bonus was provided. The motion, therefore, resolved that the Town Council should submit a by-law which would ensure the delivery of a supply of water for the town and the settlement of the adjacent land. A copy of the resolution was sent to Town Council.

Early in December 1898 the AICo. gave the citizens the opportunity to be included in their system when it offered to deliver at the town limits a stream of water flowing continuously from 15 April to 1 November of each year at the rate of 5 cfs. It also offered to build branch line canals of sufficient capacity to irrigate 20,000 acres within a radius of

Ditch system along along Stafford Drive South which served North-West Mounted Police barracks. (Sir Alexander Galt Museum & Archives, Lethbridge, Alberta)

Lethbridge ditch system along the north side of Galt Gardens.(Glenbow Archives, Calgary, PD-310-51)

10 miles of Lethbridge and to use every means to settle the territory as speedily as possible. In return for these undertakings the AICo. asked the sum of $50,000 payable in debenture bonds bearing 5 per cent interest redeemable in 25 years with the interest only on the bonds to be paid for the first 10 years.

In January and February 1899, the Board of Trade sent a letter to the new council requesting that it take action to ensure that the canal was completed to Lethbridge and organized a petition from the citizens of Lethbridge to have Council pay a bonus of $30,000 to the company to build a canal for town water supply and irrigation in the area. The *Lethbridge News* encouraged support by giving complete coverage of the irrigation promotion activities of the Board of Trade and publishing editorials and letters supporting irrigation. One such letter of support came from John Reid, the Colorado

State Engineer who commented on the idea of bringing water into a town and surrounding area. He indicated that:

> almost all of our small towns and many of our larger ones have water running on the curb of the street, for the purpose of irrigating the trees planted in front of the residences, and also for the purpose of irrigating lawns, making our towns beautiful and homelike and at the same time furnishing an invaluable drainage and sewerage for the town.[54]

Reid painted a picture of the development of a bountiful agricultural hinterland around the towns of Colorado because of irrigation:

> In the immediate vicinity of the towns truck gardens of an acre to ten acres are the rule, with chicken ranches and dairies sufficient to supply the needs of the town: beyond these come the farms proper, where alfalfa or lucerne forms the principal forage crop: wheat, oats and corn the principal grain crop. In some localities specialties are made of musk melons and water melons in others potatoes, and still others fruit: all of which crops are bountiful and of a superior quality, which we consider due to irrigation. Beyond these and above the ditches the cattle are ranged, and this forms quite an industry: all, as you see, centered about the town and tributary to it. . . . I believe that without irrigation such settlements would be impossible in our State, and as I said in my letter to you, irrigation has made the farm products of Colorado of greater value than her mineral output.[55]

The initial success of these efforts came on 6 March 1899 when first reading was given to By-law 101 which authorised the granting of a $30,000 bonus to the AICo. and the issuing of debentures to cover its cost. The by-law incorporated the terms as proposed by the AICo. in December 1899 with the added provision that the company be required to spend a further $5,000 under the direction of the council on the construction of a ditch system throughout the town. One thousand trees were also

Lethbridge ditch system south of Galt Gardens along 6th Street South. (Sir Alexander Galt Museum & Archives, Lethbridge, Alberta)

to be planted along the ditches of this distribution system. In regards to the area around Lethbridge, the AICo. agreed to deliver 135 cfs for the irrigation of 20,000 acres of land situated within 10 miles of the town. A distribution system was to be built and the land sold "to a desirable class of settlers" as speedily as possible. The work of building the canal to the town boundary was to be completed by 1 May1900 with the distribution system to be completed three weeks thereafter. The tree planting program was to be directed by the Town Council.

The introduction of the by-law intensified the ongoing debate as to the value of irrigation to Lethbridge. While there was no organized opposition to its approval, the *Lethbridge News* and the Board of Trade, as its leading advocates, had to respond to the concerns of individual ratepayers. One ratepayer questioned the viability of the AICo. given the failure of similar companies in the USA. This concern was answered by the publication of a letter to the editor from Thomas Farrar who concluded:

> those across the line which failed had not the money nor the brains to carry out their schemes . . . While our Irrigation Co. have got the money. They have bought the brains which bear a continental reputation, and if these are not enough, they have the money to buy more.[56]

Many ratepayers, particularly those living north of the railway tracks, also expressed some concern that the proposed town ditch system would not be extended to their area. The *Lethbridge News* dealt with this concern by sending a reporter to interview Company Engineer George Anderson and Mayor Mewburn. Both men assured the citizens of Lethbridge that no section of the town would be overlooked.

The additional information provided by the *Lethbridge News* resulted in the overwhelming approval of the by-law on 3 April, 1899. The vote was 96 to 6. The six people who voted against it were dismissed by the *Lethbridge News* as representing that group of people resident in every community who resisted any project no matter how beneficial. The paper was pleased to find that in Lethbridge the percentage of such people was so small.

The celebration, however, was premature because it was later discovered that the wording of the by-law had not been approved by the Attorney-General and the legal adviser to the Territorial Government. The result was the submission of two new by-laws, one of which covered the supply of water to the town, while the other covered the supply of water to the area around Lethbridge. These two by-laws were approved on 29 June 1899 with a vote of 85 to 1 for By-law 107 and 84 to 2 for By-law 108. With the passage of the by-laws, agreements were signed on 17 July 1899 between the City of Lethbridge and the CNWICo.

Fishing in the town ditch system 4 September 1900.
(Sir Alexander Galt Museum & Archives, Lethbridge, Alberta)

THE BIG DITCH

We're a-diggin' a ditch in Alberta.
See the plowmen a-turnin' the sod
An' the scrapers a-scrapin' the dirt-eh!
Oh, the pay is three dollars a rod.
Yes, she runs like a snake through the prairie
Or a tunnel turned over on top
Or a twenty league furrow of fairies,
Or a grave for a cholera crop!
There was mud in the rains,
An' the youngsters complained:
Now they growl a their lot
'Cause the sun's gettin''ot.
But the ditch will give us a farm,
An' plenty fill the plain!
I think of that an' think of thee
My lovely Mary Jane!

You've no money in 'and till you make it,
An' you don't 'ave a wife while you batch:
You can't eat your bun till you bake it,
And you can't raise a chick till 'e's'atched!
An'so my philosophy teaches
That prosperous days are in store
For the man that will work as I preaches-
There'll be wealth when there's water galore!
There was mud in the rains, etc.

So we work with a view to the future.
An' a missus and kids of our own:
An' if this 'ere job doesn't suit yer,
W'y jump it and skip and be blown,
But the fellow as stick to 'is scraper.
An' carefully 'usbands 'is pay,
Will rejoice long before 'e is grey, sir.
In a 'ome of 'is own, as I say.
There was mud in the rains, etc.

Times we're grumpy at dawn of the sky light,
An' we grovel in the 'eat of the day
But the evenin's so cool in their twillight
Drives all our ill temper a way!
See this ditch in the dawn of 'er bein',
An' the busy noon-tide just before,
W'en 'er evenin''as come, w'at a seeing'
Of bloom in the blessin's she'll pour
There was mud in the rains,etc.

Then swing clear the pick an' the shovel,
Hurrah! for the scraper an' plow!
An''ere's to the bloke with 'is level
That shows all the rookies just 'ow!
The 'eads that direct ain't contrary,
Gawd bless'em that shows us the way!
Three cheers for our scratch in the prairie
That brings in a glorious day!
There was mud in the rains
'Ow the rookies complained
"Ear 'em growl at their lot
Cause the sun's bloomin 'ot!
But the ditch will give us all a farm
An' plenty fill the plain
I think of that an' think of thee
My lovely Mary Jane!

C. Jay
Lethbridge, Alta., July 1, 1899

This poem celebrating the virtues of irrigation was published in the *Lethbridge News* on 12 July 1899.

The successful effort to bring irrigation into Lethbridge was, in the opinion of Methodist Minister, T. Jackson Wray, an act of faith by those people who believed in the future of irrigation over those people who were willing to depend on the benefits of the traditional industries of Lethbridge which were coal mining and ranching. In his view:

> Lethbridge itself [had] no great faith in the enterprise. Some trust in cattle and some in ditches, said they, but we trust in the payroll at the mines, that's good enough for us. Some of the old timers said they had been in the country going on twenty years and had never raised cabbages and that settled it. They were not all that way fortunately and there was enough public spirit in the town to put up a good bonus for the sake of having the water in the streets, but nevertheless there is still a type of townsman who prophesies terrible things and disheartening to the newcomer, unwise enough to seek advice around the bar-room. A man that is great on whisky has naturally no place for water either inside or outside.[57]

Wray further noted that given this negative attitude, the only course open to the company was to bring in Mormon settlers who could "take up land and give the thing a fair trial."[58]

Work on the Lethbridge branch canal began in the fall of 1899. Construction was similar to that on the Kimball to Stirling portion. The Mormons, under the direction of Hammond, provided the labour and Nine Mile Coulee was used to reduce the need to excavate canals. A 25 mile-long canal had to be built across the height of land between the Magrath Plains and the Lethbridge Plains known as the Rolling Hills. The construction of the 16 miles of town ditches began in the spring of 1900. The town ditch system and the tree planting program was intended to transform Lethbridge from a "very ordinary and rather unattractive town into the neatest, cleanest and most picturesque place in the West."[59] The improved drainage provided by the ditches was another advantage of the arrival of irrigation in Lethbridge. Maintenance of the town ditch system, however, was not continued and by 1908 it had fallen into disrepair. It was replaced by a system of underground pipes as part of the civic improvements in preparation for the Dry-land Farming Congress of 1912.

The system was put into use when the first irrigation water arrived in Lethbridge on 4 September 1900 at three o'clock in the afternoon. This event was first observed at the town boundary by a group of civic, government and company officials composed of Mayor Mewburn, Councillor Hyssop, Secretary-Treasurer Bowman, William Pearce, George Anderson, and P. L. Naismith. A. J. Fraser, a local land agent, and several citizens were also present. The opening of several bottles of champagne and congratulations all around were followed by the gauging of the amount of water which had arrived. The water continued its way slowly into the town where it flowed into the trenches dug along the streets. The rest of the citizens gave it another reception which was described by the *Lethbridge News*:

> There was rejoicing among the citizens, and the small boys of Lethbridge found the flowing water a source of much delight, little water-wheels being set going, and sundry fish that had injudiciously come in from the main canal to see the town being caught, a specially large one, of the "sucker" species, falling prey to the prowess of a youngster armed with a potato bag, with which he whipped the fish out of the runlet in front of the residence of Dr. Mewburn, the Mayor of Lethbridge. Mayor Mewburn was the first person in Lethbridge to plant trees on the lawns about his house and to lay pipes about his property from the irrigation trench running down the street on which he lived.[60]

OFFICIAL OPENING

The completion of the system north to Lethbridge was marked by the visit to the project by Governor-General, Lord Minto and Lady Minto on 14 September, 1900. Their tour of the project was a grander affair than that of Clifford Sifton in 1899. The vice-regal party, including representatives of the CNWICo., the State of Montana, the Great Falls and Canada Railway, the North-West Mounted Police and the *Manitoba Free Press,* left Lethbridge in the morning on a special train bound for Stirling. The group was then conveyed by carriages and horses to Magrath under less than desirable weather conditions. Addresses were presented at Stirling and Magrath by Presidents Cannon and Smith of the LDS church and lunch was provided at Magrath. Lord and Lady Minto departed from Lethbridge the next day following an address to the vice-regal couple on behalf of the Town of Lethbridge by Councillor Stovall. His address summarized the expectations of the community for their new irrigation system. Minto's visit, Stovall noted, was timely:

> in as much as you came amongst us at a season when the desires of long years are being crowned with fulfilment and a new era is opening for the Lethbridge district, which if our confident expectations are realized will gradually cause our now vacant plains to be studded with settlers, who cultivating our fertile soil under such favourable conditions as our temperate climate and an ample water supply can afford, will in time cause it to be regarded as one of the most beautiful and favoured portions of Canada.[61]

FOUR
OPERATIONS AND LAND SETTLEMENT 1901 - 1907

The construction of the main canal system through to Lethbridge was only one step in putting the St. Mary Project into operation. Building a system of laterals, obtaining a permit to divert water and the drafting of a standard water contract to govern the purchase of water rights were also required. The railway system had to be expanded specifically to serve the needs of the new irrigation-based agricultural community between Stirling and Magrath. Once the irrigation system was put into operation, its maintenance and modification were ongoing tasks. The operation and maintenance carried out during the initial period of operation included the rebuilding of portions of the system after the record amount of precipitation in 1902, the widening of the main canal and the addition of new structures. Control of the system also passed to a new company in 1904 with the creation of the Alberta Railway and Irrigation Company (AR&ICo.). It consolidated the four separate companies created since 1893 by Alexander and Elliot Galt to develop their coal and land holdings in the Lethbridge area.

Efforts to promote land settlement and assist the new water users in their adaption to irrigation agriculture included the publication of a number of pamphlets and the establishment of a model farm under the direction of William Fairfield. This farm was intended to encourage proper farming methods and to conduct irrigation crop research. The establishment of the Dominion Experimental Farm on land provided by the AR&ICo. created a permanent institution dedicated to this idea. The company's efforts to settle the land east of Lethbridge were assisted after 1901 by increased immigration, particularly from the USA, which created new urban communities such as Coaldale and rural communities such as Sunnyside. The company's co-operation with the LDS church was continued through a joint effort to start the sugar beet industry, the first example in Alberta of industrial development based on irrigation agriculture.

NEW CONSTRUCTION

The development of a network of laterals and sub-laterals off the main canal from Kimball began with the construction of the Magrath Lateral in the summer of 1900. This lateral ran from a point on the Pothole Coulee a mile west of the Magrath headgates to a location approximately three miles north west of the Town of Magrath. It served the land west of the town as well as Magrath itself. The original canal from Nine Mile Coulee east to Stirling was designated the Raymond Lateral. Extensions to the Raymond Lateral included a canal built in 1902 to the new Knight Sugar Factory which went into operation in 1903. The new Magrath and Raymond Laterals were designated part of the Southern Division of the St. Mary Project. The construction of a distribution system

Construction of Chin One Lateral, 1904.
(Glenbow Archives, Calgary, PD-310-24)

in the Lethbridge-Coaldale area, which was designated the Northern Division, continued after the construction of the canal into Lethbridge. The primary laterals in this system were Chin 1 and 2, Coaldale, East, Cameron, North East and North.

Operation of the St. Mary Project required a permit to divert water from the St. Mary River and the transfer of the federal government's water right to the CNWICo. A permit had been issued in 1897 when the federal government was planning the construction of an irrigation system using the St. Mary River (see Chapter 2). On 3 May 1899 the CNWICo. was authorised to divert 500 cfs of water from the St. Mary River during the low water stage and 1000 cfs during the high water stage. In March 1900, this permit was amended to permit removal of water flowing in Pinepound and Pothole Creeks as well as in Nine Mile, Middle and Kipp Coulees.

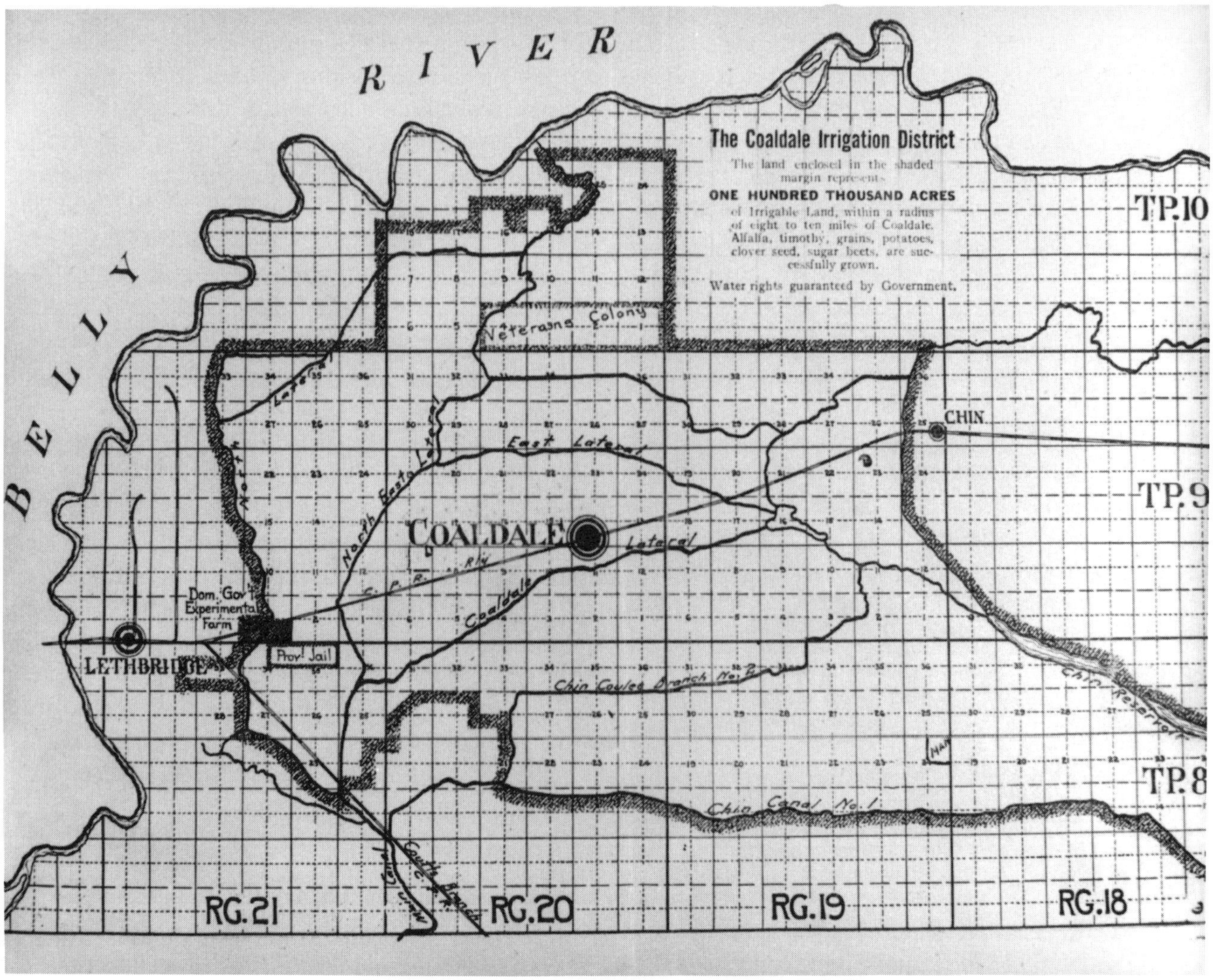

Map Two Lethbridge-Coaldale distribution system as it developed to 1914. (Coaldale: Gem of the West)

WATER CONTRACTS

The first contract to cover the rental of this water to the project's farmers was drafted by 1901. This document began with a statement of the CNWICo.'s right to divert water from the St. Mary River followed by the legal description of the land which was to receive the water. The total number of acres in the parcel were indicated but no distinction was made as to what portion of this land was irrigable and non-irrigable. For the payment of a one time charge of one dollar per acre plus the annual water rental fee, the company agreed to supply a stream of water throughout the irrigation season flowing at the rate of one cubic foot per second except when prevented from doing so by unusual storms, floods or other disasters over which it had no control. In modern terminology, this one dollar fee would have been considered the water right which was the farmer's contribution to the construction of the system. The fee for the rental was the water service charge which was the farmer's contribution to its annual cost of operation.

The annual fee was $150 for 1 cfs, deemed sufficient under the North-west Irrigation Act to irrigate 150 acres or approximately a quarter section. The rate of flow of water each farmer received and the rate he paid was dependent upon the amount of land he owned and the amount of land he wanted to irrigate. For example if the farmer owned a half section, he was entitled to receive water at the rate of 2 cfs or if he owned a quarter section but only wanted to irrigate 80 acres, he had the option to receive a stream of water flowing at the rate of ½ cfs. The amount due to the company was payable on or before 1 May of each year. Failure to take the water and pay the rental for two consecutive years made the contract null and void. The contract made no provision for an increase in the water rental rate. This lack of flexibility was an advantage to the water users but would prove to be an encumbrance for the company since it could not make any changes to reflect increased costs of operation.

Other provisions in the contract included the right of the company to shut off the water supply at any time during the irrigation season when canal repairs were required. The farmers were to be given reasonable notice, whenever practical, of any shut down for maintenance purposes. The water supplied was only to be used to irrigate the land described in the water agreement and no other land and for no other purpose. Under no circumstances was the water to be used for "mining, milling or mechanical power, or for any purpose not directly connected with or incidental to the irrigation of the said land." The company determined the location of all diversions from the canals and controlled the headgates which regulated the flow of water. The lessee was not permitted to allow water to run to waste or make any claims against the company for loss or damage to his property caused by seepage from its canals. The company was permitted to make rules for the operation of the system. Written notice to the company was required for the transfer of water rights to new owners of land. In a drought situation, when insufficient water was available to meet the contractual requirements, the company had the right to distribute the available water on a pro rata basis to all contract holders.

Water contract 1 was between the CNWICo. and the Municipality of Lethbridge for 5 cfs of water which the company was required to supply under the terms of the contract which brought irrigation to Lethbridge in 1900. Water contracts 2 to 850, which were signed between 1902 and 1908, covered the rest of the St. Mary Project from Magrath to Coaldale. The contracts for water supplied to land in the Magrath to Stirling area were distinctive because of the relatively small amounts of water supplied. Extensive subdivision of land during the creation of these communities created small parcels, which even with a full water right, received considerably less than one cfs. The decision to take out a partial water right on these small acreages reduced even further the amount of water to be supplied. The result was that the average amount of water delivered was 1/5 cfs with the lowest being 4/75 cfs for water contracts 624 and 700. The only exception were the Towns of Magrath and Raymond which received 3 and 7 cfs, respectively. Given the use of open ditches and carriage losses due to seepage and evaporation, it was physically impossible to accurately deliver such small amounts. The fact that these contracts were impractical because of the ridiculously small amounts of water desired by the farmers was noted by the CPR when it was requested to split a water right between two properties one of which was to receive 1/36 of a cfs while the other was to receive 5/36. In response to this request, the CPR indicated that it:

> could not, under any circumstances, issue water agreements for such amounts. 1/36 of a cubic foot of water would not commence to even

wet the ditch, leave alone run through it to anybody's place. We certainly cannot undertake to do the impossible.[62]

THE ST. MARY'S RIVER RAILWAY

The AR&CCo. expanded its railway network with a view to serving the needs of the settlers located along the canal west of Stirling. In June 1900 the St. Mary's River Railway Company (SMRRCo.) was incorporated by an Act of the Dominion Government to build a line from Stirling to some point on the international boundary between Ranges 24 and 30, south of Cardston. The shareholders of the company as identified in the Act, were Elliott Galt, Charles Magrath and Peter L. Naismith of Lethbridge, William Ramsay of Montreal and Alexander Ferguson of Ottawa. The operation of the line was provided for in Section 9 which authorised the new company to enter into an agreement with the CPR or the AR&CCo. This was the third railway built in the area but the first which was designed specifically to serve the agricultural settlement of the region. The first two railways built by the Galts were designed to meet the needs of their coal mining operations. The railway to Spring Coulee went into service on 28 November 1900 with an inaugural run by a special train carrying company officials, Mayor Mewburn, members of the Lethbridge Town Council and representatives of the Lethbridge Board of Trade. The group had a pleasant trip because of the quality of the construction which was appreciated particularly by Mayor Mewburn who remarked that when he went over the Crow's Nest line during construction, the train went off the track 10 times. When thanked for his hospitality, Naismith replied, "Gentlemen I thank you. It [was] a pleasure to have you along. I trust that you will pass the further extension of the road we are to make as satisfactorily as you have that which you have seen. If you are satisfied. I am gratified, and if you are gratified, I am satisfied. Have another cigar."[63] The railway was eventually extended southeast to Whiskey Gap and around the south end of the Blood Reserve to Glenwood by the mid-1920s.

CORPORATE REORGANIZATION

Given the commencement of operations it was necessary for the CNWICo to have a separate office. This was acquired in the summer of 1899 when a building described as the Builders' Co.'s Hall was converted from a "place of amusement" to an office for the irrigation company. It was used after 1904 by the Alberta Railway and Irrigation Company (AR&ICo.) The AR&ICo. was created in 1904 by Elliott Galt by merging the AR&CCo., the CNWICo., the SMRRCo. and the Lethbridge Land Company. The Lethbridge Land Company had been created in 1891 to administer the sale of land in the Lethbridge townsite. The head office of the new company was in London, England.

Office of the Canadian North-West Irrigation Company, 1904. It was located on the east side of Galt Gardens and was demolished to make way for what is now the Garden Market IGA.
(Sir Alexander Galt Museum & Archives, Lethbridge, Alberta)

OPERATIONS AND MAINTENANCE

Damage to main canal caused by heavy rains in 1902.
(Sir Alexander Galt Museum & Archives, Lethbridge, Alberta)

The task of rebuilding the system began much earlier than expected when floods in 1902 caused by unusually high amounts of rainfall destroyed the diversion structure at Kimball and portions of the main canal. When the canal was originally built, the headgates were constructed on the old river channel without any diversion dam across the river. The 1902 flood scoured out the river bed so that a dam was required to divert the flow into the canal. About 500 feet upstream from the original

Damage to main canal caused by heavy rain in 1902. (Sir Alexander Galt Museum & Archives, Lethbridge, Alberta)

Close-up view of new headgates at Kimball, 1912. (Glenbow Archives, Calgary, NC-2-160)

Repair work on canal in 1902.(Sir Alexander Galt Museum & Archives, Lethbridge, Alberta)

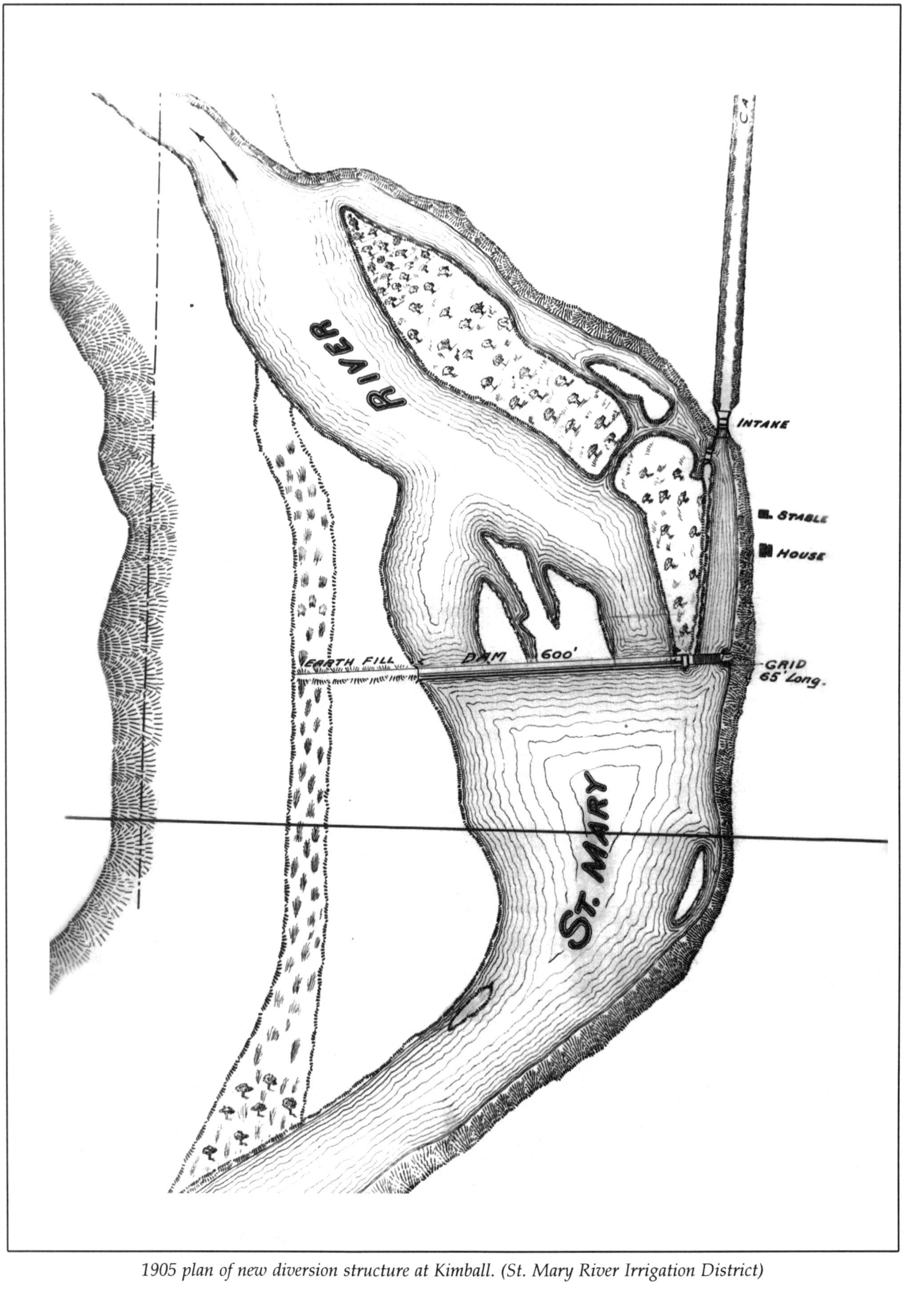

1905 plan of new diversion structure at Kimball. (St. Mary River Irrigation District)

New diversion structure at Kimball built after the flood of 1902.(Glenbow Archives, Calgary, NC-2-156)

headgate, a grid ice breaker and an overflow dam about 130 feet long were built. Shortly after these modifications were made, the St. Mary River overflowed its banks again. The flood waters scoured a new channel west of the new dam leaving an island about 300 feet wide between the end of the dam and the new channel. In 1905 a dam about 170 feet long was built across the new channel and a cut-off wall across the island connected the two

First dredge used in widening of the main canal, approximately 1906.
(Glenbow Archives, Calgary, NA-1323-16)

dams. Eventually, floods overflowed this cut-off wall and scoured out the gravel below. In 1908 the two old dams were connected across the island to make a continuous dam 600 feet long. Concrete headgates replaced the original timber headgates in 1908. The changes to the main canal system included widening the main canal to a capacity of 1,200 cfs. with the use of floating barges.

WATER USERS' COMPLAINTS

The CNWICo. had entered into these contracts with the expectation that the water users would play a significant role in the operation of the irrigation system through their management of the lateral system. The fact that the water users were not working together to achieve this objective became evident as early as 1906 when John B. Stocks, the Commissioner of Irrigation, received a letter, dated 16 February of that year, from the Raymond and District Board of Trade. The Commissioner was advised that, at the insistence of a large number of water users in the district, "they wanted to bring to the attention of the Government the very unsatisfactory state of affairs that prevailed in the matter of the distribution of water." According to the Raymond Board of Trade:

> the farmers themselves have no protection against one another and the man near the canal may if he sees fit take all the water he wants while those below him on the same lateral have no redress. The matter is one of great importance to the well being of the whole community and a matter that cannot be allowed to drag on through another season without some redress. We are confident that should you make an investigation of the management of the distribution of water you cannot fail to see that drastic remedies must be introduced at once to relieve the prevailing causes of discontent and loss to the farmers.[64]

To deal with this situation, the Board of Trade suggested a convention where the farmers could lay before the Commissioner some of the causes of their complaints and suggest remedies. Stocks replied that the *North -west Irrigation Act* of 1898 made full provision for the remedying of abuses, such as the illegal diversion of water by one water user to the detriment of another. Rather than call a convention, which would be legally powerless to act against these abuses, Stocks suggested that those suffering from such abuses apply for redress to the Department of the Interior in the prescribed manner and he would see that justice was done.

Stocks also wrote a personal letter to Frank Oliver, the new Minister of the Interior, in which he further outlined his position that if a specific complaint was received, the Department would use the mechanisms provided in the North-west Irrigation Act to deal with the situation. "We never hesitate to step in and enforce the Act when such is necessary" Stocks continued "but it had not been considered wise to take up other people's quarrels, and even now . . . we must insist upon a bona fide complaint."[65] Stocks concluded his letter with a comment which clearly indicated that at least some officials of the Department of the Interior, as well as the CNWICo. expected the water users to solve their own problems with respect to the operation of the lateral system. Stocks indicated that:

> there is no branch of the service where more petty complications arise between neighbours than in connection with the distribution of water under the irrigation system during the dry periods; that they steal from one another I have not the slightest doubt; but it is desirable that they should settle their quarrels among themselves, unless in the case of the Company discriminating, when it would certainly be our duty to interfere.[66]

The farmers in the Raymond area made a second attempt the following year to influence the operational procedures of the company. On 8 February 1907 the Irrigation Committee of the Alberta Farmers' Association made a number of recommendations which would have given the company a greater role in the distribution of water. The two most important being the maintenance and operation of all laterals by the company and the distribution of the water at the rate of 2½ cfs rather than 1 cfs. The committee elected to bring these concerns to the company's attention had no success in making any changes.

While the company was not prepared to intercede between individual water users, it took action to protect the interests of the farmers who had water contracts from those people who simply took water without an agreement with the company. John Stewart, who succeeded Stocks as Commission of Irrigation in the fall of 1906, reported in 1906 that the illegal diversion of water had reached such proportions that the company was forced to patrol the canals with men on horseback to protect the interests of those who bought water from the company.

RAYMOND: THE SUGAR CITY

By 1900 the first three steps in the creation of the Alberta irrigation industry had been completed with the construction of the main canal to Stirling and Lethbridge, the construction of the St. Mary's River Railway as far west as Spring Coulee and the establishment of the towns of Magrath and Stirling. After 1900 Charles Magrath and Mormon entrepreneur Jesse Knight took the process one step further with the construction of a sugar beet processing plant at a third new townsite which would become Raymond.

The extraction of sugar from beets was first accomplished by German chemist Andrew S. Marggraf in 1747. The first commercial extraction of beet-sugar was undertaken by one of his students, Karl Franz Achard, the son of a French refugee in Prussia. The beet-sugar industry was established in Europe on a paying basis as a result of the Napoleonic Wars. Napoleon's blockade against Britain cut off Europe's chief source of sugar cane, forcing the cultivation of large quantities of sugar beets as a substitute for sugar cane. Following

Construction of canal into the Knight Sugar Company, 1903. (Provincial Archives of Alberta, Edmonton, A 7998)

the Napoleonic Wars, the industry continued to grow with Germany becoming the world's largest producer by 1878.

Sugar beets were first grown in America about 1830 at Ensfield near Philadelphia. The first attempt to bring the crop to the American West was made by John Taylor. He studied the beet-sugar industry in Europe during a mission to France for the LDS Church, and in 1852 he purchased a complete outfit of machinery for making beet-sugar and shipped it to Provo, Utah. The machinery was later relocated to Salt Lake City. Production difficulties, however, forced the project to be abandoned in 1855. The first successful operation was established by E.H. Dyer in Alvarado, California who was able to put it on a paying basis by 1879. The LDS Church was also successful with a sugar factory it built at Lehi, Utah in the 1890s.

The cultivation and processing of sugar beets in Canada began in the late 1870s and early 1880s. The first plant built to extract sugar from beets was built in 1881 by La Compagnie de Sucre de Betterave de Quebec at Farnham, Quebec. Based upon the expectation that the beet sugar industry would be important to the entire country, the newly created Central Experimental Farm in Ottawa began a systematic seed testing program in 1888. Three of the four types of seed tested that year were provided by Wilfred Skaife, President and Manager of the Berthier Sugar Beet Factory of Berthierville, Quebec. The fourth type of seed tested was the Wanzleben variety, obtained from Haage & Schmidt of Enfurt, Germany. These programs continued throughout the 1890s with seeds from Germany and France being used in most of the experiments. The experimental farm at Indian Head, North-West Territories, became involved in the testing program in 1890 when it sowed three varieties of beets along with two varieties of sugar cane. This is the first documented instance of sugar beet cultivation in the prairie West. The sugar cane was the more successful of the two crops, but the testing of sugar beets rather than sugar cane continued through out the 1890s.

The first documented instance of sugar beet growing in the Lethbridge area was in 1894. An article in the 3 October 1894 issue of the *Lethbridge News* described a sample as "syrup crudely made

Knight Sugar Company factory at Raymond, 1903. (Raymond Museum)

from the juice of the sugar beet grown here." From this small beginning, the paper confidently predicted the establishment of sugar factories in the North-West Territories especially in those regions fortunate enough to have irrigation. The *Lethbridge News* indicated that these experiments were being continued and that samples were being sent to the Central Experimental Farm in Ottawa for analysis. While the newspaper accounts do not identify who was responsible for these initial experiments, it may have been Charles Magrath. In June 1892 John W. Taylor had suggested to Magrath that the area was suitable for the cultivation of sugar beets. He may have also provided seeds since sugar beets were being introduced into Utah at that time.

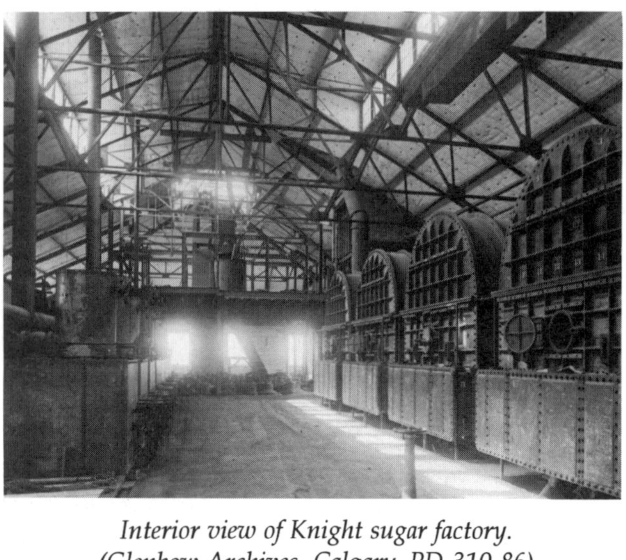

Interior view of Knight sugar factory. (Glenbow Archives, Calgary, PD-310-86)

In March 1900, the CNWICo. distributed a large quantity of sugar beet seed free to farmers in the area. Farmers living within 15 miles of Lethbridge picked up their sample packages at the company office while samples were sent to outlying area farmers upon request. The *Lethbridge News* immediately got behind the venture by encouraging the full cooperation of the farmers. It pointed out that the seeds had been provided "with the hope that each will sow the seed and give it a thorough test, and thus see whether or not sugar beets can be satisfactorily grown in this district. Extreme care, therefore, should be taken by all in the matter in order that the very best results may be obtained. A great deal depends upon the outcome of this experiment." [67] The *Lethbridge News* continued to pass on advice to the farmers. In a letter to the editor published in the 5 April 1900 issue, J. J Gibb of Magrath, a nine year veteran of growing beets in Utah, advised on cultivating the land and planting the seeds. In November, samples were sent to the Dominion Experimental Farm in Ottawa and to the Utah Sugar Company at Lehi for analysis. The Central Experimental Farm also received samples

Knight Sugar Company Barn. (Raymond Museum)

for testing from Manitoba and Prince Edward Island. The Alberta sample consisted of four large plants grown at Magrath and nine smaller plants grown at Stirling. The plants had all been grown on sod rather than on cultivated land and without the benefit of irrigation since the system was still not in operation. The difference in size was accounted for by the fact that Magrath received more rain than Stirling. Based on the analysis of these samples, Frank Shutt, the chemist at the Central Experimental Farm, concluded that the results were exceedingly good despite the fact that they grown on sod. He did not think, however, that "it would be safe to consider them as necessarily indicating that the Lethbridge district would always yield beets with a high sugar content."[68] He recommended that further data be obtained before any further conclusions were drawn. The test results from Lehi also indicated that the sugar beets were very satisfactory.

The results of these tests combined with the completion of the irrigation system and the growth of the industry in Utah encouraged Charles Magrath to pursue the establishment of the sugar industry in southern Alberta. He repeated the approach that had proved so successful in 1897 when he and Elliott Galt negotiated the contract with the LDS Church to build the canal system. Magrath returned to Salt Lake City in 1900 to expand the partnership with the LDS Church to include the growing and processing of sugar beets. His visit with John W. Taylor, who had not lost any of his enthusiasm for the future of Alberta, led to a meeting with Jesse Knight, a wealthy entrepreneur of Provo, Utah. Knight was responding to the call of the LDS church leaders to use his fortune to provide financial assistance to the Church and its members through corporate ventures.

Magrath and Taylor succeeded in interesting Knight in southern Alberta as a place to invest. Following an inspection of the area by his two sons, Knight's first investment was the purchase of 30,000 acres of land near Spring Coulee. The expansion of this investment to include a sugar factory and additional land was first discussed during his visit to Stirling in 1901. During this visit, Charles Ora Card, John W. Taylor and Charles Magrath joined in promoting such a venture. Knight's concerns about its success were overcome by the enthusiasm of John W. Taylor and his own desire to use his wealth for philanthropic purposes. The result was an agreement signed on 10 July 1901 with the CNWICo. by which Knight agreed to buy more than 200,000 acres of land including the Kirkcaldy Ranch and build a sugar factory to be operational by 1903. The agreement also required him to keep the factory in operation for at least 12 years.

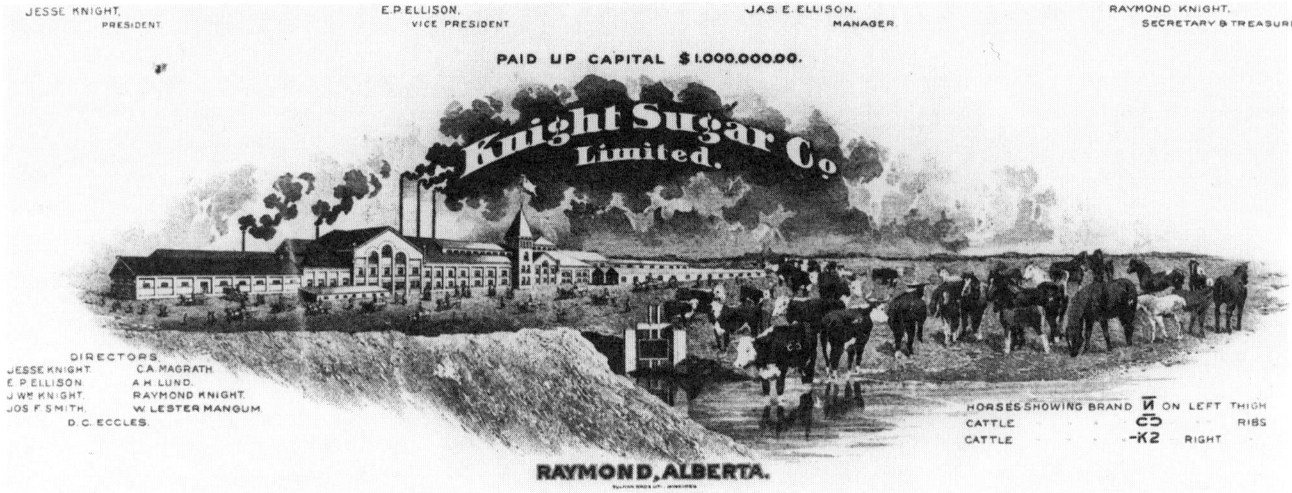

Knight Sugar Company stationary. This letterhead illustrates the sugar and ranching interests of the company. (Glenbow Archives, Calgary, M 1574 File 1916)

The implementation of the agreement involved establishing a town, creating a company to oversee the growing and processing of the beets, constructing the processing plant and organizing the production of sugar beets for processing in the plant. The development of the townsite, which was named after Jesse Knight's eldest son Raymond, was begun immediately. The town grew rapidly and by the spring of 1902 had over 400 residents. The movement of Mormon settlers to Raymond and area was not driven by religious persecution or by missionary calls as had been the case with Cardston, Magrath and Stirling, but by the economic opportunities created by the construction of the sugar beet plant. Immigration was also encouraged by the general prosperity of western Canada which was experiencing a significant economic boom caused by railway construction and a large increase in immigration from Europe and the United States.

The settlement of Raymond was also distinct because of the incorporation of Mormon principles, which were strongly supported by Knight, in the land purchase agreements. The titles included a caveat that in the case of the establishment at any time of a saloon or drinking place on any part of the property, it would revert to the original owner.

The Knight Sugar Company was created in the fall of 1902. Jesse Knight and his friends held three-fourths of the stock while E.P. Ellison and friends held one-fourth. The stock holders met at Will Knight's residence in Raymond on 15 October 1902 to elect the officers for the company and complete the arrangements for its organization. Jesse Knight was elected President and Ephriam P. Ellison was elected Vice-President and General Manager. The company was incorporated for the purpose of processing sugar beets but in 1903 it also acquired the Knight Ranching Company and the Bar K2 Ranch thus making it a combined sugar beet and ranching company. Knight returned to Utah shortly after the creation of the company, leaving the management of the company in the hands of Ellison. Ellison was a very successful entrepreneur in his own right from Layton, Utah, who had established or served on the board of directors of a number of successful companies related to the processing of agricultural products including sugar beets. In 1893 he raised 5 acres of beets for the new processing plant at Lehi. In 1897 he became a director of the Ogden Sugar Company and in 1902 a director of the Amalgamated Sugar Company. The choice of Ellison to oversee the plant's construction and initial operation provided a bonus for the economic development of the Raymond area since he established the Ellison Milling Company.

Ellison's initial task was supervising the construction of the plant, which began in October 1902 under the direction of E.H. Dyer Company of Cleveland, Ohio, and continued through to September 1903. Organizing production began in October 1902 with the decision to sell land at $6 per acre in 40 acre and 80 acre plots. The land purchase contract included a commitment to grow not less than 10 acres of sugar beets each year. Land to be used for ranching was located south of the townsite. Farmers found that growing sugar beets was backbreaking toil compared to producing wheat. They had to plow, level, and seed the land. Then, labourers had to thin them by hand, constantly hoe and cultivate between the rows and irrigate several times. Harvesting was equally arduous. The workers had to plow up the beets, knock off the hardened soil, pile the beets into windrows, cut off the tops by hand, load the beets by hand into wagons, haul them to the factory, unload them and haul back the dirt.

The first beet crop was planted in the spring of 1903 and successfully harvested and processed by November. The first year of operations was concluded on a very successful note, according to both the *Deseret Evening News* and the *Lethbridge News*. In 1904 production was down to 3,600,000 pounds but recovered to 4,622,900 pounds in 1905. The success of the Knight Sugar Company from both its sugar beet and ranching operations permitted the payment of an 8% dividend in 1904, 1905 and 1906. During these years the Knight Sugar Company also brought in the first Japanese labour.

LAND SETTLEMENT IN THE NORTHERN DIVISION

The contracts with the LDS church and with Jesse Knight were the first means used by the CNWICo. to occupy their irrigated land. The Mormons who arrived under the terms of these contracts settled primarily in the Southern Division which was now made up of the Magrath, Raymond and Stirling areas. In 1899 the company began its efforts to settle the land in the Northern Division with a sales campaign in Ontario. It undertook this task with the help of Lethbridge Presbyterian Minister, the Rev. Charles McKillop who went to Hamilton in the spring of 1899 as a delegate from the Presbytery of Calgary to the General Assembly of the Presbyterian Church. He began his sales campaign while still in Hamilton when he was able to arrange for one of the speakers at the Assembly to make a presentation on the value of irrigated land in the Lethbridge area. The speaker "extolled it highly . . . and in conclusion strongly advised the ministers to use their influence to induce any who intended to emigrate to settle around Lethbridge."[69] This presentation, according to McKillop, was listened to by hundreds of clergymen and laymen, besides a large number of the citizens of Hamilton. During the three months following the Assembly, he was able to get statements from:

> some twenty-seven of Ontario's brightest farmers, to the effect that they would become settlers of the Lethbridge district during the next few months, and from many others as well who will come in a course of a year or so. In addition to this he was able to send far and wide throughout this American continent literature and verbal tidings of the vast area of first-class land to be had in this country and the bright prospects now opening up for all who wish to take advantage of them.[70]

Model farm of William Fairfield.
(Glenbow Archives, Calgary, PD-310-138)

The company also produced traditional promotional literature which included three pamphlets entitled "Irrigated Lands in Southern Alberta." These publications encouraged investment in the area by stressing the advantages of its climate, the value of its agricultural land, the extensive markets for agricultural produce, railroad facilities and its community services. These publications also discussed methods of putting water on the land to give prospective farmers an early start on how to farm the land using irrigation. One of these fact sheets referred to the irrigated lands of southern Alberta as "the Colorado of Canada" and included a letter from Levi Hacker addressed to the CNWICo. which praised the value of the project.

The efforts of the CNWICo. to complete the settlement of the land east of Lethbridge also involved the creation of a model farm. The individual responsible for suggesting the idea of such a farm was William Harmon Fairfield, another American whose talents were made excellent use of by the CNWICo. Fairfield grew up on an irrigation farm in Colorado. After obtaining Bachelor of Arts and Master of Science degrees from the College of Agriculture at Fort Collins, he became an assistant professor of agriculture at the University of Wyoming.

Dominion Experimental Station.
(Glenbow Archives, Calgary, NA-3092-44)

Fairfield's long and significant association with the development of southern Alberta agriculture began as a result of a visit to Lethbridge by his brother, who wrote to William suggesting that he negotiate an agreement with the CNWICo. whereby it could help them establish a farm in the area. Fairfield acted upon his brother's suggestion and wrote Charles Magrath offering to start an irrigation demonstration farm if the company would provide the necessary equipment and seed. The idea appealed to Charles Magrath who arranged to meet Fairfield in Salt Lake City to negotiate an agreement. The memorandum of agreement was signed on 1 October 1901. The preamble of the agreement indicted that the CNWICo. was "desirous of establishing a model farm for the benefit of persons settling on their irrigated lands." [71] The company agreed to erect buildings, build fences and advance moneys for the purchase of a team of horses, harness and such machinery and other chattels as might be necessary for the cultivation of the land. Fairfield agreed to take charge of the farm, devoting his whole time to his duties for a three year period during which time he would receive all the profits derived from the sale of the produce of the farm. At the end of this period, he had to repay the money advanced and had the option to purchase the farm at the rate of $8 per acre. Fairfield had to lay out the land in a practical way so that the farm was a model for practical farmers. He was required to raise field and garden crops together with trees and small fruits which were likely to do well in the Lethbridge area. He was to give special attention to the growing of alfalfa. Five acres had to be sown the first year of operation with a total acreage of 20 acres to be sown by 1904. In addition to operating the farm, he was required to instruct settlers in the proper use of water.

The farm developed rapidly under the enthusiastic efforts of Fairfield. An inspection of the farm in October 1903 by a reporter from the *News and Alberta Irrigationist* revealed that he had more than lived up to the terms of the contract since 300 acres had been broken to grow fodder crops such as alfalfa and grains such as wheat and flax. Apple and plum trees were grown on land near the house. The protection provided by the house from the wind was Fairfield's explanation for the success of these fruit trees. Fairfield also successfully grafted the Wealthy variety of apple onto a crab tree. Gooseberries and strawberries along with vegetables and flowers were grown. Elm, Russian poplar, ash and maple trees were grown to serve as windbreaks. An additional 2000 to 3000 trees were also being grown for use in the shade tree planting program in Lethbridge. The farm's livestock included pigs, chickens and cattle.

The crop which received special attention from Fairfield and the one which he would continue to promote throughout his career was alfalfa. Fairfield's enthusiasm for this crop was noted by the reporter who indicated:

> although naturally pleased with the abundant grain crops and minor success in fruit and other lines, it is when pointing out the luxuriant growth of that wonderful fodder plant, alfalfa, that Mr. Fairfield becomes enthusiastic and dilates at length upon the effect the successful growth of this plant will have upon farming operations in the future. [72]

The success of cultivation of this crop, as Fairfield explained, was the presence of a microbe on the root of the plant which helped in the storage of nitrogen. The spread of this microbe would ensure the spread of alfalfa production. His view of the importance of this crop was further indicated by his plans to shift towards its increased production at the expense of grain since cereals he said, only depleted the soil.

Fairfield also took an active role in the community and in the promotion of irrigation. He spoke at meetings of the agricultural society and wrote articles for the *News and Alberta Irrigationist* on irrigation and the crops and agricultural economy which could be based on it. In one such article he wrote about the potential for growth in the cattle industry given the nature of southern Alberta's climate and the production of alfalfa using irrigation. He said irrigated farming was the ideal way to

John Hamilton was one of the first of many successful irrigation farmers in the Coaldale area. (Glenbow Archives, Calgary, NA-2919-5)

furnish winter feed to the cattle industry. Fairfield also participated in various organizations created by the farmers in the area to deal with problems which arose in the operation of the irrigation system. He was a member of the Irrigation Committee of the Alberta Farmers' Association, the group which tried to persuade the CNWICo. to operate all laterals and distribute water at the rate of 2.5 cfs rather than 1 cfs.

DOMINION EXPERIMENTAL FARM

The concept of the model farm was made a permanent feature of the Lethbridge area with the creation of the Dominion Experimental Station which was well equipped and financed to deal with the issues facing the development of irrigation agriculture in southern Alberta. Agitation for the creation of such an institution in Lethbridge had begun as early as 1893 but no decision was made until 1905 because of competition amongst the various Alberta centres. The great debate over where the first experimental station in Alberta should be located was resolved in 1905 when the AR&ICo. donated 400 acres of land along with water rights to the federal government. The operations of the Experimental Farm were directed by Fairfield who was appointed director on 1 August 1906 and by Asael Palmer who joined the staff in 1921 as assistant director. Fairfield retired in 1945 and was succeeded by Palmer who remained until 1953. Fairfield divided the operation of the farm into irrigated and dryland portions and conducted parallel experiments in the growing of crops on both farms. Under the leadership of these two individuals the Experimental Farm investigated a number of issues related to irrigation agriculture including the duty of water for specific crops, the best method of crop rotation and ways to counter soil drifting.

LETHBRIDGE: GARDEN CITY OF THE WEST

With the arrival of irrigation in Lethbridge, the town council provided for its distribution, the implementation of the tree planting program and continued its efforts to promote irrigation. A decision on the rates to be charged for the water was made in the fall of 1900. The monthly charge ranged from 25 cents for a lot less than 33 foot in width to $1 for a property between one half and one acre in size. An additional dollar was charged for every acre thereafter. The charge for supplying a livery stable was fixed at $50 for the season. This charge was made deliberately high in order to deter "the said business from using the water so as to conserve it for other purposes."[73]

The town council helped promote the development of irrigation by placing advertisements in the *News and Alberta Irrigationist* throughout 1901. The first ad simply drew the attention of people to the availability of irrigated land in the Lethbridge area while the second was more dramatic. It advised its readers that there was "No drought here!" and that "every man [was] his own rainmaker." It invited people desirous of changing their location to investigate the grand crops grown under irrigation in the district. All enquiries were to be promptly answered by town Secretary-Treasurer C.B. Bowman. A number of copies of the newspaper were sent by the council to Canadian government immigration agents in the USA.

The promotional efforts of the company were assisted by the *Lethbridge News*. The *Lethbridge News* changed its name, beginning with the 29 November 1900 issue, to the *News and Alberta Irrigationist*. This change in name indicated that it was now its intention to "present the news of the irrigated district to our readers and to provide material bearing on the subject generally; and by making *The News* a welcome addition to the many new homes which will soon be dotting our prairie . . .." The *News and Alberta Irrigationist* also carried as a regular promotional feature after 20 February 1902 a summary of the town's development which now included its economic and physical transformation because of irrigation. According to the *News and Alberta Irrigationist,* Lethbridge was on its ways to becoming the "Garden of the West."

IDEAL FARMERS AND IDEAL FARMS

The development of the model farm by Fairfield southeast of town near Six Mile Coulee was followed by the development of similar farms in the immediate vicinity. The "ideal" farm of D. J Whitney was located to the north and west of the model farm while Edgehill Farm of the Rev. Mr. White was located to the south. Both individuals operated similarly to Fairfield. Whitney was particularly ambitious in his efforts to use irrigation. He had originally come to the Lethbridge area in the 1880s at which time he purchased a farm eight miles from town. In 1902 he purchased a 470 acre farm 3 miles from town for $20 per acre in a deliberate effort to create an irrigated farm. Irrigation for Whitney was "insurance against failure" and he believed it was a "sound investment to pay a little premium for this insurance." In the year he purchased the farm, he built a brick house which has survived and is located south of the Sportsplex.

Whitney's efforts to demonstrate the potential of irrigation were featured in a special 1908 promotional issue of the *Lethbridge Herald.* The article on Whitney entitled "An Ideal Farm And An Ideal Farmer" began by pointing out that "while farmers in the West have been able to produce splendid results by following crude methods, Mr. Whitney belongs to the class who makes a study of his profession and puts brains into the soil." Whitney operated a very diversified farm which grew wheat, timothy hay, alfalfa, potatoes, vegetables and small fruits. Spring wheat using Turkey Red seed was the most successful wheat crop. The yield on his timothy hay was 2 tons to the acre which he sold for $14 a ton in the stack. He cut two crops of alfalfa in June, August and September and made enough money to pay the interest on the entire farm. He had equal success with his vegetables. His farm yielded 300 bushels to the acre of "The Irishman's Delight." An acre of berries on his farm in 1907 yielded a revenue of $1,600. An eighth of an acre of strawberries produced 100 quarts of strawberries. His garden also included water melons, squash and pumpkins. He planted 13,000 elm, cottonwood, maple, ash and Russian poplar seedlings as part of his tree cultivation program which he started in 1904. Some of these trees by 1908 were 20 feet high and 6 inches in diameter. His farm included hogs, chickens, ducks and bees. As a "poultry fancier" he raised barred Plymouth Rocks which were "spoken of all over the Province" and purchased by the government for the Provincial Poultry Farm. He sold his chickens for $5 to $6 each and eggs for $1.50 a dozen. Whitney's 15 colonies of bees were the only domesticated bees in southern Alberta. Whitney's success was not only in financial terms but he was consistently awarded prizes for his poultry and ducks. The Inspector of Forestry for the provincial government awarded Whitney a prize for his trees which made the best shelter belt in the province.

COALDALE: GEM OF THE WEST

The settlement of the land east of Lethbridge was facilitated by the immigration of many Americans from North Dakota and Illinois to the area after 1904. One of these people was the Rev. B.F. Cokley from Illinois who purchased 28 sections of land from the AR&ICo. in the area of Coaldale, the first station east of Lethbridge. A company was formed to administer the land with Cokley's son-in-law, H. A. Suggit, as secretary-treasurer. Before the plans of the company could be put into operation, however, Cokley died and a new company was formed which included Mrs Cokley, H.A. Suggit and W.H. Fairfield. A group of settlers from North Dakota brought to the area by Suggit in 1905 were the first purchasers of land in the area. The first individual in this group to actually locate on the land was S.A. Hjelmstad of Leeds, North Dakota, who broke 160 acres in 1906.

The townsite of Coaldale, which was to be the service centre for this new area, was surveyed in 1906 and put on the market shortly thereafter. Advertisements for town lots focused on what irrigation could do for a community. It was going to make Coaldale into a thriving little city. Irrigation meant that there was no chance of crop failure thus there was no chance of a business depression or hard times. It meant intensive farming based on dairying, gardening, stock feeding, sugar beet and potato cultivation. Irrigation meant a more thickly settled rural community because the farms would be small. The need to pay the water rates also ensured that the land would be kept busy. Irrigation-related industries such as a canning factory, sugar factory, potato starch factory, and flour mills would also be established.

FIVE
DIVIDING THE WATERS
1900 - 1921

Irrigation in Whoop-Up country began in the 1870s with American projects on the lower Milk River. Canadian government support for the St. Mary Project was driven largely by the desire to ensure the use of the St. Mary River by Canada before the Americans accessed it as a source of additional water for their Milk River projects. The successful launch of the St. Mary Project in 1898, with the assistance of a Colorado irrigation engineer and the Utah based LDS Church, gave Canada a prior claim to water from the St. Mary River. Having won the race, the Canadian government issued a permit to the CNWICo. to divert a volume of water from the St. Mary River which precluded American use of the river for irrigation. The USA had effectively laid claim to the Milk River and Canada had taken possession of the St. Mary River. Accommodating these claims to reflect geography and the needs of the two countries was the challenge after 1900.

The seriousness of the American threat to the water supply of the St. Mary Project depended, to a large extent, upon the willingness of the US federal government to help Montana implement its water transfer scheme. In the 1890s, the Americans had just begun to develop a national program for irrigation development in the West and thus the federal government could offer only limited support. The situation changed dramatically by the turn of the century as the Americans began to implement an extensive land reclamation program in the West under the provisions of the *Reclamation Act* of 1902.

Canada met the renewed American threat after 1900 in the same way it had in the 1890s, by supporting the plans of a private company to make use of the water while at the same time pursuing a diplomatic solution which would ensure the equitable partition of the water supply. Support for the CNWICo. took the form of the sale of 500,000 acres of land northeast of the Milk River Ridge and an authorisation to divert water from the Milk River. In 1902, the diplomatic approach to resolving this issue began when Canada requested the USA not to proceed with the diversion of the St. Mary since it would be injurious to Canadian interests. The negotiation of the terms of an understanding was completed 19 years later by the International Joint Commission.

IRRIGATION EXPANSION IN THE AMERICAN WEST

The adoption of a comprehensive national plan for irrigation development in the USA began in 1897 when Captain Hiram M. Chittenden, of the American Army Corps of Engineers, made a report on the potential reservoirs in Colorado and Wyoming. The Chittenden report was the first official advocacy of government participation in irrigation development in the USA. He proposed that the federal government construct the reservoirs, administer the distribution of water and leave the states to manage the irrigation. By 1900 the national political parties in the USA supported federal aid to reclamation of land through irrigation. The result was the passage of the *Reclamation Act* in 1902 which provided for the creation of a reclamation fund from the sale of public lands in Arizona, California, Colorado, Idaho, Kansas, Montana, Nebraska, Nevada, New Mexico, North Dakota, Oklahoma, Oregon, South Dakota, Utah, Washington and Wyoming. The fund was to be used in the examination and survey for and the construction and maintenance of irrigation works for the storage, diversion and development of waters for the reclamation of arid and semiarid lands in these states. The Secretary of the Interior was authorized to use this fund to undertake this work.

Even before the passage of this legislation, the United States Geological Survey began the investigation of the Montana proposal to divert the St. Mary River for possible inclusion in this national program. The results of their investigation were presented to the hearings of the House of Representatives, Committee on the Public Lands, in January 1901 by F. H. Newell, Chief of the Hydrography Division of the United States Geological Survey. The American government version of the St. Mary Project, as explained by Newell, was similar to the plan developed by the Great Northern Railway in the early 1890s which intended to make use of the north branch of the Milk River to transport water from the St. Mary River to the lower Milk River via Canada. In his testimony, Newell indicated it was not possible to divert the water on the Canadian section of the Milk River for use in Canada.

CANADIAN MILK RIVER CANAL

The government of the North-West Territories initiated the Canadian response to the revival of the American version of the St. Mary Project. In 1901 it surveyed a 500 cfs canal to take water from the Milk River to irrigate approximately 350,000 acres east of the Milk River Ridge. Suds, Weston, Tyrrell and Verdigris Lakes were identified as potential storage facilities which would allow the retention of the total high water or flood flow of the Milk River. The plan made use of the coulees in the area to maintain a year-around flow of water thus raising the value of the adjacent territory for stock-raising. On completion of the survey, the territorial government issued a permit to itself to divert water from the Milk River into this canal in order to ensure that it had priority in any future use of this river. This procedure was a duplication of the federal government's actions

Construction crew of Adelbert Cazier on the Canadian Milk River Canal, 1903-1904. (Glenbow Archives, Calgary, NA-323-2)

Construction of the Milk River Canal, 1903-1904. (Glenbow Archives, Calgary, PD-310-21)

between 1895 and 1897 when it surveyed the St. Mary Canal and issued itself a permit to divert the water.

Completed portion of the Milk River Canal. (Glenbow Archives, Calgary, PD-310-19)

In August 1902, the territorial government advised the federal government of developments in the United States and the action it had taken to counter the threat. Based on this information, J.S. Dennis, Deputy Commissioner of Public Works for the North-West Territories suggested that:

> the facts should be intimated to the American government simply from the standpoint of the injustice that will be done to the Canadian North West Irrigation Company if the water of the St. Mary River is diverted without any reference to the record which they have made against the stream on our side of the line. At the same time it should, of course, be intimated to the American Government that if it is necessary to do so the rights which we claim to have in this stream can be protected to a certain extent by preventing the use of the channel of the Milk River in so far as it is situated in Canada for the conveyance of any water which they propose to divert from the St. Mary River into that stream for the purpose of again diverting from the Milk River after it crosses the international boundary . . ..[75]

Dennis suggested that the Americans might build a diversion canal which would eliminate the need to use the Canadian portion of the Milk River but he felt that it was a "scheme which will cost such a large sum as to be entirely outside reasonable financial limits until land values in the West have increased enormously."[76]

CANADIAN NORTH-WEST IRRIGATION COMPANY PLANS

From the moment the Americans announced their plans, Elliott Galt and Charles Magrath were deluged with information about them along with counter-strategies from William Pearce and George Anderson. Pearce initiated this correspondence in December 1901 when he sent a letter to Charles Magrath enclosing an extract from the 1901 report of the Secretary of the Interior which described the American diversion plans. According to Pearce, the report did not make clear what the Americans were going to do with the water once it reached the north branch of the Milk River. If they let it go north via the Milk River, Pearce suggested, "we might take it out." Pearce further speculated that:

> possibly the United States government imagines that having once put it into that stream, they are entitled to take out as large a quantity as they put in, or in other words, that we Canadians should not take out any of that is thus put in, having once adopted the principle of diverting water from their natural channels, I cannot very well see how they logically can object to our diversion of the Milk River waters. However the whole question of utilization by another country of international waters is in a very unsatisfactory condition.[77]

In January 1902, Pearce sent Magrath a letter he had received from F. H. Newell, Chief Hydrographer, United States Geological Survey. Newell concluded his letter with the statement that they had not arrived at a decision as to the feasibility and desirability of the project. Despite Newell's comments, Pearce felt that the Americans were already committed to the project despite its high cost because for "a wealthy country like the United States nothing is prohibitory if they make their mind to do it, particularly if by so doing they prevent Canada benefiting to the extent she would were it left undone."[78] If the project was built, Pearce suggested, the Belly and the Waterton Rivers could be diverted to compensate Canada for its loss of water to the Americans on the St. Mary and Milk Rivers. Pearce was the first to make this suggestion which was ultimately the direction of the St. Mary Project.

George Anderson provided Elliott Galt with an extract from a speech he had given to the Denver Chamber of Commerce as well as copies of background papers on the American proposal. In these documents, he concluded that the American version of the St. Mary Project was expensive but that it would probably be built given the level of public interest. Anderson's suggestion on how to counter the plan was called the "high level canal from the St. Mary." This plan involved the construction of a dam on the St. Mary River one mile north of the international boundary. According to Anderson, it would be possible and reasonably cheap to construct a solid masonry dam at this point because there was a abundance of rock in the canyon immediately adjacent. A rock fill dam or a crib dam of timber and rock were also possible. The construction would be facilitated by its close proximity to the route of the proposed extension of the St. Mary's River Railway thus permitting the delivery of construction materials and labour at reasonable cost. The dam was to be 113 feet above the bed of the stream with a spillway incorporated into its design. The reservoir created behind the dam would be three miles long and average about 3/4 of a mile in width. It would thus extend for 1.5 miles into the United States. The capacity was estimated at 60,000 acre-feet. A canal system would take the water further east to the Milk River where it would be removed by another canal for Canadian use.

Based on this information, the CNWICo. made a proposal to the federal government in August 1902 for the extension of the St. Mary Project by 500,000 acres to the north and east of the Milk River Ridge. The proposal had the full support of J.S. Dennis who indicated to Clifford Sifton, Minister of the Interior that:

> it is, of course, possible that if representations are made, as originally recommended, to the American Government through the proper channels, that they may re-consider the decision to proceed with this undertaking, but in view of the political pressure which it would seem is being brought to bear as indicated by the American newspapers to have this project proceed with I am of the opinion that it would perhaps be well to protect the rights which have been granted to the Canadian North West Irrigation Company by encouraging them to

> proceed with the construction of the Milk River canal on the understanding that we will in that way utilize any water delivered into that stream by the American Government.[79]

Dennis also provided a blind copy of this letter to Elliott Galt in order "aid him in his negotiations" with the federal government.

The result of this close cooperation between CNWICo., the territorial government and the federal government was an agreement dated 12 December 1902 which provided for the expansion of the system already built as well as the sale of 500,000 acres east of the Milk River Ridge at $3.00 per acre. The company was given a credit to the amount of $300,000.00 for the expenditures it already made on construction of the irrigation works from Kimball. The company was given a further credit of $700,000.00 on account of the future expenditures it was to make in enlarging and extending the irrigation system in the west. The remaining $500,000.00 was to be paid in 10 equal instalments of $50,000.00 each, the first of which was to be paid on 1 December 1907 with interest at 5% from the first day of December, 1905. This agreement resulted in a third revision of the CNWICo. permit to divert water. The company was permitted to divert an additional 500 cfs from the St. Mary River as well as 500 cfs at the low water stage and 1,500 cfs at the high water stage of the Milk River. Rolph or Willow Creek, a minor stream located on the west side of the Milk River, was also added to the permit.

AMERICA WILL PROTECT

Initial discussions between Canada and the USA over dividing the water took place between 27 October and 29 December 1902. The exchange was initiated at the request of and with information supplied by the territorial government. The Canadian position was that Canada had built a canal capable of diverting the total amount of water in the St. Mary River available for irrigation and that this amount could be beneficially used in Canada. Canada also indicated that it would remove any water from the Milk River which the Americans put into it by their diversion canal.

The American response was that Canada did not have at present nor would it have in the future the capacity to divert the total water supply of the St. Mary River. The American plan, in their view, was an advantage to Canada since it not only protected the Canadian supply but would also prevent further damage from floods such as those of 1902. Both countries indicated that the plans of the other country would be disastrous for their own citizens.

The federal government referred the matter back to the territorial government for clarification rather than forcibly restating Canada's intention to maximize its use of the St. Mary River as indicated by the commitments it had made to the CNWICo. This failure to press its stated position reflected at least two factors: the view held by the federal government since 1895 that there was no principle in international law which could be used by Canada to prevent the Americans from carrying out the diversion of the St. Mary River; and support within the Department of the Interior for the American position. The Surveyor General felt that since the Americans proposed to store and utilize only the flood waters of the St. Mary River which were not used in Canada, the rights of the CNWICo. were not affected.

The Canadian response, which was sent on 6 January 1903, simply made reference to an earlier dispute concerning Boundary Creek. The Canadian government hoped that the Americans would show as much consideration to Canada in the present situation as Canada had shown to the Americans over Boundary Creek. The Americans dismissed the Boundary Creek dispute as being totally irrelevant and repeated their view that their plan for the St. Mary River "would avoid all interference with the amount of water to which the Canadian canal was entitled."[80] This view was unfair to Canada since the prior vested rights recognized by the Americans were to be strictly limited to the carrying capacity of the Canadian canal at the time of construction which was 300 cfs. No recognition was made by the Americans of the planned expansion of the canal by the CNWICo. as provided for in its 12 December 1902 agreement with the federal government.

INFORMAL DISCUSSIONS

The failure of diplomacy at this stage to resolve the issue led to negotiations between October 1903 and June 1905 by senior staff from the United States Geological Survey and Elliott Galt and Charles Magrath who were acting on behalf of the CNWICo. without any formal authority from the government of Canada. These informal discussions were initiated by Cyrus C. Babb who was the official in charge of the American surveys in connection with the diversion of St. Mary Lakes. In October 1903 he made a trip to Lethbridge to meet Elliott Galt and Charles Magrath. During his trip, the Canadians and Americans exchanged information on their plans and discussed an agreement between the two countries.

Following Babb's visit to Lethbridge, F. H. Newell, Chief Engineer of the United States Geological Survey, indicated in a letter to Charles Magrath that these informal discussions should be continued, based on the recognition that any correspondence and general understanding which might be reached was of course to be considered entirely unofficial and a wholly personal matter between themselves. Despite the unofficial nature of these talks, Newell hoped that they could be part of the process leading up to a formal consideration of the matter by the proper authorities of the two countries.

In this same letter, Newell suggested that a certain minimum flow be allowed to cross the international border and that the rest be diverted into the Milk River. He also suggested that Canada could take some water out of the Milk River while letting the rest go to the USA. Magrath's response to Newell indicated that he could not deal with the situation without passing it before the directors of the company in England but that he wanted to talk to him informally. Various other suggestions were made by both parties at two subsequent meetings which were held in Washington.

CANADIAN CONSTRUCTION AND AMERICAN PROTEST

While these informal discussions were taking place, the CNWICo. began construction on the Canadian Milk River Canal which was part of an overall plan prepared by George Anderson to use water from the St. Mary and Milk Rivers to irrigate the land east of the Milk River Ridge. The plan included the construction of a dam just north of the international border and a diversion canal to carry the water east to Whiskey Gap. A second dam was planned for the Milk River near Whiskey Gap. From the reservoir created by this dam a canal was to be built which transported the water to what is now called Shanks Lake. The capacity of Shanks Lake was to be increased through the construction of dams. The water was then to be carried further east by natural streams eventually ending up back in the north branch of the Milk River. A system of canals and reservoirs was planned to take water out of the Milk River and convey it around the Milk River Ridge to what is now Ridge Reservoir. The intake was located west of the town of Milk River. From the Milk River Canal a distribution system took the water further east to irrigate land in what is now the Warner area. Brunton and Verdigris Lakes were to be developed as reservoirs in this system. The planning for the Milk River main canal system of the St. Mary Project by Anderson also included his examination of what he called "pak-ogh-kee Lake" known today as Pakowki Lake.

The route was surveyed in November 1903 and construction began in December when a contract was let to Adelbert Cazier for the first seven miles of the canal. A second contract for the construction of the next seven miles was let to H. S. Allen of Raymond in the summer of 1904. Construction continued through to November 1904 at which time tests of the new canal revealed major seepage problems. These engineering difficulties and the concerns of the Americans prompted the company to stop construction. Contrary to the popular view, the CNWICo. had every intention of completing the project. It concentrated its efforts, thereafter, on the system which had been built west of the Milk River Ridge. The destruction of the headgates by a flood in 1908 rendered use of the system even more remote. The CNWICo. spent $250,000.00 on the construction of the 16 miles of canal before it abandoned the project. The company was able to retain the 500,000 acres which this system was designed to irrigate.

The construction of the Canadian Canal resulted in a protest by American irrigation farmers at Chinook, Montana in May 1904. This protest was passed on by the American government to the Canadian government which merely stated that authorization granted to the company allowed them to divert water from the Milk River.

INTERNATIONAL BOUNDARY WATERS TREATY

In the fall of 1904 Elliott Galt and F. S. Newell decided that they should make an effort to have their respective governments resume negotiations. In a letter to the Minister of the Interior dated 22 September 1904, Galt indicated that he was:

> anxious that some arrangement be made as to the proper division of these waters for we are convinced that if nothing be done the United

> States will build one of . . . the two canals . . . which will lead to difficulty later and meantime it will become public property that the United States are going to divert our water which although in a measure impracticable, would undoubtedly cause uneasiness among our settlers as to their future water supply and also disturb the minds of our proprietors who are supplying the funds for the prosecution of our very extensive enterprise.[81]

Galt ended his letter with the suggestion that a treaty be arranged that would give Canada two-thirds of the waters of the St. Mary River and the Americans one-third.

The American attempt to initiate negotiations involved included the suggestion in a 30th September letter that a meeting be held to discuss the issue. In March 1905, the Americans made a further attempt to initiate negotiations when Dr. Charles D. Walcott, Director of the United States Reclamation Service made a personal visit to Ottawa. This visit had the opposite effect since the government of Sir Wilfred Laurier decided that the resolution of the Milk and St. Mary Rivers issue would be dealt with as part of an effort to deal with all boundary water issues.

The negotiations for this treaty were subsequently carried on by Dr. W.F. King, Dominion Astronomer and F.H. Newell. The treaty which was ratified by both countries in 1909 included article 6 which read as follows:

> The high contracting parties agree that the St. Mary and Milk Rivers and their tributaries (in the state of Montana and the Provinces of Alberta and Saskatchewan) are to be treated as one stream for the purposes of irrigation and power, and the water thereof shall be apportioned equally between the two countries, but in making such equal apportionment more than half may be taken from one river and less than half from the other by either country so as to afford a more beneficial use to each. It is further agreed that in the division of such waters during the irrigation season, between the 1st of April and 31st of October inclusive, annually the United States is entitled to a prior appropriation of 500 cubic feet per second of the waters of the Milk River or so much of such amount as constitutes three-fourths of its natural flow, and that Canada is entitled to a prior appropriation of 500 cubic feet per second of the flow of the St. Mary River so much of such amount as constitutes three-fourths of its natural flow.
>
> The channel of the Milk River in Canada may be used at the convenience of the United States for the conveyance, while passing through Canadian territory, of waters diverted from the St. Mary River. The provisions of article 2 of this treaty shall apply to any injury resulting from the conveyance of such waters through the Milk River.
>
> The measurement and apportionment of the water to be used by each country shall from time to time be made jointly by the properly constituted reclamation officers of the United States and the properly constituted irrigation officials of His Majesty under the direction of the International Joint Commission.

The treaty also created the International Joint Commission (IJC) which had the responsibility to interpret the meaning of the various sections.

INTERPRETING THE MEANING OF ARTICLE SIX

Implementation of the treaty proved to be difficult when the accredited officers of the two governments responsible under the treaty were unable to agree upon the interpretation of Article 6. With the approval of the IJC an informal conference was held in Washington in April 1915 between the reclamation and other officials of Canada and the USA for the purpose of outlining a basis of settlement. The main points of contention which remained after the meeting related to the water covered by the treaty and priority. The Canadian position was that the waters to be divided included the St. Mary and Milk Rivers and all their tributaries or the total flow of the combined watershed down to the mouth of the rivers. The United States held that the waters to be divided were those waters only which flowed across the international boundary. The Canadian position on the priority issue was that in the division of the waters of these streams the priorities amounting to three-quarters of the natural flow up to 500 cfs should be set aside and the remainder of the flow then divided equally. The American position was that the equal division should include these priorities.

Meetings were held through to 1921 when the

IJC met at Chinook, Montana, on 15 September and at Lethbridge on 17 September. The Chinook meeting was the occasion for a demonstration by American farmers which included a special train which carried people from the town of Glasgow and points west. They had a band and banners, one of which read, "We want water." A group of 350 school children also put on a demonstration which included more banners one of which read, "Canada has rye, we want water." At this meeting, F.P. Stranahan of Fort Benton used a story to illustrate his view:

> Once upon a time an Indian and a white man went shooting. They agreed to make an equal division of the spoils at the end of the day. On their return their bag contained a turkey and a crow. Said the white man to the Indian, 'Now, either I'll take the turkey and you take the crow, or you take the crow and I'll take the turkey.' The Indian pondered this equal division for a minute then said, 'You no talk turkey to me at all.' That's about the situation in this controversy," Mr. Stranthan concluded. "Canada has the turkey and we have the crow." [82]

According to the *Lethbridge Herald*, the Chinook meeting left a solution to the problem as far away as ever. The only positive item to come out of the meeting, it said , was Montana Governor Dixon's suggestion that the two rivers should be considered as one, not only in the matter of the average flow but in the possibility of impounding all flood water in reservoirs on the St. Mary and Milk River Basin. This suggestion was endorsed by Charles Magrath, the former Land Commissioner of the AR&ICo. who was now a Canadian representative on the IJC.

The Lethbridge meeting of the commission was organised by the Lethbridge Chamber of Commerce and the CPR. Magrath began the session by declaring that if the commission failed to settle the question in the near future it would have to return the treaty to the two governments for clarification. The Lethbridge meeting did not include the hoopla which was organized by the Americans at Chinook. The IJC was, however, pressed by all the Canadian delegates for a decision to be made. This advice was also provided by Premier Greenfield and George Marnoch of the Irrigation Development Association.

The results of these meetings was an order issued by the IJC on 4 October 1921 respecting the measurement and apportionment of the water of the St. Mary and Milk Rivers. With respect to the St. Mary River this order had the following provisions:

(a) During the irrigation season when the natural flow of the St. Mary River at the point where it crosses the international boundary is 666 cfs or less Canada shall be entitled to three-quarters and the United States to one-fourth of such flow.

(b) During the irrigation season when the natural flow of the St. Mary River at the point where it crosses the international boundary is more than 666 cfs Canada shall be entitled to a prior appropriation of 500 cfs and the excess over 666 cfs shall be divided equally.

(c) During the non-irrigation season the natural flow of the St. Mary River at the point where it crosses the international boundary shall be divided equally between the two countries.

The order had the following provisions for the Milk River:

(a) During the irrigation season when the natural flow of the Milk River at the point where it crosses the international boundary for the last time is 666 cfs or less the United States shall be entitled to three-fourths and Canada to one-fourth.

(b) During the irrigation season when the natural flow of the Milk River at the point where it crosses the international boundary for the last time is more than 666 cfs the United States shall be entitled to a prior appropriation of 500 cfs and the excess over 666 cfs shall be divided equally between the two countries.

(c) During the non-irrigation season the natural flow of the Milk River at the point where it crosses the international boundary for the last time shall be divided equally between the two countries.

In addition to making a ruling on how Canada and the United States would divide the water, the International Joint Commission on October 6, 1921 also recommended that the governments of the United States and Canada enter into an agreement for the construction of a reservoir at St. Mary Lakes in Montana to control the flow of the St. Mary River and that the Reclamation Service of the United States proceed with the construction of the proposed Chain-of-Lakes Reservoir in Montana and the Canadian Reclamation Service build the proposed Verdigris Coulee Reservoir.

SIX
ACQUISITION AND OPERATION BY THE CANADIAN PACIFIC RAILWAY 1908-1918

In1908 the CPR purchased a controlling interest in the AR&ICo. thus giving it control of the St. Mary Project. This purchase culminated a long period of cooperation between the two companies in the development of southern Alberta and gave the CPR a virtual monopoly of the irrigation industry in Alberta. The Bow River Project, then under development by the Southern Alberta Land Company, was the only major irrigation development in Alberta not under its control. This change of ownership affected the St. Mary Project through to 1946 in terms of the way the land was settled as well as the way the system was expanded and maintained. The commitment of both the AR&ICo. and the CPR to the operation of the sub-laterals by the farmers and the terms of the water agreements ensured continuity in the day to day operations of the project.

The CPR became involved in irrigation with the expectation that the increased productivity and diversity of irrigation agriculture would generate a sufficient increase in traffic to justify the expense. The company hired knowledgeable people such as William Pearce, J.S. Dennis and George Anderson to design, build and operate one of the largest irrigation projects in North America. Unlike the AR&ICo., the CPR constructed their entire system over a short period rather than developing it incrementally. Such an approach increased the chance of mistakes and increased the cost and inconvenience of their subsequent correction.

The CPR encountered a number of problems in their attempt to use irrigation to develop a mixed farming agricultural economy in the area east of Calgary. One of the reasons for this failure was the lack of cooperation from the water users in implementing mixed farming practices and the growing of such crops as alfalfa. Disagreements over land classification and the way the CPR delivered water led to a full scale revolt by a number of the farmers in the Gleichen and Strathmore area after 1911. Similar problems occurred in the St. Mary Project, but the relationship between the company and the water users did not deteriorate to the point of rebellion. The water agreements, which were insurance against water rate increases were an important reason for the relative peace in the St Mary Project. The difficulties, particularly in the western division, reduced the willingness of the CPR to invest in the expansion of their entire irrigation system and ultimately led to its decision to end its involvement in the industry between 1935 and 1946.

IRRIGATION AND THE CPR

The CPR's association with the development of irrigation in Alberta began in 1894 when it considered the development of a 3,000,000 acre irrigation project between Calgary and Medicine Hat. As discussed in Chapter One, legislation was passed in 1894 to create a solid block of land to facilitate this project. The CPR, however, did not proceed with these plans at that time because of concerns it had with sections of the North-west Irrigation Act which could have led to the CPR losing their land if they failed to complete the project on time. The CPR directors also wanted to complete the Crow's Nest Pass Railway before starting any new projects.

In the autumn of 1899, the CPR and the federal government resumed their negotiations for the development of an irrigation project east of Calgary. The success of the St. Mary Project was one of the factors that led to the successful conclusion of these negotiations. Construction of the western section of the development began in 1903 and on the east in 1910.

The efforts of the CPR to settle its irrigated land included the publication of a series of brochures. The first brochure, published in 1906, indicated that the western section of the CPR block would become the "Future Home of the most closely settled and prosperous mixed farming, stock raising and dairying community in Canada." The publication of brochures with the same message continued through to the 1920s.

In addition to the production of brochures, which described the type of agricultural economy they wanted to establish, the CPR took other steps to ensure that the land was settled and not used for land speculation. A program of assisted settlement was initially conceived by Thomas Shaughnessy, the President of the CPR during a trip to England in 1904. During this visit, he was advised that the better class of farmers in the British Isles could only be induced to emigrate if they knew that they would have a home immediately on arrival. The CPR did not pursue the idea until 1909 when the Salvation Army of England indicated to the CPR that it was interested in sponsoring a program of assisted land settlement in Canada. The Salvation Army proposed that they should obtain English families of small means to be located on Canadian Pacific lands with aid from the CPR. A proposal for assisted settlement was also made at the same time by Father Van Aken who had been engaged in colonization work for the company in Holland. He also indicated that many Dutch farmers could be attracted to western Canada if they could be put on farms improved to the extent of a house to move into and a certain amount of land under cultivation.The idea was enthusiastically endorsed by the CPR management which immediately began planning its implementation.

In the fall of 1909, recruiting of British settlers for the program began with a publicity campaign. At the same time, the company developed 24 quarter sections which had been selected for the Salvation Army Colony. The CPR built a house and barn on each farm, fenced the land, dug a well, and prepared a portion of the land for seeding in the spring of 1910. Named Nightingale, this community became the first of the ready-made farm colonies of the CPR. A total of 1,622 farms were created between 1909 and 1914 under this program. While the original intent was to concentrate on recruiting British settlers, American settlers were settled on these farms as well.

As a further aid to settlement, the company established demonstration farms within the irrigation block starting in 1905 with the first located at Strathmore. These farms were similar to the model farm of Fairfield and the Dominion Experimental Farm established at Lethbridge in 1906. The staff undertook experiments on the growing of a variety of crops under irrigation and provided advice to farmers throughout the irrigation block. The company made a special effort to encourage the growing of alfalfa. It imported alfalfa seed and retailed it at cost to the farmers and provided an alfalfa expert to give advice on its cultivation. Farmers were encouraged to include livestock in their operations and the company provided seed grain to prevent the growth of noxious weeds.

When the western section of the CPR irrigation block went into operation in 1911, the CPR based its administration on cooperation between the water users and the company. Their view of the importance of cooperation was outlined in the 1912 issue of its rules and regulations. This document noted that irrigation projects in America succeeded in exact proportion to the development of the cooperative spirit amongst the water users and between them and the company. Cooperation in agriculture, it continued:

> is manifestly a sound principle, and the nature of irrigated communities is such that they lend themselves particularly well to co-operation by virtue of the denser settlement and the many common interests that exist. It is a wellknown fact that there is less friction on irrigation systems where the farmers themselves enter actively into the business of the delivery of water

and maintenance of the distributary system than where such is wholly under the corporation.[83]

CONSOLIDATION OF CONTROL

The AR&ICo. was purchased by the CPR in order to ensure that the AR&ICo. would continue to provide traffic to the CPR mainline. This change in ownership had no immediate impact on the operation of the irrigation or railway system. The *Lethbridge Daily Herald*, however, regretted this change since, in its view, it made Lethbridge a one railway town thus reducing its status in an era when economic growth was measured by the number of railways which passed through a community.

The integration of the AR&ICo. into the overall operations of the CPR came in 1912 with the creation of a Department of Natural Resources. This new department consolidated all the company's land management offices in Calgary. Peter Naismith, the former General Manager of the AR&ICo, was appointed manager of this new department which administered the coal mining, timber, irrigation and land settlement operations of the CPR. The administration of the CPR's three irrigation projects was placed under the Engineering Branch. The superintendent of each project reported to the Chief Engineer. The other branches of the Department of Natural Resources directly involved in the ongoing management of the St. Mary Project were Agriculture and Animal Industry, Colonization & Development, and Accounts.

ORGANIZATION OF THE ST. MARY PROJECT UNDER THE CPR.

When the newly created Engineering Branch of the CPR's Department of Natural Resources took over the management of the St. Mary Project in 1912, it inherited the administrative structure established by the AR&I. Given the fact that the CPR and the AR&ICo. had similar management philosophies, it is unlikely that the new owners made many changes. The CPR, therefore, continued to depend on the water users to maintain the system of sub-laterals, a large contingent of seasonal workers to carry out maintenance work in the spring and fall and a small permanent staff.

In 1918, the permanent office and field staff consisted of Superintendent of Operations and Maintenance, Sam Porter, along with an Assistant Engineer, Chief Clerk, a stenographer, a chauffeur-mechanic, two canal superintendents, eight water masters, a ditch rider, a gate keeper and a caretaker. The seasonal staff, which totalled 192 in 1918, was paid as much as $4.50 per day because of the scarcity of labour caused by the First World War.

The permanent staff was located in a number of offices and buildings distributed from Kimball north to Lethbridge and east to Chin Coulee. The gate keeper was stationed at Kimball in order to operate the diversion structure on the St. Mary River. A water master and the superintendent for the Southern Division were stationed at Magrath. Their duties included the operation of the diversion structure south of Spring Coulee, the Magrath diversion structure and the Magrath Lateral. The water master at Welling operated the Raymond Lateral in cooperation with the farmers. The headquarters for the St. Mary Project continued to be the office building in Lethbridge acquired in 1899 by the CNWICo. The company also had a garage in downtown Lethbridge for servicing its fleet of five cars and a one-ton truck. The day to day operations of the Northern Division and the overall maintenance of the system were directed from the Home farm. Located east of Lethbridge and south of the Dominion Experimental Station, the 300 acre Home farm housed the company's horses and grew food for their use as well as for the work camps operated by the company. A caretaker was the only staff member permanently located at the farm. In 1918 the home farm housed 46 horses many of which were described as being "old and not able to do steady heavy work."[84] Their condition prompted the company veterinarian to recommend that 20 head be sold and replaced with younger animals. The remaining staff were stationed east of Lethbridge at Chin headgates, Royal View and south of Coaldale. Communication between the various offices was provided by a copper circuit telephone system which dated from the turn of the century.

OPERATIONS AND MAINTENANCE

The maintenance of the system up to 1918 included the replacement of existing structures. Among the structures included in this program was the flume on the main canal over Willow Creek (also called Rolf), the first major structure east of the Kimball headgates. It was a very large structure, 656 feet long and 27 feet wide with a water depth of 6.5 feet at full supply level designed to permit the passage of the floating dredges used in

Concrete diversion structure built in 1916 south of Spring Coulee to replace a wooden structure. (Sir Alexander Galt Museum & Archives, Lethbridge, Alberta)

the maintenance of the canal. Three carpenter crews were kept busy repairing and renewing the wooden drops and checks on almost all the canals. The 1918 *Annual Report* indicated that "practically all the structures are of timber and many have reached the end of their life."[85]

In a limited number of cases the CPR replaced wooden with concrete structures. New concrete diversion structures were built at Kimball and south of the town of Spring Coulee where the water was diverted to the east of the town into the west arm of Pothole Creek and where water was diverted into Nine Mile Coulee.

Wooden drop structure in Pothole Creek which was built to slow down the speed of the water and thus reduce erosion. (Sir Alexander Galt Museum & Archives, Lethbridge, Alberta)

New structures added to the system included an inclined flume with a baffle chamber at the bottom which was constructed in 1914 from the end of the Northeast Lateral into a coulee leading to the Oldman River. This new structure was intended to reduce the erosion of the coulee. The most important new construction work was the modification of Chin Coulee to permit its use as a storage reservoir. That stretch of Chin Coulee located northeast and southeast of the Readymade area originally consisted of five lakes which on the early maps were designated from north to south Lakes One, Two, Three and Four with the last, Upper Willows. Prior to the modification of Chin Coulee, the AR&ICo. had used the natural channel from Malloy Lake to waste water from its Chin Two and Coaldale Laterals into the south end of Lake One. Chin Two Lateral originally tailed out into the south end of Lake Four.

In 1911 the AR&I began the construction of a dam between Lake One and Two and a diversion canal which redirected the water coming from Malloy Lake via the natural channel around the height of land into Lake Two which had now become the reservoir behind the dam. The AR&I also built a dam on the height of land between Lakes Two and Three at Chin Butte. This work involved excavating a cut 25 feet deep and several hundred feet long. Two 36 inch pipes were then placed in the cut and a dam constructed over the pipes with the earth excavated from the cut. The water in this reservoir was controlled at the dam by means of mechanically operated valves through a reinforced concrete valve tower on the top of the dam. The reservoir behind the dam included the former Lakes Three and Four. The area around the south end of Lake Four where water flowed in from Chin One lateral was also modified to permit its redirection into Upper Willows Lake as the need arose. This gave the AR&I complete control of their water supply.

A number of timber drops were built in the natural channels to reduce the harmful effects of erosion where grades were steep. Contrary to the expectations of George Anderson, the natural channels did not erode to solid material and stabilize. The high rate of erosion of the bottoms and sides of the natural channels was caused by the high velocity of the water which had to drop about 900 feet as it went from the intake at Kimball to the Coaldale area. The eroded material was carried to the main canal and laterals where it was deposited because of the water's slower speed. The 1918 *Annual Report* indicated that this was "the basis of most of the trouble which has been experienced with the canals silting up and the consequent shortage of water. This condition will recur repeatedly until the cause is removed by building more drops and protection works in the natural channels."[86]

Cleaning out laterals of the accumulated silt and widening the main canal were ongoing tasks. In the fall of 1918 between 60 and 130 teams were at work cleaning out the Coaldale, Chin 1, Chin 2, Six Mile Coulee, Magrath, Raymond and Cameron Laterals. Enlarging the main canal from Nine Mile Coulee to Chin Reservoir, completed in 1918, gave the main canal a capacity of 1,000 cfs from Kimball through to Chin Reservoir.

Weed control was another ongoing maintenance task. In 1918 this problem was made worse by the drought. As the 1918 *Annual Report* indicated,

Later type of dredge used to clean out and widen the main canal. This dredge is working near Kimball. (Sir Alexander Galt Museum & Archives, Lethbridge, Alberta)

Dredge working on the Magrath to Welling portion of the main canal. (Sir Alexander Galt Museum & Archives, Lethbridge, Alberta)

"many fields seeded to grain did not produce enough crop to handle and were permitted to go to weeds. Russian thistles especially are bad as they seem to thrive in dry years when they are not crowded out by growing crops."[87]

Despite the amount of work required and the interruptions of service in the Coaldale area, Superintendent Porter was confident that the maintenance program was adequate to deal with all problems encountered in the operations of the system in 1918. He predicted that the St. Mary Project would "enter the next season with the system in much better condition than it was last year and barring accidents will be able to deliver all the water required."[88]

1918 IRRIGATION SEASON

The 1918 irrigation season was affected by drought. There was no rain in May, June or July. The result was a total failure of practically all non-irrigated crops and reduced yields on well irrigated land. The demand for water was the highest in the history of the system. The increased demand began in May when the use of water went from 125 cfs to over 500 cfs by the end of the month and remained high until July. In August the demand declined as the farmers shifted their activities to harvesting. The water demand increased again in October as the farmers began to fall irrigate. The system was shut down on 11 October in preparation for maintenance work. Water diversions averaged 350 cfs for the entire irrigation season. The only operational difficulty despite the increased demand was a temporary water shortage in the Coaldale area caused by the silting of Chinland 2. No problems, however, were encountered because of an inadequate supply of water from the St. Mary River.

SETTLEMENT PROGRAMS

With the inclusion of the St. Mary Project in the irrigation operations of the CPR after 1912, its settlement policies were also now applied to the Lethbridge region. Lethbridge was thus included in the pamphlets which the CPR produced as part of its effort to attract settlers. Its 1912 brochure entitled "Get Your Canadian Home from the Canadian Pacific" provided details on the length of

Standard house plan No.1A and standard barn No. 7 (Glenbow Archives, Calgary, Na-2829-6)

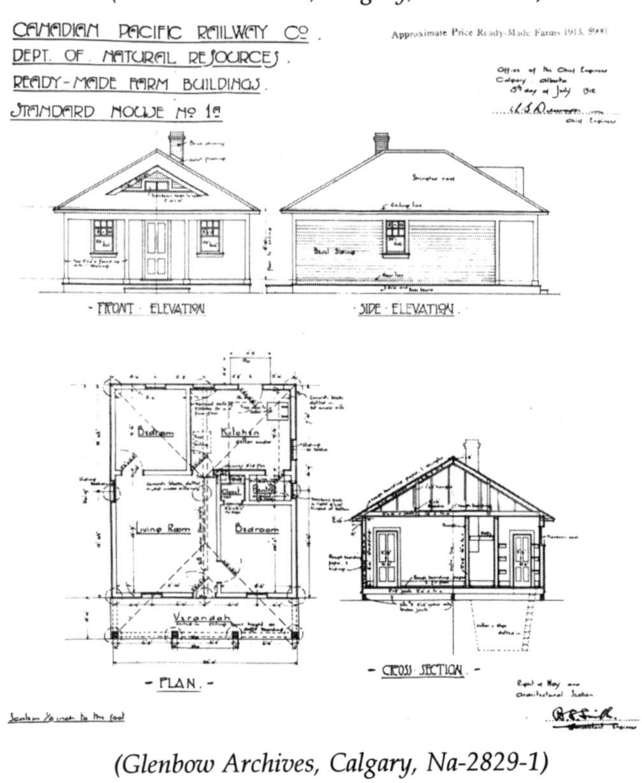

(Glenbow Archives, Calgary, Na-2829-1)

the canal system, the amount of land under irrigation, which was overestimated at 100,000 acres, and the fact that the area was well served by railway facilities.

The irrigation section of this pamphlet also dealt with the type of farms and crops which the CPR felt were appropriate to the irrigated regions of southern Alberta. Because irrigable and dry land was generally mixed together, the CPR encouraged the development of what it described as combination farms which would include both types of land. The land above the canal was to be used for the production of winter wheat and the grazing of livestock while the irrigable land below the canal was to be used for the production of crops which required abundant moisture such as alfalfa, barley, timothy, sugar beets and vegetables. Combination farms were "one of the best agricultural propositions on the North American Continent." The idea of combination farms had been suggested first by William Pearce in the 1880s as the best way to maximize the use of the land.

The readymade farm program was introduced into the Lethbridge area in 1912. The intention of the CPR to do this was announced on 9 May 1912 by J.S. Dennis, Assistant to the President. He indicated that 17 readymade farms of 160 acres each with buildings and improvements were to be

Standard house plan No.6 and standard barn No.1
(Glenbow Archives, Calgary, Na-2829-18)

Barn No. 7
(Glenbow Archives, Calgary, NA-2829-22)

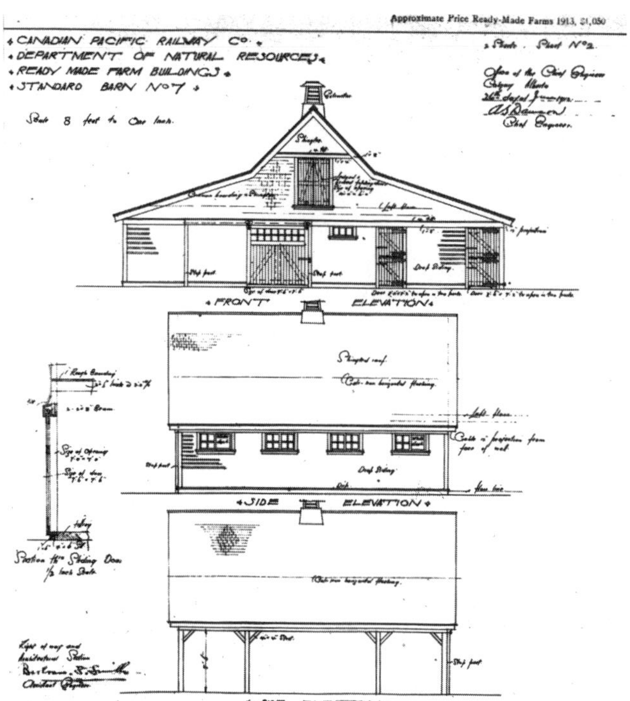

(Glenbow Archives, Calgary, Na-2829-31)

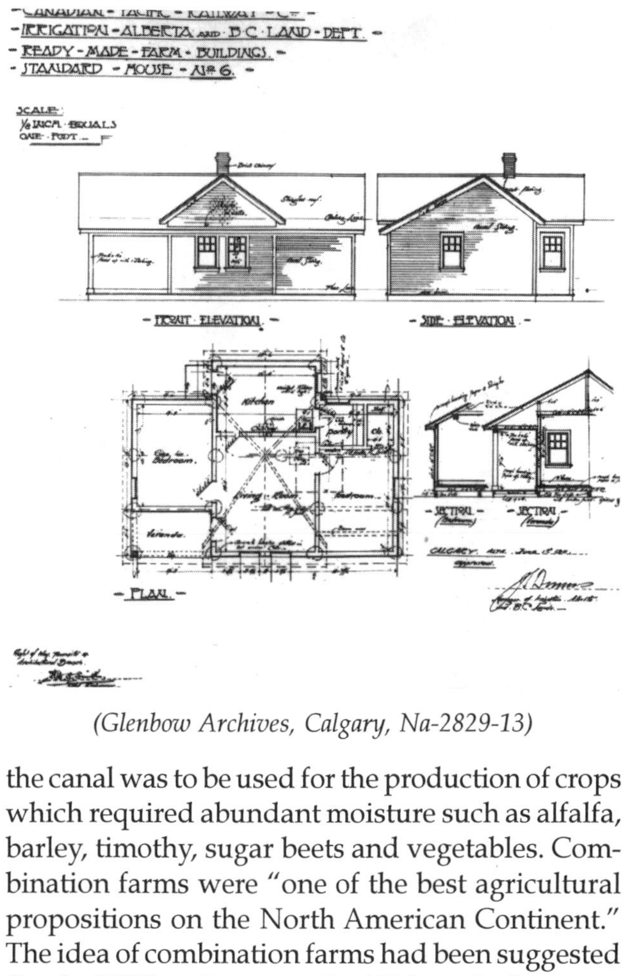

(Glenbow Archives, Calgary, Na-2829-13)

placed in 1912 with the occupation of the farms to follow the following year. The ready-made farms in the Coaldale colony followed Plan 7. The land development program was under the direction of the Agriculture and Animal Industry Branch of the CPR's Department of Natural Resources.

In addition to the farms on the Coaldale Colony which were intended for sale, the CPR retained four sections of land for use as a company farm to grow hay for shipment to the CPR lumber camps in British Columbia. The buildings on the farm included a bunk house for the men, barn, machine sheds and the farm house for the cook. The equipment on the farm included 8 rakes, 8 bull rakes, 2 overshot stackers and 11 mowers. The manager was also expected to provide advice and assistance to the farmers who had purchased the readymade farms in the area.

Standard house plan No.4 and standard barn No.1 (Glenbow Archives, Calgary, Na-2829-12)

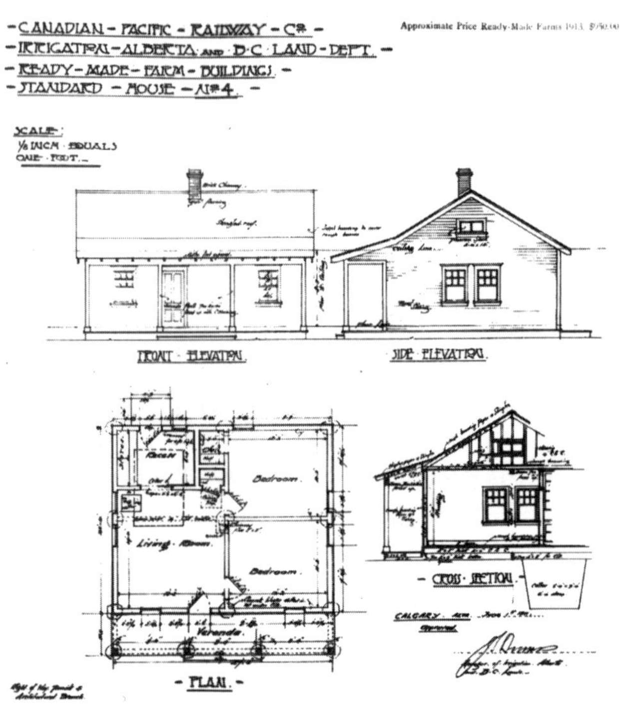

(Glenbow Archives, Calgary, Na-2829-7)

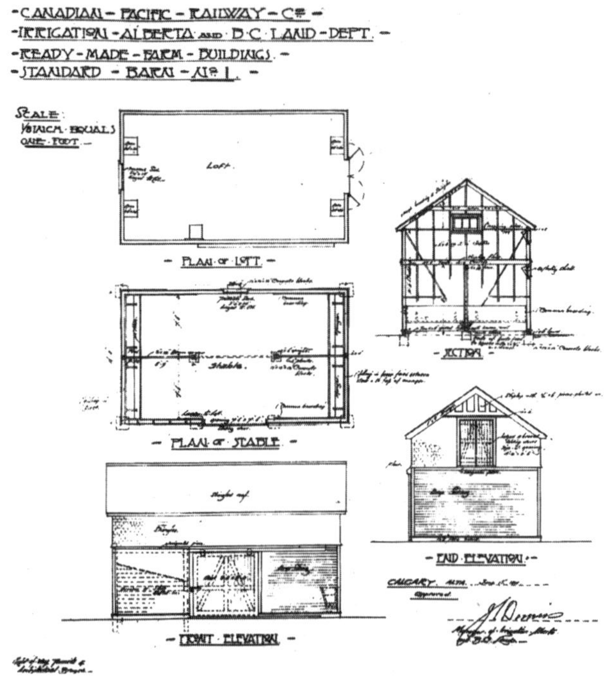

(Glenbow Archives, Calgary, Na-2829-30)

A veteran's colony along the same lines as the ready-made colony was also established by the CPR during the First World War. The intention of the CPR to create veteran's colonies was announced on 17 March 1916 by Lord Shaughnessy with the complete details of the program announced in November 1916.The Lethbridge Board of Trade immediately took action to ensure that the Lethbridge area would be the location for one of these colonies. In a telegram to Lord Shaughnessy, the group expressed its satisfaction that the CPR proposed to do something practical regarding the settlement of returned veterans on Canadian lands and its hope that "one of these colonies [would] be established on your company's irrigated lands near Lethbridge, where intensive farming is being successfully carried on . . . ."[89]

The CPR responded to the request and began the construction of a veteran's colony in May 1916 when it let a contract for the construction of the first buildings. According to the *Lethbridge Daily Herald* the contract called for the construction of 25 farms to be called the Van Horne Colony after former CPR President Sir William Van Horne. As with the ready-made farm development southeast of Coaldale, the CPR built and operated a company farm as part of the Van Horne Colony. The central farm was operated by a farm superintendent whose duty it was to look after the welfare of the farmers of the colony.

The ready-made farm idea was copied by the M.A. Wilson Land Company of Minneapolis. The *Lethbridge Daily Herald* for 23 January 1915 indicated that this company had purchased lumber to build six houses and barns and posts enough to fence that many half sections. The company's plan was to put up the buildings as soon as the land was sold. Like the CPR readymade farm plan Wilson also assisted the farmers with the purchase of horses, livestock, seed, and feed with long term loans.

THE CPR AND THE DEPARTMENT OF THE INTERIOR

The CPR acquisition of the St. Mary Project occurred when the federal government was becoming increasingly concerned with the administration of western Canada's water resources. The International Waterways Treaty of 1909 along with the rapid settlement of the West made the government aware of the need for the efficient use of what was now seen as a limited resource. Another reason for this increased concern was the number of problems which had arisen in the western section of the CPR block and in the Bow River Project then under development by the Southern Alberta Land Company.

The federal government became directly involved in the development of the western section of the CPR irrigation project in 1911 when the Department of the Interior received a number of requests from the water users to investigate the charge that the company's canals were inadequate to supply water for the entire area of allegedly irrigable land. A second grievance of the water users in the western section was the way the company had classified the land as irrigable or dryland. Some land had been classified as irrigable although it was above the level of the supply canal, or could not be irrigated because of topographical conditions. The third concern was that the farmers disputed the view of the federal government as originally proposed by William Pearce that the area east of Calgary was a semi-arid region requiring irrigation to be productive. The water users claimed that the rainfall was generally adequate for satisfactory crop production without irrigation and that the impervious character of the subsoil, combined in some localities with the presence of alkali in considerable quantities, precluded the satisfactory practice of irrigation. The water users were thus challenging the reason for creating an irrigation system and the CPR's way of carrying out that objective.

The federal government responded to these concerns by undertaking its own land classification survey of the western block in 1913. The net result of this investigation was a reduction in the irrigable acres from 367,000 to 225,000. The special investigation to determine whether or not the climatic conditions on the western section were such as to make the practice of irrigation agriculture practically impossible, as claimed by certain farmers, was conducted between 1913 and 1915. The results of this investigation were published in 1915 in a document entitled "Report on the Climatic and Soil Conditions in the Canadian Pacific Railway Company's Irrigation Project, Western Section." The report, which dealt with the climate, the temperature and the content of the water in the canals and the soil, concluded that there was no basis for any of the farmer's concerns.

The need to more closely monitor the construction of the various systems was also illustrated by the unexpected collapse of the intake of the Southern Alberta Land Company on the Bow River in 1912. The collapse was described as an accident by company officials but W. G. Bligh, Inspecting Engineer for the Department of the Interior concluded from his inspection of the works that the collapse was "due, not to some unforseen accident or the treacherousness of the river bed, but simply to the disregard, or crass ignorance of the engineering principles which had to be applied on the part of the engineer who designed the works."[90]

The impact of the federal government's new approach on the CPR's administration of the St. Mary Project was first evident on 5 January 1912 when Commissioner of Irrigation, Peters indicated to Naismith that the plans which the CPR had provided showing the projected reservoir system in the vicinity of Chin Coulee were inadequate. The CPR was asked to provide the necessary plans, profiles and cross sections on tracing linen in duplicate. The Commissioner further indicated that in the past, the Department of the Interior had not been very strict in the matter of the submission of the plans called for by the North-west Irrigation Act but the policy now adopted by the Department required that the plans be submitted and approved before any work of this kind was undertaken.

In addition to enforcing the rules regarding the submission of plans more closely, the government required the CPR to comply with section 52 of the North-west Irrigation Act despite the fact that the final license for the St. Mary Project had not been issued. This clause in the North-west Irrigation Act required that licensed companies operating under the authority of the Act had to submit an annual report to the Minister of the Interior outlining the year's operations in terms of the expenditures made and areas irrigated. Because the St. Mary Project had not received its final license, this particular section had not been applied. The Minister of the Interior decided, however, that since the permit to divert water gave it the same privilege as a licensed company, the CPR should be required to comply with all the provisions of the law respecting such companies including the submission of the annual return called for by Section 53.

The CPR submitted its first annual return as requested in 1914 for the period ending 31 December 1913. The document indicated that 676 water users in the St. Mary Project were irrigating 76,200 acres. These statistics were, however, at variance with the government's own records. P.M. Sauder, Acting Commissioner of Irrigation, notified the CPR of this discrepancy. According to Sauder, the figures should have been 683 water users and 48,961 acres. Peter Naismith explained these discrepancies by stating that the company really had no idea of the exact amount of land irrigated since it merely delivered water to its users and did not keep records of its use.

Lack of accurate information on the amount of water being used was a problem because that information was required before a permanent license could be issued for the diversion of water. The CPR ended the anxiety within the Irrigation Branch over this issue when it applied for a five year extension to the 1922 deadline for completion of the project. In making its application, it cited the effects of the First World War as the main cause for the delay.

DISTRIBUTION SYSTEM

The St. Mary Project was jointly operated by the Engineering Branch and the water users with the farmers being responsible for all sub-laterals through to the farm gate. The CPR did, however, provide assistance to the farmers in laying out these sub-laterals. The area covered by water agreements in the southern and northern divisions was 102,900 acres by 1918. This was an increase of 9,600 acres since 1917 and included 26,000 acres of fractional water rights in the Magrath and Raymond areas. This statistic was derived from the water contracts and did not represent the actual amount of land irrigated each year. The company believed there was the potential to double the irrigated land in the region. The 1918 *Annual Report* noted that:

> most of the Company's land is now sold but there are a number of large holdings which should be divided up and quite a few tracts in non-resident ownership which are not being farmed at all or are very indifferently farmed. As a whole the section as a farming community has been fairly prosperous but there is room yet for a vast improvement. It has not yet reached more than 50% of the development which can be reasonably expected.[91]

The first association of farmers organized to operate the lateral system which served their land was created in 1910 under the *Societies Act*. The charter members of the Pawson Lateral Association were H.A. Suggitt, Syver Crest and W.H. Pawson, Jr, all very prominent members of the Coaldale community. Suggitt not only owned land in the area but had been active in the company that established the town of Coaldale in 1906.

SEVEN
CRISIS IN THE SUGAR BEET INDUSTRY 1908 - 1917

The Knight Sugar Company went into operation in 1903 and enjoyed success up to 1907. The output of the factory increased and the company paid a substantial dividend each year. The company, however, faced two problems in making the plant a long term success. The first problem was competition from cane sugar which was supplied to western Canada by the Vancouver based BC Sugar Company established in 1890 by B. T. Rogers. Rogers' initial response to the opening of the Knight plant was to try to use the company as a domestic source of raw materials. The Knight Sugar Company, however, was not in a position to provide unprocessed beet pulp because the plant was designed only to produce the finished sugar. With the failure to obtain a supply agreement, BC Sugar tried to drive the Knight Sugar Company out of business by reducing the price of sugar sold in Alberta. In an effort to obtain protection from BC Sugar, Knight requested that tariffs be placed on cane sugar. While the federal government did not give sugar beets any tariff protection, it did pay a bonus of 50 cents per 100 pounds of sugar which was divided equally between the beet growers and the factory. It also eliminated the taxes on the factory for the 12 years for which is was required to be operated.

The attempt of Rogers Sugar to put the Knight Sugar Company out of business generated a "buy Alberta" campaign led by the *Lethbridge Daily Herald* which was eventually supported by the City of Lethbridge, the Lethbridge Board of Trade and businessmen in other communities. It began as early as 1906 when the *Lethbridge Daily Herald* pointed out that there were two reasons for everyone in the western half of the prairie country and especially every Albertan should be interested in the future of the sugar industry. The first reason was that it provided

> security against monopolistic prices being levied upon the purchasers of sugar by the big coast refineries. Competition between local manufacturers and the coast combine protects us against this. In the second, there is the value to the country of thriving industry, employing large numbers of men, providing a market for an additional crop, and putting in circulation large sums of money that would otherwise be sent out of the province.[92]

The disadvantage of the Knight Sugar Company in its effort to increase its market share was the perception that cane sugar was superior to beet sugar. According to the *Lethbridge Daily Herald*, this preference extended even to Lethbridge where "the housewife and others who buy sugar at retail are under a misapprehension in regard to the quality of Raymond sugar." [93] In an effort to counter this view, the *Lethbridge Daily Herald* published a letter from James Ellison, Manager of the Knight Sugar Company. He noted the tests made by the provincial government and his own company showed that Raymond sugar was 99.8% pure while the cane sugar was 99.7% pure. He also described how beet sugar was being used successfully in the canning of fruits in California. He even promised to send a company representative to Lethbridge to demonstrate what could be done with beet sugar.[94]

LACK OF SUPPLY

A second and much more serious problem, was the difficulty of obtaining a sufficient quantity of beets to keep the factory in full operation. The beet shortage problem was evident as early as 1908 when E. P. Ellison noted in his annual report that the farmers were slow to sign contracts.[95] The farmers in the Cardston area had quit growing beets altogether and those in the Stirling area had also stopped and switched to the production of wheat and livestock when the Knight Sugar Company refused to increase the price it paid for the beets.

The company's efforts to solve the supply problem included growing their own beets on company land. The cultivation of company lands was first done by Japanese labourers brought in especially for that purpose. When many of the Japanese workers acquired their own farms, the company shifted their recruiting efforts to Europe. The result was the arrival of a number of Belgian families who were considered by the *Lethbridge Daily Herald* to be very suitable because they understood the sugar beet industry and all had wives and children who could assist in weeding and thinning[96]. By 1913, the Knight Sugar Company was cultivating 3,000 acres of beets per year which represented approximately three quarters of the total crop for that year.

The company also borrowed an idea from the CPR and began its own program of ready-made farms in 1913.[97] The plan was to provide each settler with 40 acres on which the settler had to raise at least 20 acres of beets to be supplied to the factory. Ten acres also had to be seeded to what the company called garden truck or hoed crops. The settlers were free to use the remaining 10 acres for whatever purpose they wanted. The settlers were required to pay $2.50 an acre every year. The company provided each farm with horses, cows, pigs, chickens, a house and barn which the settler paid for over time According to an article in the *Lethbridge Daily News*, the company was anxious to encourage their Belgian workers to occupy many of these farms.

In addition to these efforts to encourage local production the company also imported beets from Europe and sugar cane from the West Indies. The first shipment of imported beets, consisting of 25 carloads or 2,000,000 lbs arrived from Germany in June 1908.

CLOSING THE FACTORY

The Knight Sugar Company first considered closing the factory at a meeting of the board of directors on 22 December 1911. At this meeting, E.P Ellison was authorized to sell the entire holdings of the company including the sugar factory and its farms and ranches. The board of directors were willing to sell the land at $17 an acre and give the sugar plant free of charge to anyone who would buy the land. Since no buyer could be found, the sale of the plant was deferred.

The future of the sugar plant was again in jeopardy in July 1913 when area farmers were warned at a public meeting that their continued failure to grow beets would result in the closure of the plant and its relocation to another community. The warning was not delivered by a company official but by Bishop Charles W. Nibley of Salt Lake City, who was described by the *Lethbridge Daily Herald* as being in charge of the business affairs of the LDS Church. According to the 31 July issue of this paper, "he went exhaustively into the principles of business success and in the course of his address very bluntly told the Raymond district people that they were falling down hard on the job of supporting the sugar factory and unless more interest were taken in it, the factory would be taken away."[98]

The company's point of view on the situation was provided by Ellison in an interview with the *Lethbridge Daily Herald* following the meeting. Ellison outlined the value of the plant to the economy of Raymond and southern Alberta as well as the value of sugar beet growing to the individual farmer[99]. The benefits to the community included the $60,000 paid in wages. These wages were used to purchase goods in Raymond stores whose owners in turn purchased their supplies from wholesale supply companies in Lethbridge. Based on an income of $50 per acre and production costs of $30 per acre, Ellison anticipated a profit of $20 per acre for the growers. This figure was based on the production of 10 tons per acre and a payment of $5 per ton by the company. He estimated that each farmer could cultivate from 10 to 15 acres depending on the supply of labour, that is, the size of his family. Ellison concluded his remarks by placing the responsibility for the future of the plant entirely in the hands of the farmers who simply had to grow 10 to 15 acres of beets each to save the plant. According to Ellison, "a magnificent industry could be built up if the people of the home district would only cooperate."[100]

The claim that the farmers were entirely responsible for the problems of the Knight Sugar Company did not go unchallenged. On 6 August the *Lethbridge Daily Herald* published a letter from an "interested farmer" which appeared in a column entitled "Why Farmers Don't Sow Sugar Beets."[101] The author disputed the claim that a farmer could make a $20 per acre profit by growing beets. "If raising sugar beets really is a paying proposition" he concluded "a quick way to convince the farmers would be for the company to show them tabulated statements of their own operation." The comments of the "interested farmer" were immediately responded to by J. W. Evans, the agricultural superintendent of the Knight Sugar Company in an article in the *Raymond Gazette* and by George Budd whose article appeared in the *Lethbridge Daily Herald*. Budd put the blame back on the farmers who he felt did not have enough resources or the right attitude to grow beets properly. According to Budd:

> we expect too much of the country, too much of the climate, too much of nature and are not willing to do our share in the same manner we have been accustomed to do in the towns where we came from. Does this "interested farmer" put the same energy into his work here on the same sized piece of ground or is he like the majority of others trying to farm five times as much land with the same outfit and with no more help than he used on his farm where he came from.
>
> Its usually the case of a thousand dollar man, with a five hundred dollar outfit trying to cope with a thirty-five hundred dollar proposition which spells failure on the face of it. [102]

The "interested farmer"and Evans continued their very public debate about the profitability of growing sugar beets in subsequent issues of the *Lethbridge Daily Herald*. This public exchange of correspondence came to an end on 4 September when J. W. Evans wrote another letter to the editor of the *Lethbridge Daily Herald* in which he provided production statistics from local farmers and the Dominion Experimental Station. He also invited everyone, including the interested farmer, to visit the company's beet fields. He felt that the visit would be particularly beneficial for the interested farmer "since he desires so much that the Sugar Co. demonstrate that they can grow a large tonnage and he may after seeing, decide to use his talents in boosting a good idea."[103] Despite the efforts to increase the participation of local farmers in the production of beets, the majority of the beets for the 1913 campaign came from company land.

On 10 February 1914, James Ellison announced that he had been given the privilege of making one more effort to interest the people of Alberta in this enterprise and that the factory would run another year.[104] This announcement launched the second effort to save the Raymond plant. The campaign was begun by James Ellison himself who coupled his announcement with a review of the advantages of growing beets. Growing beets gave the farmer sure cash at a contract price, provided work for his children since it was especially adapted to child labour, improved the soil, eradicated weeds and produced valuable by-products such as the tops. The *Lethbridge Daily Herald* contributed to the effort by including statements from area farmers who endorsed the growing of beets in an article which announced the possible closure of the plant. The reporter covering the story went further by interviewing farmers from the area such as Arthur Dahl who said that the $500 he made each year was the easiest money he earned.

Lethbridge city officials and the Lethbridge Board of Trade also got involved in the effort to increase beet production by encouraging the growing of beets in the city and surrounding area[105]. The city's contribution was the growing of 10 acres of beets in Henderson Park. The committee on rural relations of the Lethbridge Board of Trade made arrangement with the Knight Sugar Company whereby the same price would be paid for sugar beets delivered at Lethbridge as for those delivered to the plant at Raymond. The Board of Trade also obtained a supply of seed from the Knight Sugar Company and made it available to city residents at a reduced price. The *Lethbridge Daily Herald* felt that the growing of sugar beets would have a special appeal to urban residents because:

> it would be a profitable as well as a pleasurable one, and anyone who sniffs the fine spring air these mornings with a longing, vague desire to get closer to mother earth will have the

> greatest opportunity to satisfy his longing if he follows the directions herein contained and grows sugar beets in his spare time.[106]

The net result of this effort was again very disappointing since only 600 tons of beets were delivered to the plant when at least 35,000 tons were required. The shortfall in the local supply was compensated for by two shipments of raw sugar from Europe which left Antwerp just before the outbreak of World War One on 4 August 1914. The commencement of World War One eliminated the one way which the company had been able to keep the plant open since 1908. Given these circumstances, the Knight Sugar Company had little choice. The decision to relocate the plant was made public on 19 December 1914.

The third community effort to save the plant began two days later at a large meeting of Raymond citizens and farmers. J. W Evans and Arthur Dahl were elected as a committee to visit the various towns in the south including the cities of Lethbridge and Medicine Hat where they were to ask for help from the Boards of Trade. On this occasion only Coaldale responded to the request for help. The effort was given further support on 12 January when Raymond Knight joined the campaign.

This campaign had no effect on the plans of the company which announced on 14 January 1915 that the plant was to be moved to Layton, Utah. This decision was confirmed by James Ellison speaking at the Lethbridge Board of Trade. He identified the labour problem as the most important reason for the company's lack of success. Ellison made a special point of emphasizing the hard work required in cultivating beets. "Let an average man thin [beets] for half a day" he said "and he will be ready to go home and go to bed."[107]

The dismantling of the plant in preparation for its relocation to Layton began 24 February 1915 but this activity was suddenly stopped when the CPR obtained an injunction based on the argument that the terms of the original contract with the AR&CCo. had been broken. Under the circumstances the Knight Sugar Company simply built an entirely new plant in Layton and left the Raymond plant to be disposed of as soon as the injunction was lifted. A canvass of the area in the spring of 1915 indicated that only 800 acres of beets would be grown by area farmers, providing little incentive for the company to reverse their decision. The plant was finally dismantled and moved to Montana in 1917.

EIGHT
BIRTH OF THE LETHBRIDGE SOUTH EASTERN PROJECT 1909 - 1918

The decision of the CPR to maintain the St. Mary Project largely as they found it in 1908 and the International Waterways Treaty of 1909 had the effect of bringing the federal government back into active involvement in determining the future of the St. Mary Project. These two events combined to produce this result because Canada then faced the need to use its share of the water from both the St. Mary and Milk Rivers or face its permanent loss. The Lethbridge South Eastern Project (LSEP) was the name given by the federal government to its plan to use Canada's share of this water. The development of this plan began with a review of the proposals to divert water from the St. Mary and Milk Rivers which had been developed since 1895. These plans included those prepared by the government's own irrigation surveys in 1895 and 1901 as well as the plans prepared for the CNWICo. by their engineering consultant, George Anderson. The LSEP envisaged the expansion of the St. Mary Project on a scale first attempted by the CNWICo.

THE CALGARY IRRIGATION OFFICE

Prior to 1909, the federal government had left the development of irrigation in the West in the hands of private corporations or the territorial government to which it transferred the administration of the *North-west Irrigation Act* and the general irrigation surveys. In requesting this transfer in 1898, the territorial government had argued that irrigation was a local question which need not concern the federal government and that the legislation could be more efficiently administered from Regina. To carry out this new responsibility, the territorial government established an Irrigation Branch within the Department of Public Works which was administered initially by J.S. Dennis who was followed in this position by B. J.Saunders and John Stocks.

With the creation of the provinces of Alberta and Saskatchewan in 1905, the federal government transferred the administration of irrigation back to its office in Calgary rather than permit the new provinces to take over the territorial government's responsibility to administer water resources. This new irrigation office was placed under the administration of the Forestry Branch of the Department of the Interior, the director of which was designated the Superintendent of Forestry and Irrigation. In April 1906, John Stewart was appointed Commissioner of Irrigation and was succeeded in 1911 by Frederick H. Peters. In 1912 a separate Irrigation Branch was created within the Department of the Interior. This new branch was placed under the direction of E.F. Drake who was given the title of Superintendent of Irrigation. Peters, along with the bulk of the staff of the branch, remained based in Calgary while Drake and his much smaller staff were based in Ottawa. Peters and Drake soon acquired a dislike for one another which at times threatened to bring the work of the branch to a standstill. On the eve of the First World War, the permanent and seasonal staff had increased to 180 employees. The budget appropriation for the operation of the branch increased from $12,000 in 1907 to $100,000 in 1913. The expansion of the Calgary office during this period also included the creation of a separate branch to collect hydrographic data. The creation of this office was suggested by P. M Sauder who became its first director in 1909.

Planning the future of the St. Mary Project was not an entirely new departure for the Department of the Interior, since it had made the first survey of an irrigation canal to make use of water from the St. Mary River in 1895. Its planning activities after 1909 were, however, far more extensive. The Department of the Interior initially referred to their plan for the expansion of the St. Mary Project as the Milk and St. Mary Rivers Irrigation Project but changed its name to the Lethbridge South Eastern Project to distinguish it from the Lethbridge Northern Irrigation project which was being planned at the same time. The primary goal of the federal government after 1909 was to collect the necessary data to permit the expansion of the St. Mary Project within the terms of the International Waterways Treaty and to compensate the AR&ICo. for the loss of water to the Americans. This action was not taken in the interests of the company but in the interests of those Canadian farmers who depended on the company for water. As Drake pointed out:

> these surveys were not made in the interest of the Irrigation Company, as might perhaps be assumed, but for the development of irrigated farming in the Lethbridge district and to demonstrate the possibility of fulfilment by this government of certain obligations to supply water to the company in lieu of the proportion of the flow of the St. Mary River which will be taken by the United States in accordance with the provisions of the Waterways Treaty.[108]

SURVEYS

Planning the LSEP involved a significant increase in survey activity in order to collect additional information on a wide range of subjects. The first task was to determine the water supply available from the watersheds of the Waterton, Belly, St. Mary and Milk Rivers. It was also necessary to determine the amount of land which was being irrigated both in Canada and the United States and the potential amount of irrigable land which could be developed using this water supply. The best locations for canals, reservoirs and dams which could be used to expand this irrigable area had to be investigated. The federal government also wanted to know the effect on the Canadian portion of the Milk River of its use by the Americans to transport approximately 1,000 cfs of additional water for use in Montana as provided for in the Treaty.

Hydrographic investigations in the foothills to determine the supply of water for irrigation purposes were begun in 1895. The systematic collection of this type of information was not carried on all the major rivers until after 1903 when P. M. Sauder was hired and a separate hydrographic branch had been created under his direction in 1909. As part of this program, gauging stations were established on the Waterton River at Waterton Mills at the north end of Lower Waterton Lakes and at Standoff on the Belly in 1909.

The hydrographic service was able to give special attention to the Milk River because of a separate appropriation made after the ratification of the International Waterways Treaty in 1909. This special survey took stream measurements along the entire length of the Canadian portion of the Milk River and evaluated various locations for permanent gauging stations so that in succeeding seasons records of all stages of flow could be made. A permanent station was established east of Whiskey Gap in July of 1909. The surveys continued in 1910 during which time frequent stream measurements were taken throughout the year on the Milk River as well as on the St. Mary River and all their tributary streams which could be reached.

The Milk River was also evaluated to determine the effects of its use by the Americans. It was concluded that this extra volume of water would increase the river's tendency to erode its banks and change its course, thus increasing the overall width of the valley. The extra water would thus make the river less fordable during the summer months thus necessitating the construction of highway bridges at several points. In 1915 a detailed traverse survey of the river was made to evaluate potential damage to the river channel caused by the extra water which did not begin flowing down the north branch of the Milk River until 1918.

Information on the extent of irrigation development using the St. Mary and Milk Rivers in the United States and Canada was first collected in 1904 when Arthur Wheeler prepared a report on American irrigation development on the Milk River. In 1913, Ralph Burley inspected the Frenchman River from its source at Cypress Lake in south western Saskatchewan to its confluence with the Milk River north of Saco, Montana. The irrigated hay flats above Beaver Creek near Saco were inspected along with other projects on the lower Milk River which included the Dodson diversion dam near Malta and the Vandalia project near Glasgow.

Surveys to determine the size of the irrigable area southeast of Lethbridge were conducted as early as 1895. No information was available, however, of what portions of the immense area from the Belly River to the Milk River and east and west from the St. Mary River to Forty Mile Coulee and Pakowki Lake was irrigable given a sufficient quantity of water. Surveys to design a system to irrigate this land began in 1912 when Ben Russell investigated the feasibility of the construction of a canal from the Waterton and Belly Rivers to a location on the St. Mary River upstream from the AR&ICo diversion structure at Kimball. The result was a canal which followed a route very close to that of the current main canal of the Mountain View Irrigation District from the Belly River to a location south of Beazer. From Beazer, the proposed route ran southeast via Boundary Creek to a point on the St. Mary River one mile north of the international boundary. A canal system to take water from the Waterton and the Belly to the same location on the St Mary River was surveyed in 1914 by T.A. Montague but was located further north on the Blood Reserve.

The survey of a canal system to take water east of the Milk River Ridge began with a review of the plans already prepared. The 1895 plan of the Department of the Interior required the construction of a canal from Kimball around the western side of the Milk River Ridge to Ridge Reservoir. In the plan suggested by George Anderson, water from the St. Mary River was to be carried through Whiskey Gap to the Milk River. Both these plans were abandoned by the fall of 1914 and the Irrigation Branch decided to plan the expansion of the St Mary Project based on the use of the existing AR&ICo. canal system.

The reservoir and dam site surveys of 1914 and 1915 included the investigation of three possible sites on the Waterton and one on the St. Mary Rivers. The first two Waterton River sites were at Upper Waterton Lake while the third was three miles inside the Waterton Park boundary. The first two reservoirs were to store water and the other was to divert water into the proposed canal from the Waterton River to the St. Mary River, which had been surveyed by Montague. It was determined that the third site would require a dam 215 feet long and 10 feet high in order to divert the required amount of water. A dam site on the St. Mary River south of the Kimball intake and a mile north of the international boundary was also investigated At the time this site was considered to be the only site on the St. Mary River in Canada where any amount of water could be stored. A dam at this site was part of George Anderson's plan to divert water to the Milk River via Whiskey Gap.

Reservoir sites investigated north of the Milk River Ridge included Horse Fly Lake and Chin Coulee. Horse Fly Lake was shallow with a relatively large surface area, thus losses to evaporation would be very high in proportion to the quantity of water which could be stored. Despite the potential losses to evaporation, the Irrigation Branch wanted to determine the possibilities of draining the lake and the amount of additional land that would be flooded if it was used for irrigation purposes. It was found that although the drainage channel was not well defined and was very flat for a considerable distance from the lake, it could be drained by cutting a canal from the northeast end of the lake to Purple Springs. Chin Coulee which was already in use as a reservvoir was considered to have a potential storage capacity of 110,000 acre-feet of water.

Brunton and Verdigris Lakes were also investigated as possible reservoir sites east of the Milk River Ridge. Both had been identified by Anderson and the 1901 government plan for the Milk River canal. The Brunton site in Verdigris Coulee had an estimated capacity of 20,000 to 25,000 acre-feet. The Verdigris site further south in Verdigris Coulee, was considered one of the best sites in which to store a large quantity of water estimated as at least 116,000 acre-feet.

IRRIGABLE AREA

Surveys of use of water from the Milk and the St. Mary Rivers up to 1915 indicated that the AR&ICo. was irrigating 100,000 acres in the Magrath, Raymond, Lethbridge, and Coaldale areas and the company's system could be expanded to another 71,296 acres. These acres along with other potential additions to the St. Mary Project were subdivided by the Irrigation Branch of the Department of the Interior into eight separate tracts.

Tract 1A was near Magrath, between Pothole Coulee and the St. Mary River. It included 36,480 acres, with 20,440 considered irrigable some of which was already irrigated by the Magrath Lateral. Tract 1B consisted of 4,096 acres south of Pothole Coulee in the Bradshaw area. The last tract in the Magrath area, 1C, consisted of 9,600 acres on table land south and east of the present location of Jensen Reservoir. The 31,700 acres of potential irrigable land in the Raymond area was designated Tract 2. Tract 4 was 6,770 acres south of Raymond at a higher level than Tract 2. In the Coaldale area, Tract 3 had an estimated irrigable area of 104,450 acres.

Tracts 5, 6, 7 and 8 were beyond the existing system of canals. Tract 5 was north and east of Chin Coulee, south of the Oldman River and west of Forty Mile Coulee. The 2700 foot contour was used to separate this area into two sections. North of this contour 123,000 acres were designated as Tract 5A, could be commanded by a canal built from the north end of Chin Coulee. To the south Tract 5B required a flume across Chin Coulee and a high line canal further south along the 2,900 foot contour. The 63,000 acres of irrigable land south of Chin Coulee, north of the CPR between Etizkom and Swift was designated as Tract 6. It was to be irrigated by water taken from the Milk River by a canal from Verdigris Lake across Etzikom Coulee.Tracts 7A and 7B included the land between Etzikom and Verdigris Coulees. Tract 7A north of Verdigris Lake had 29,146 irrigable acres, Tract 7B north of lower Verdigris Coulee had 35,232 acres and Tract 7C west of Pakowki Lake had 26,622 irrigable acres. Tract 8 included 20,000 irrigable acres between the proposed main canal from the Milk River and Verdigris Coulee.

CANAL AND RESERVOIR PLANS

Following the completion of survey work for 1914, the Irrigation Branch prepared plans to irrigate Tracts 1 through 8. The first plan outlined a system of canals to provide water to the entire 548,456 acres. It included canals surveyed by Montague to take water from the Waterton and Belly Rivers to the St. Mary River at a point south

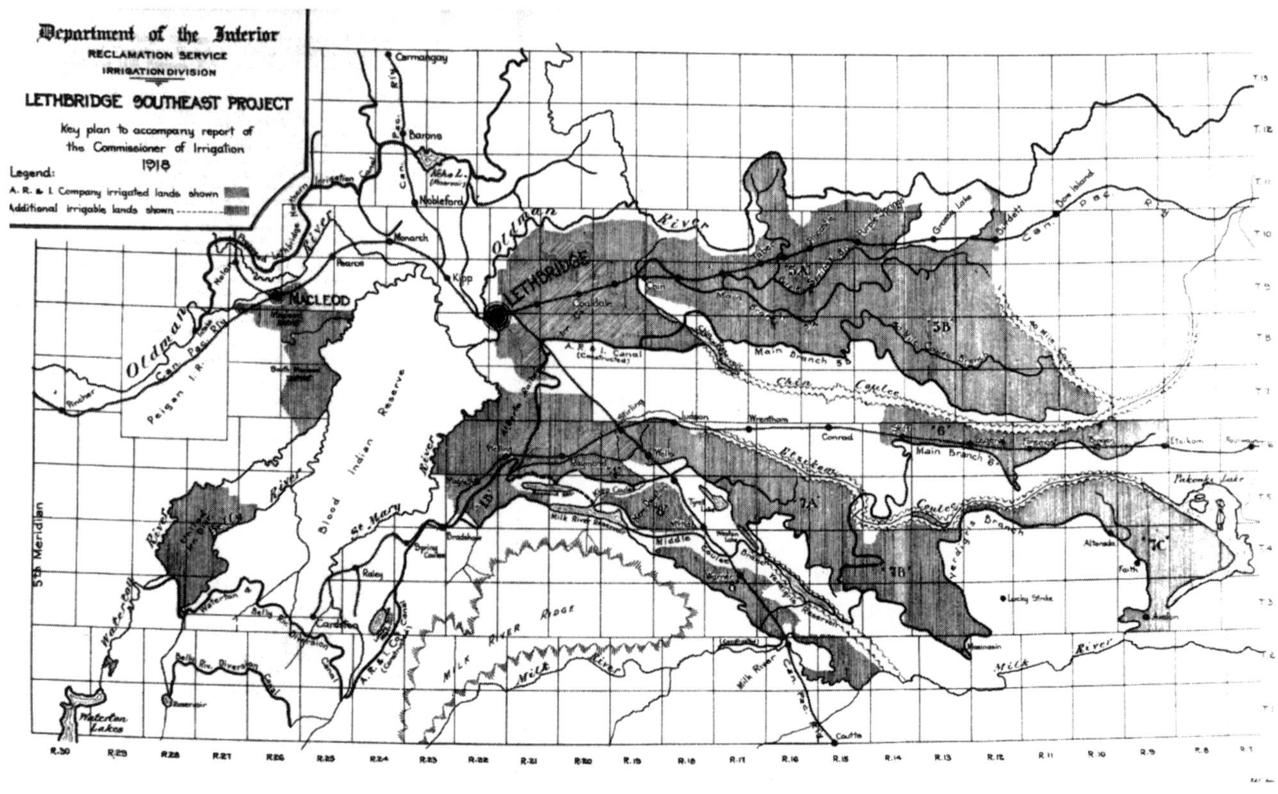

61

Plan of the Lethbridge Southeast Project(Annual report of the Department of the Interior, 1918-1919)

of the diversion at Kimball, four miles north of the international boundary. The Lumpy Butte Reservoir midway between Kimball and Spring Coulee, was the only addition to that part of the existing system. This location had been suggested as a reservoir site by George Anderson. The Magrath lateral was to be extended north of the town to irrigate Tract 1A. A second canal was to be built east from the system at Bradshaw to the Milk River Reservoir to irrigate Tract 1B and to fill the reservoir. Water from the Milk River Reservoir was to be taken in two directions: to the proposed Raymond Reservoir, from which canals were to be built to irrigate land south of Raymond; and from the Milk River Reservoir east to the proposed Brunton and Verdigris Reservoirs. Water from the Milk, Brunton and Verdigris Reservoirs could then be taken to irrigate the land south of Chin Coulee and east of the Milk River Ridge in Tracts 6, 7, and 8. Additional water was to be provided by the completion of the canal from Milk River.

In the Northern Division of the St. Mary Project, various extensions of the existing system were suggested to irrigate Tract 3, which included the land between Lethbridge and Chin Coulee. Two canal extensions east of Chin Coulee were planned, one from the north end to the Bow Island area would irrigate the 123,000 acres in Tract 5A and a second extension from a dam on Chin Coulee east to Forty Mile Coulee would irrigate the 84,000 acres in Tract 5B.

The second plan was designed to irrigate a considerably smaller area, 354,834 acres. The plan to take water from the Waterton and the Belly Rivers was replaced with the canal taking water from the Belly to the St. Mary River surveyed by Ben Russell in 1912. The Milk River diversion canal was to be linked only to the Milk River Reservoir and not to continue to the Verdigris or Brunton Reservoirs. Water from the Milk River Reservoir was to irrigate only Tract 7A and 7B. The canal from Chin Coulee Reservoir to irrigate Tract 5B was also eliminated.

In 1916, the Irrigation Branch prepared a much more detailed set of plans for the expansion of the St. Mary Project which included cost estimates. The plan was divided into seven variations, each a progressive enlargement of the area to be irrigated and higher capital expenditures. These seven variations extended the existing system of the AR&ICo. into Tracts 5 to 8. Expansion of the canal system to irrigate additional land in the Magrath, Raymond and Coaldale areas, Tracts 1 to 4, was left to the CPR and water to irrigate 110,000 acres was reserved for this land.

The first variation involved the construction of a small canal from Chin Coulee Reservoir to irrigate 17,000 acres in Tract 5A in the Barnwell and Taber areas. The cost of this extension was estimated in 1916 at $11.50 per irrigable acre. The only new construction required was the canal since Chin Coulee Reservoir provided sufficient storage.

The second variation was designed to irrigate the entire 78,000 acres in Tract 5A plus 15,000 acres in Tract 5B. This plan necessitated enlarging the main canal of the AR&ICo. to a gross capacity of 1,450 cfs, the renewal of most of the structures

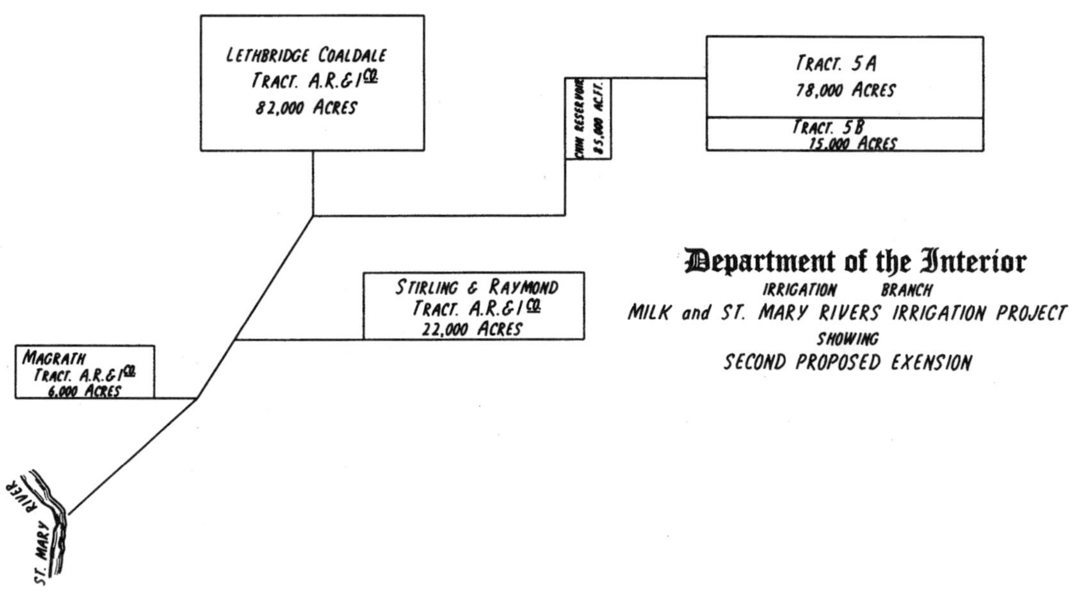

Second proposed extension (Annual report of the Department of the Interior, 1918-1919)

and building new structures to control the flow through the natural watercourses of that system. Chin Coulee Reservoir would also have to be enlarged to a capacity of 85,000 acre-feet. The water supply for this extension was to be obtained entirely from the St. Mary River. The estimated cost of this extension was $2,596,267.85, with $1,085,600.00 to enlarge the AR&ICo. canal the largest single item. This extension would expand the irrigable area by 93,000 acres at a cost of $27.90 per acre.

In the third variation, 85,000 acres in Tract 5B were added to the St. Mary Project. The existing main canal would have to be increased to give it a maximum gross capacity of 2,440 cfs and a canal, with 750 cfs capacity built from the Belly River along the canal route surveyed by Ben Russell in 1912. All the structures of the main canal were to be renewed and drops and chutes be added to carry water through the natural channels of the system. The Chin Coulee Reservoir was to be enlarged to a capacity of 102,000 acre-feet. Costs for the expansion were estimated at $5,113,168.10 for 178,000 acres, $28.75 per acre.

The fourth extension required the construction of additional canals and the creation of storage to irrigate the 49,100 acres in Tract 7A and the 17,400 acres in Tract 7B. It required the addition of four new reservoirs as well as 102,000 acre-feet of water provided by an enlarged Chin Coulee Reservoir: 15,500 acre-feet in the Belly River Lakes Reservoir, 79,900 acre-feet in the Milk River Reservoir, 17,000 acre-feet in the Raymond Reservoir and 9,000 acre-feet in Verdigris Reservoir. The canal from the Milk River was also needed to increase the water supply. This canal would permit filling of the Verdigris and Milk River Reservoirs from either the Milk or St. Mary Rivers. The variation would irrigate 244,500 acres at a cost of $7,045,952.35 or $28.80 per acre.

The fifth extension involved the irrigation of all

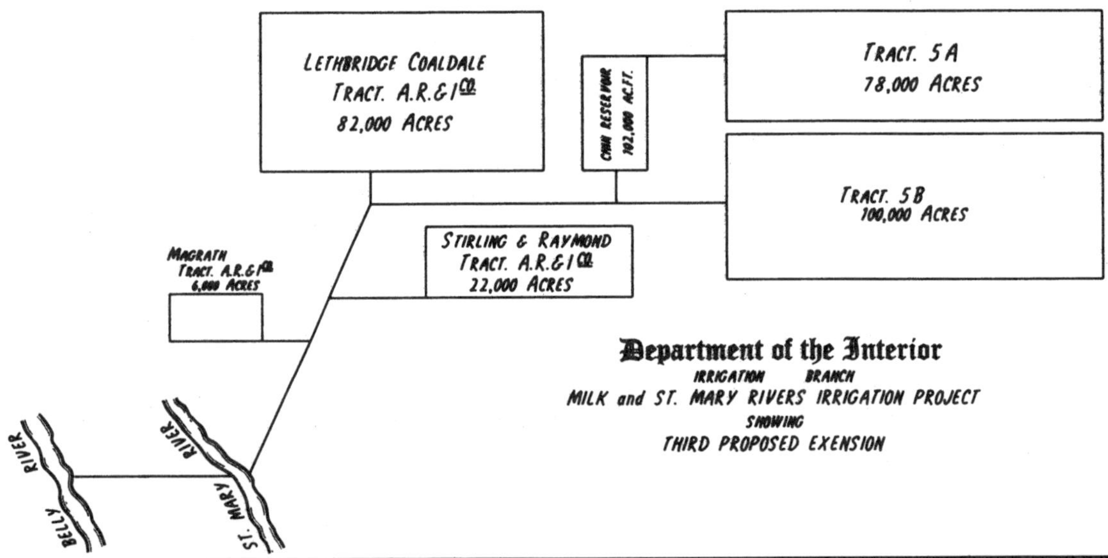

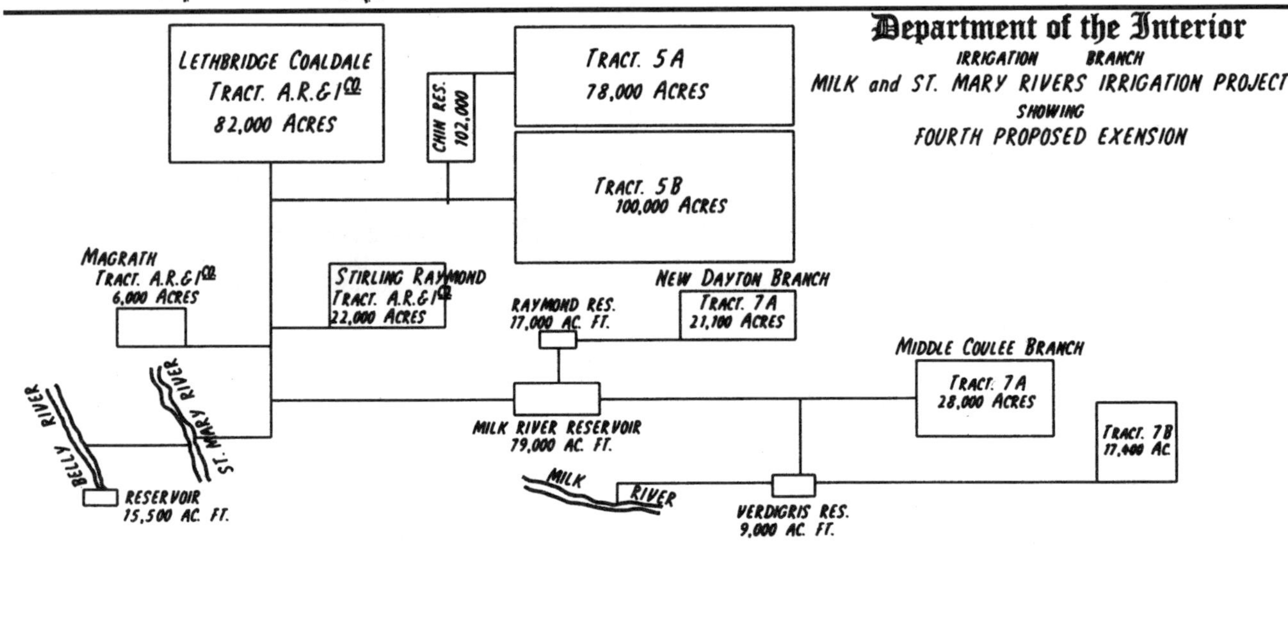

20,000 acres in Tract 7B. Irrigation of these acres required increased storage in Verdigris Lake Reservoir. Its capacity was to be raised to 21,000 acre-feet from 9,000 acre-feet. Costs were quite similar to those of the previous option at $7,275, 530.40 but 256,100 acres could be irrigated at a cost per acre of $28.45.

Extension six included the addition of the 24,500 acres in Tract 7C. The capacity of Verdigris Lake was to be increased to 37,000 acre-feet and a reservoir on the Belly River at Wests' ranch, which had a capacity of 7,700 acre-feet, was added to the system. The additional cost of this extension was $1,174,679.25.

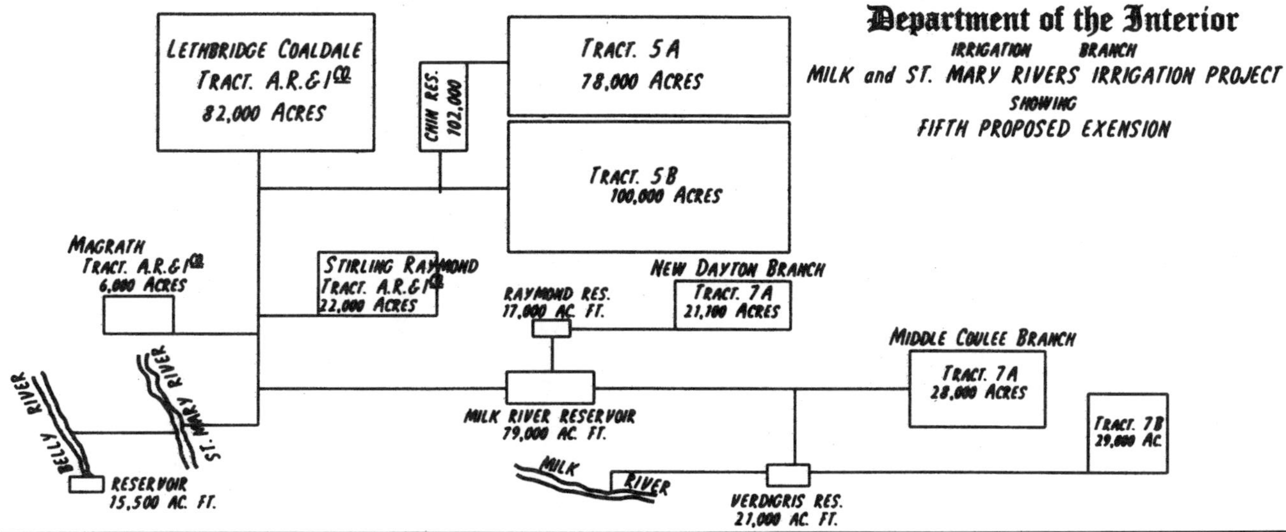

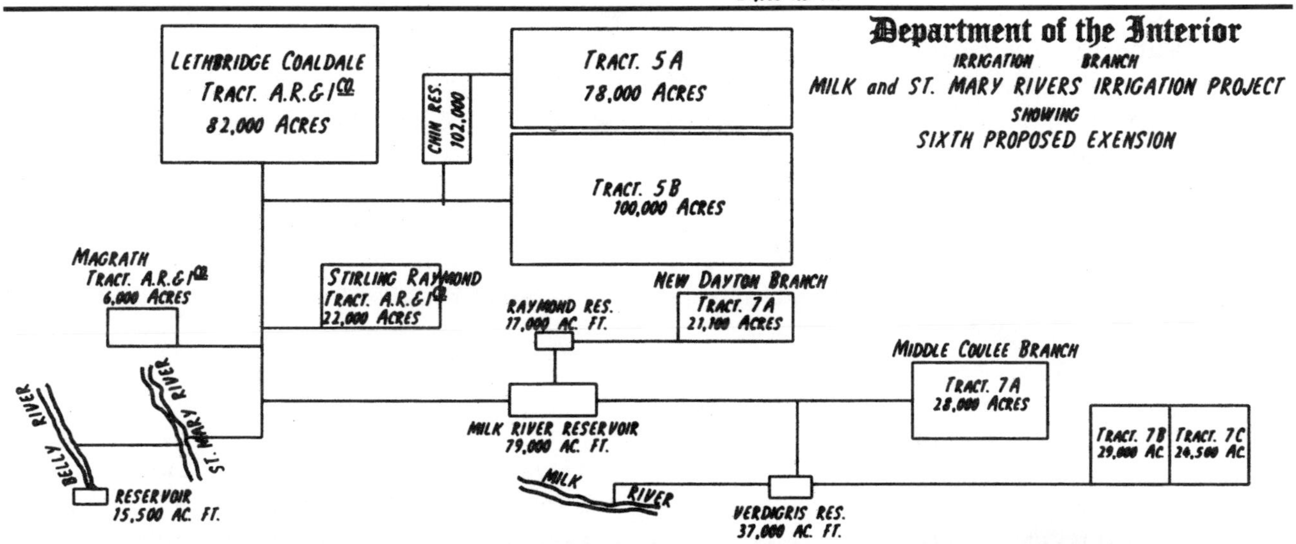

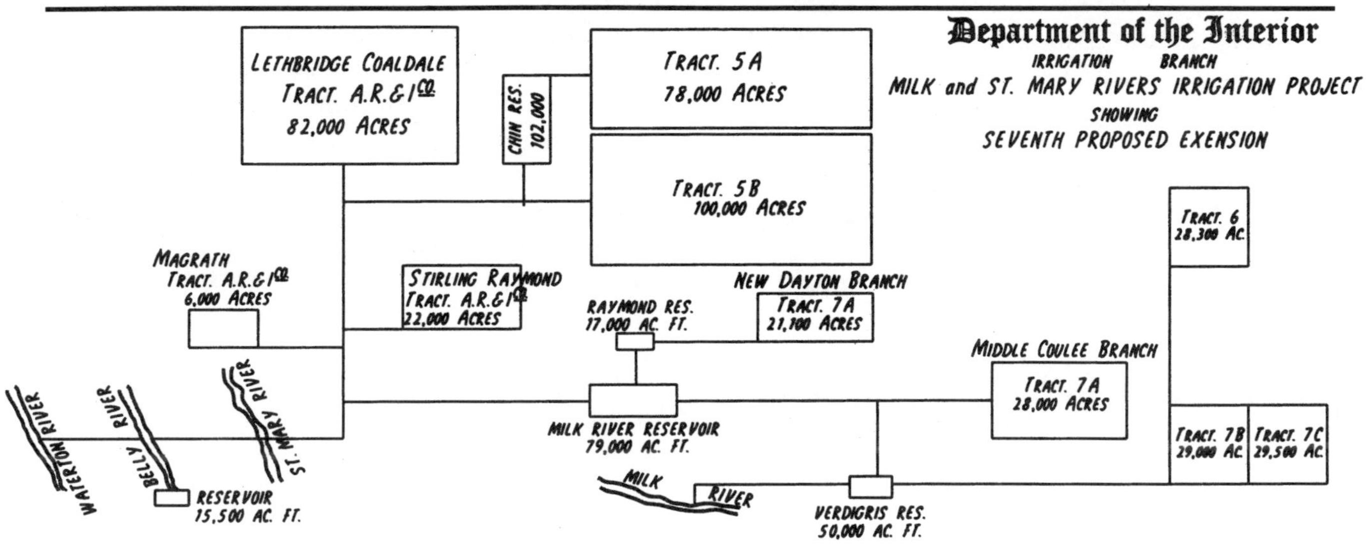

The seventh and last extension proposed in 1916 incorporated the 28,300 acres in Tract 6 into the project. This expansion required water from the Waterton and the Belly Rivers using the canal surveyed by Montague in 1915. This canal required a capacity of 400 cfs. In all previous extensions when water from the Belly River was required, it was planned to carry it via the Mountain View and Boundary Creek route in 1912 by Russell. Use of water from the Waterton, however, necessitated a more expensive route via the Blood Reserve. The construction of such a canal system was estimated at $645,917.00, the largest component of the $1,311,420 for this option which would bring the total area of the St. Mary Project to 313,900 acres. The total cost of irrigating this area was $9,780,661.55 or $31.15 per acre.

IMPLEMENTATION PROBLEMS

Despite the detailed construction cost estimates prepared as early as 1916, more planning was required. The estimates for enlarging the AR&ICo. canal for example were very uncertain because there was little information on the amount of earth to be excavated throughout the graded portions of the canal. The Irrigation Branch had practically no information from which to estimate the cost of enlarging those portions of the main canal which used the natural channels of Pinepound and Pothole Creeks. The "Report on Irrigation Surveys and Inspections for 1916-1917" indicated that "the enlargement of the natural water courses was roughly estimated from any scattered information which could be used."[109] The erosion problems created by the increased volumes of water were also a concern.

DUTY OF WATER

In addition to costing problems, the Irrigation Branch did not know the actual amount of water required to irrigate the area proposed for development. Research to determine the correct value for the duty of water had only just begun. The "duty of water" was legally defined as the area of land which could be sufficiently irrigated by one cubic foot of water per second flowing constantly throughout the irrigation season, or the equivalent of such flow applied to the land during a shorter period of time. The amount of water which would maximize plant growth did not necessarily correspond to this amount.

Research into determining a duty of water appropriate to the soils and climate of southern Alberta was first begun by Dominion Experimental Farm in Lethbridge in 1908 and continued on a much larger scale after 1911. In that year the annual convention of the Western Canada Irrigation Association held in Calgary passed a motion which called on the government to undertake a "thorough system of investigation to determine the duty of water in the different provinces and for the different crops, so that such duty may then be determined with approximate exactness." Investigations began in 1913 in the Coaldale area under the direction of G.D. Walters who worked with area farmers D. J. McArthur, A. F. Stewart, H. A. Suggitt, A. Haley and W.H. Pawson. Accurate measurements of the water applied, rainfall, the water wasted and the amount of crop produced on each of the fields of each farmer. Walters came to various conclusions based on this first series of measurement with the most obvious being that alfalfa required more than twice the amount of water needed by wheat. He also pointed out in his report that the data agreed very closely with the results obtained for nine large irrigated districts in the United States which had climate conditions and soils somewhat similar to those of southern Alberta. He concluded that during the period from April to September 25 inches of moisture was very desirable for good plant growth.

The third problem was the lack of accurate information on the amount of water available. In some cases, rivers had not been surveyed to determine the feasibility of diversion and utilization. Stream flow records, particularly for the Waterton and the Belly were inadequate for accurate determination of the quantities of water available.

The LSEP would have required the cooperation of the Canadian Parks Branch and the American government since Waterton Lakes was an international park. No contacts with either the Canadian Parks Branch or the American government were made during the planning process. By the end of the First World War the LSEP was a general plan which could only be implemented after a number of engineering and political questions had been dealt with in greater detail. It would become a beacon of false hope in the troubled times ahead.

NINE
FROM THE BOUNTIFUL CANAL COMPANY TO THE TABER IRRIGATION DISTRICT 1908 - 1917

As the federal government planned the expansion of the St. Mary Project in the years up to 1916, the farmers in the Barnwell and Taber areas were engaged in a similar endeavour, albeit on a more modest scale. The construction of a canal which could take water east from Chin Coulee was initiated by a coal company which required it for use in its mining operations. The use of this canal for irrigation purposes attracted the attention of the Taber Agricultural Society which provided the leadership necessary to fully exploit the irrigation potential of this supply of water. The interest of the residents living east of Chin Coulee in wanting to expand the St. Mary Project would eventually lead to the passage of the Irrigation Districts Act in 1915 which in turn set the St. Mary Project on a course which would see farmer run irrigation districts take over complete operation of the system by 1968. The opportunity to use stored water by the farmers in the Taber area made possible by their proximity to Chin Coulee gave them an advantage not shared by the other farmers in the St. Mary Project until the completion of the St. Mary Dam in 1951.

CANAL COMPANY

In October 1908 the Bountiful Canal and Irrigation Company began the construction of a ditch to take the water wasted into Chin Coulee by the AR&ICo east to the town of Coal City which had been established in October 1907. The water was required at Coal City to operate steam engines at a mine operated by B.K. and W.E. Bullock. Farmers along the route were also interested in development of the ditch since it would provide water for their stock. The ditch was laid out by surveyor J.F. Hamilton and built by the farmers along the route. Water was turned into the canal in June 1909. By 1911, at least four other landowners had constructed canals to take advantage of this water supply. The federal government had issued water permits to some of these people on the understanding that the supply was temporary. Their supply of free water ended with the construction a dam in Chin Coulee and the modification of the natural drain into Chin Coulee from Malloy Lake in 1911.

TABER AGRICULTURAL SOCIETY

The unofficial beginning of irrigation east of Chin Coulee followed shortly after the establishment of the townsite of Taber on 22 February 1906 when a subdivision plan was registered by Ransom Van Orman, William Probert and W. R. Russell. Van Orman had arrived in Canada as part of the Mormon group which came from Utah to build the St. Mary Canal. Taber was not a company townsite established by the AR&ICo. or the CPR.

Like that of Lethbridge, the early economic development of Taber was based on coal mining. Also in common with Lethbridge, the local business community was anxious to diversify the town's economy through the development of agriculture. The creation of the Taber Agricultural Society, in May 1908, and the organization of the first Taber Fall Fair in September of 1909 were intended to achieve that objective. Taber businessman Ted Sundal was the secretary of the Taber Agricultural Society.

Sundal, on behalf of the Taber Agricultural Society, initiated efforts to construct a new canal in cooperation with the CPR. In December 1913, he arranged a meeting between some area farmers and Peter Naismith Manager of the CPR's Department of Natural Resources. The Taber delegation included Hans Hansen, John Anderson, Lawrence Peterson, W. E. Bullock and Ted Sundal. Peterson and Anderson had been involved in the construction of the Bountiful Canal. At this meeting, Naismith indicated that the CPR would extend the system if the farmers to be served contracted for water sufficient to irrigate 10,000 acres. The CPR estimated that it would cost $150,000.00 to build the 25 miles of canal required to irrigate the 17,000 acres of irrigable land in the area of Barnwell and Taber. The cost to the farmer was to be approximately $10 per acre to cover construction costs and $1 per acre to cover the annual water cost. The CPR also indicated that it would take first or second mortgages on the land being served as security.

On 20 December 1913, the terms of the CPR proposal were discussed at a meeting of area farmers in Taber. Following their acceptance of these terms, the CPR prepared a formal agreement which was sent to the organizers in March 1914. The agreement as drafted by the CPR, indicated that the farmers agreed to pay not less than $10 per acre. The Taber Agricultural Society and the Taber Board of Trade, who were now cooperating to bring irrigation to the area, requested that the wording be changed to "not more than $10." With this change, the organizers sent out contracts for signature by farmers who had agreed to take the water. Surveying work was to begin as soon as a sufficient number of farmers had signed up. CPR surveys began in June of 1914 and construction was expected to begin in the fall. The survey work included ditches to bring water into the Town of Taber for the irrigation of gardens.

The mood of optimism came to an end with the beginning of World War One on 4 August 1914. Despite assurances from the CPR that they intended to proceed, construction did not start in the fall of 1914. The fear that the CPR was no longer interested in the project, or was at least prepared to delay construction, came true in January 1915 when the Taber Agricultural Society received a letter from Peter Naismith indicating that the "company was anxious to await developments in connection with the Oldman River irrigation proposition."[110] The citizens of Taber and Barnwell could not see the connection between their project and the Lethbridge Northern Project. According to the *Taber Times* of 14 January 1915, the Lethbridge Northern Project was "to be undertaken by the government while the Taber system would merely be an extension of the CPR."[111]

The delay led to what was described as an indignation meeting by the farmers in the Barnwell area in January 1915. They expressed the view that the CPR had not "kept faith with the farmers in the matter of irrigation."[112] The organizers of that meeting subsequently met with the directors of the Taber Agricultural Society who passed a resolution urging the CPR to complete their proposed project as soon as possible. They indicated further that many farmers were prepared to leave the country if the water was not forthcoming very soon. The resolution was also sent to the Minister of the Interior with the request that he use his influence with the CPR to hasten the project along.

TABER IRRIGATION DISTRICT

Naismith's response to the protest was to indicate to the organizing committee that the CPR in March 1914 had decided not to deal with the farmers east of Chin Coulee on an individual basis. It wanted to work with one organization which would handle the day to day operation of the distribution system and thus function in much the same way as a lateral association. He stated it would now be necessary for the Taber and Barnwell area farmers to "secure legislation from the Provincial House to create an irrigation district on the credit of which bonds may be issued for funds to proceed with the work, as the railway company does not see it possible to take second mortgages on the farm lands."[113] The insistence that the farmers create an irrigation district provided the CPR with greater security for the expenditures required to expand the system. Naismith concluded his letter by encouraging farmers to obtain the necessary legislation as soon as possible "since nothing would stand in the way of proceeding once the necessary legislation has been passed and the bonds sold."[114]

The Taber Agricultural Society responded by immediately going to Edmonton to lobby for the passage of legislation to create irrigation districts. The idea of irrigation districts managed by the water users was not new to the Canadian West. Similar legislation had been passed by the territorial government in 1894. The *Irrigation District Ordinance*, which was based on the Wright Act of California, was passed at the request of a number of settlers in the Springbank area west of Calgary. The Ordinance required the residents of an area who wanted to establish an irrigation system to provide a survey showing the feasibility of the proposal and to prepare a petition indicating that the majority of the residents agreed with the creation of the irrigation district. Following the submission of these documents, a notice was published in the *Official Gazette*. If no objections were received, the district could be organized with the election of a board of directors. The board of directors could then apply for the right to divert water and hire an engineer to supervise the construction of the system as described in the original memorial requesting the creation of the district.

Given the support of farmers and the CPR, the Alberta Legislature moved quickly and Taber and area farmers had a draft of the bill to consider on 31 March 1915. At the farmers' request, the provision of the act requiring that two thirds of the land in a district be irrigable was changed to one half. Another change was that non-resident owners who lived within five miles of the district were allowed to vote. The Irrigation Districts Act was passed in late April 1915 and incorporated most of the features of the earlier legislation.

Along with their involvement in the passage of the enabling legislation, the organizing committee for the Taber Irrigation District also had the responsibility of determining what land was to be included in the district. One of their concerns was that 8,000 acres of school land were considered irrigable and included in the proposed district. In each township school lands were set aside by the federal government to support the development of education. The proceeds from the sale of these lands were put into the School Endowment Fund. Under the provisions of the law respecting this fund, the Dominion Government acting as the trustee of the fund was required to invest all the proceeds of the sales of school lands and to transfer the interest thereon, less expenses of management to the provincial governments for educational purposes. Title to the lands was only transferred from the Crown to the purchaser after the entire payment had been made, thus the occupants of the day could not pledge their land as security for the irrigation district's bonds.

This issue was first raised by Sundal in a meeting with Commissioner of Irrigation, Peters in March 1914. Peters suggested to Sundal that possibly the Department of the Interior could take steps which would permit these lands to be pledged as security. Given assurance that the government would deal with the matter, the organizing committee included school land within the boundaries of the proposed irrigation district.

On 11 May 1915, Ted Sundal, on behalf of the farmers of the area, sent the required petition for the creation of the Taber Irrigation District (TID) to Charles Stewart, Minister of Public Works for the Province of Alberta. In the covering letter, Sundal indicated that there were approximately 36,960 acres in the proposed district and that signatures of the owners of 18, 480 acres of this total had been secured. Sundal wanted the notice of the petition advertised at once "as we are very anxious to have the vote taken as soon as possible, so we may get our organization under way, and make a definite agreement for the construction of the ditch immediately, in order that the system may be completed this season." [115]

The petition was immediately returned to Sundal because the petition and the pages with the signatures had been sent to the Department of Public Works as two separate documents. Sundal was advised that it would be necessary for him "to have them so connected with one another that they come to us as one document otherwise the petition [could] not be recognized."[116] After a paper clip or some other device was used to deal with this problem, the necessary steps to create the district quickly followed.

The notice of the intention to create the Taber Irrigation District was published in the *Taber Times* on 27 May 1915 and in the *Alberta Gazette* on 31 May 1915. On 11 June 1915, Lawrence Peterson of Barnwell was appointed Returning Officer to supervise the vote of land owners on 29 June at the Barnwell schoolhouse. Immediately after the vote in favour of creating the TID, Lawrence Peterson, John Anderson and Carl Cook were elected trustees for the new district.

The new board of directors held their first meeting on the afternoon of the vote and Carl Cook was elected as the first Chairman of the TID Ted Sundal was appointed Secretary-Treasurer. The ministerial order creating the Taber Irrigation District was published in the *Alberta Gazette* on 15 July 1915.

The vote on June 29 was the occasion for a community celebration which began with speeches by A. J. McLean, the Minister of Public Works in the provincial government, Mayor Malo of Taber and others. A musical program featuring various people from Barnwell "consisted of songs, duets, quartets, among which was an original composition sung by four men entitled Irrigation."[117] The musical presentation was followed by a baseball game between Taber and Barnwell and a dance.

The Irrigation Branch of the Department of the Interior immediately came to the assistance of the TID preparing a land classification survey and offering to serve as its district engineer. The duties of the district engineer were to make detailed surveys of the land, prepare maps and plans, estimate costs of construction, estimate annual maintenance and operating charges and then prepare an assessment roll for the district. The Irrigation Branch land classification survey was begun on 9 October 1915 by a party under the direction of J. S. Tempest.. The willingness of the Department of the Interior to undertake this work was probably the result of their experience with the western section of the CPR block. Government surveys there had shown that over 200,000 acres of land had been incorrectly classified by the CPR surveyors.

Despite the offer of help from the Irrigation Branch, the board hired David Walker Hays in October of 1915 to serve as the district's first engineer. His services were to be retained until the work of classifying the land, preparing an assessment roll and building and testing the system by the running of water was completed. He was to be paid $20 per day with the total payment not to exceed $500.00. At the time of his appointment, Hays was the manager of the Medicine Hat-based Southern Alberta Land Company which in 1909 had begun construction of an irrigation system to take water from the Bow River at Carseland to irrigate land from the Vauxhall area eastwards to Suffield.

FURTHER DELAYS

With the design work and land classification survey well underway, it appeared that construction could begin in the spring of 1916. Beginning in December 1915, however, a number of issues derailed plans for the new canal and distribution system. All related in one way or another to the need to change the boundaries of the district. The first problem with boundaries concerned the inclusion of the school lands. The first approach was to change the law. The TID asked the Government of the Province of Alberta as the beneficiary of the school lands program to consent to the taxation of school lands to raise funds for the construction of irrigations works under the Irrigation Districts Act. On the 16 February the Province of Alberta complied with the request and passed an order-in-council approving federal legislation to give the school lands within the Taber Irrigation District the same status as patented lands. The federal legislation passed shortly thereafter provided for the inclusion of these lands in the district provided that in the event of any such lands being sold under the provisions of the Irrigation Districts Act, the sale price had to include an amount sufficient to fully satisfy the prior claims of the Crown. The practical effect of this legislation was to make the Irrigation District bonds a second mortgage and not a first mortgage on the schools lands. This method of trying to keep the schools lands within the district was found to be unsatisfactory to the CPR, who declined to accept the Taber Irrigation District's bonds. The problem created by the school lands was solved on 21 February 1917 when the Board of Directors requested their removal from the district.

A second reason for boundary changes was water supply. Given the size of Chin Reservoir the water available for expansion permitted the irrigation of 17,000 acres. The land classification survey of the Irrigation Branch indicated 19,000 acres of irrigable land were included in the TID. The third reason for the re-examination of the boundaries of the districts was the large number of applications received from farmers asking to have their land included in the district after July 1915.

The opportunity to accommodate all of these changes came in 1917 when the Taber Irrigation District failed to request an extension of the period in which to negotiate an agreement with the CPR for the construction of the system. Having failed to request this extension, the TID technically ceased to exist. The boundaries were amended to exclude all school lands and to include other lands at the eastern end of the district. The 14 April 1917 issue of the *Alberta Gazette* contained the notice of the intention to create the district for the second time. The *Taber Times* for 26 July 1917 included the notice of a meeting at which landowners would vote whether or not to create an irrigation district. The vote was followed by the immediate election of the same people to the Board of Directors.

With the boundary issue resolved, the Board of Directors anticipated construction would begin no later than the spring of 1918. The Board of directors, however, were disappointed again when the CPR decided that because of increased construction costs caused by the First World War it was no longer prepared to build the system under the terms of the original contract which was based on a cost of $11.50 per acre. The CPR was not amenable to any of the efforts by the TID to change its mind thus ensuring that construction would not start until after the end of World War One.

At the end of the First World War, farmers around Barnwell and Taber still faced uncertainties about the future of irrigation in their district because of the ongoing reluctance of the CPR to build the system. The people in the area had, however, shown themselves to be resourceful and good organizers whose problems were not of their own making. Despite the delays, they had created an institution which would be the basis for the growth of the industry in the post-war period.

TEN

DROUGHT AND EXPANSION THROUGH THE CREATION OF IRRIGATION DISTRICTS 1918 -1926

The next phase in the history of the St. Mary Project was initiated by a drought which began in 1917 and continued through to 1926. Like the 1890s drought, it resulted in a dramatic rise of interest in irrigation on the part of boards of trade and farmers. Given the spread of settlement since the 1890s, interest this time extended as far south as Milk River and as far east as Foremost, Etzikom, Bow Island and Medicine Hat. In 1918, Lethbridge received the lowest amount of rainfall since measurements had been taken. The 7.6 inches received that year was down from a high of 24.8 inches in 1916. The amount of rainfall steadily increased after 1918 reaching 23.8 inches in 1927. In the Medicine Hat area, with greater fluctuations in rainfall, the drought was more severe. In 1919 it received 7.7 inches compared to 13.6 inches in 1923; 9.9 inches in 1924; 25.3 inches in 1927; and 7.6 inches in 1928. In July of 1917, the average temperature in Medicine Hat was over 90°F compared to 77°F in 1915.

The effects of the drought were evident in the drop in wheat production. In the Lethbridge area, dryland crops averaged 35 bushels per acre in 1915 but dropped to 13, 5, 0, 5 and 6 bushels for 1917, 1918, 1919, 1920 and 1921. The 1918 CPR *Annual Report* for the Lethbridge section provided more details of the economic disaster which was overtaking farmers, particularly those beyond the salvation provided by irrigation:

> The general conditions for farming this season have been adverse. The year 1917 itself was a dry year, the driest since 1910 until the present one. There was therefore very little moisture in the ground at seed time and poor germination resulted. The cold weather with frost almost every night from May 20th to June 6th retarded the growth very materially, especially of alfalfa and timothy. Late in June and throughout July the temperature ranged very high with strong winds blowing almost continuously. These conditions caused a total failure on practically all non-irrigated land in this section of the province and very materially reduced the yield on well irrigated land.[118]

Low commodity prices made the crisis brought on by the drought even worse. Wheat dropped from a high of over $2 in 1918 to less than half that in 1922 and 1923. The price rose to over $1 in 1924 but never returned to First World War levels for the rest of the decade. There was no drop in the cost of living to compensate the farming community for this drop in income.

ORGANIZING THE FARMERS AND BUSINESS COMMUNITY

The first indication of a heightened interest in irrigation and in the government building irrigation systems as public works came at meetings held in Fort Macleod on 20 December 1918 and 28 January 1919. The first was a public affair organized by the Fort Macleod Board of Trade featuring Commissioner of Irrigation Peters as the guest speaker. It was widely advertised using hand bills which asked everyone interested in irrigation development in the Fort Macleod district to attend. Peters dealt primarily with the Lethbridge Northern Irrigation Project, but he also discussed the overall role of the federal government in irrigation development in the West. This was, he said, to carry out the necessary development surveys to prove the feasibility and estimate the cost of irrigation in the drier areas of the province where water seemed available. Construction was left to individuals, private corporations or cooperative efforts by farmers. D. E. Mathieson, who served as the secretary for the meeting, felt "it might be desirable for the Dominion Government to go further than making the development surveys only, and institute a policy that would carry the desirable projects through until they were actually constructed, and the water delivered to the land for use."[119] The meeting ended with the creation of two committees to deal with the regions north and south of Fort Macleod. The committee dealing with the land to the south of the town acted first by organizing a petition signed by the area farmers requesting that the government further investigate the idea of using the water of the Waterton to irrigate the area and, if found practicable, to undertake its construction as a government project. This petition requested, in effect, that Mathieson's idea of government investment in irrigation be implemented.

On the 28 January 1919, Peters attended a regular meeting of the Fort Macleod Board of Trade which discussed the issue of irrigation development. In his comments at this meeting, Peters went well beyond the immediate issue of irrigation development in the Fort Macleod area to discuss some of the fundamental problems of irrigation development throughout southern Alberta, one of which was financing. He noted the greatest difficulty to be overcome was :

> the question of finances.[He] pointed out that eastern financial houses were not familiar with the subject of irrigation, and that the people in the East generally were not familiar with or favourable to irrigation development. [He] pointed out that it would be necessary to go either to the Government in Eastern Canada, or to the large financial institutions in Eastern Canada to get the capital for the proposed development, and that before this capital would be forthcoming it would probably be necessary to carry on a good deal of educational work to show that the people in the West need irrigation and must have it, and further, to show that irrigation is no matter of experiment, but recognized as an established undertaking and one that will undoubtedly provide the greatest improvement to the agricultural areas in southern Alberta.[120]

In the discussion that followed, it was pointed out that the task of education with respect to irrigation was too large for one community and that the best approach would be for "all of the interested districts [to] get together and combine with a view to making the strong effort that would be necessary to place these questions favourably and strongly before Eastern interests."[121]

The creation of such an organization began when on January 1919 the Raymond Board of Trade organized a meeting attended by 300 farmers from all points between Nemiscam in the east and Magrath in the west along with Commissioner Peters, W.A. Buchanan, M.P. for Lethbridge and George Marnoch, President of the Lethbridge Board of Trade. The meeting began with speeches of encouragement by the three guest speakers. Peters outlined the plan for the LSEP which included reservoirs to assure an adequate supply of water. The question period following his presentation dealt with the division of the water in the St. Mary and Milk Rivers between Canada and the USA and the canal built by the CNWICo. to divert water from the Milk River for use in Canada. He estimated the cost of completing this canal to Etzikom Coulee and Pakowki Lake to be $70,000 to $80,000. Many attendees supported the idea that the government should go ahead and build the canal, charging the cost to the landowners served by the scheme and handing the system over to the farmers when it was ready for water to be turned on to the land. With respect to the construction of the system, Peters said the government had not yet defined a policy, but he intimated it would be negotiated between the government and the farmers.

Buchanan declared his support for irrigation. He also declared that the Minister of the Interior, Arthur Meighen was "greatly concerned over the possibility of irrigation in Southern Alberta, its need and about the various projects."[122] He argued that government support was justified because it would help keep people on the land where they had been placed by the government itself. His second justification was that Canada had a large national debt and that its payment was only possible through the full development of its natural resources. He concluded his speech with the suggestion that if the farmers convinced the government that they wanted the water for the proposed project, the government would do its best to supply it. Marnoch echoed Buchanan's view of the need for the farmers to make their views known and noted that Lethbridge was interested in irrigation because it would lead to a more stable business economy.

Following these presentations, a committee was created to organize the various towns and regions interested in irrigation. It was made up of a chairman and secretary with representatives from Lethbridge, Magrath, Stirling, Foremost, Etzikom, Milk River, New Dayton, Nemiscam and Welling. Raymond was the best represented since John Powelson was elected chairman, John Anderson was secretary and T. J. O'Brien and J.W. Evans served on the committee.

The committee immediately organized community meetings through to 12 March at Magrath, Taber, Foremost, Etzikom, Milk River, New Dayton, Bow Island and Burdett. These meetings followed the pattern established at Raymond with encouraging speeches from Peters, Buchanan, and Powelson as well as William Fairfield of the Dominion Experimental Farm and Don Bark from the Experimental Station at Brooks. Each meeting concluded with a unanimous or near unanimous motion of support for the construction of the LSEP. According to Peters, those farmers who were in favour of irrigation generally operated large farms and were the best educated and most intelligent men in the communities while the farmers who opposed irrigation were generally small farmers and "judging from appearances and speech, are men of very little education, and with no knowledge of what is being done outside of the locality in which they are at present residing."[123] A committee was often organized for the purpose of circulating a petition to create an irrigation district and work with the central committee in Raymond. In addition to attending these meetings, Peters encouraged participation in this movement to convince the federal and provincial governments to build the LSEP by distributing copies of the plan for the LSEP to all the communities southeast of Lethbridge.

INTERNAL DISAGREEMENT

The series of community meetings had a strong impact on Commissioner Peters, who kept the Superintendent of Irrigation Drake informed through detailed reports. After his second meeting in Fort Macleod, he indicated to Drake that "the interest in irrigation matters that is being displayed at the present time, is generally much greater than at any previous time, I believe, in the history of Western Canada, and certainly since 1911."[124] The current interest, in Peters' opinion, was distinguished by the extent to which Boards of Trade were becoming involved. He cited the Boards of Trade at Fort Macleod, Lethbridge, Raymond and Cardston as examples of this involvement. He also noted the recent visit to his office by two representatives of the Taber Board of Trade who indicated that their organization was going to take an active interest in irrigation development in their district. The farmers of the Taber area had always been interested but this was the first time that the people in the town of Taber had definitely and strongly become interested in irrigation development. As interest in irrigation increased so did Peters amazement and willingness to provide assistance. In a letter to Drake dated 20 February 1919, he indicated:

> it is quite impossible for me by correspondence, to make you understand the complete change which has come over all the south country in reference to its attitude towards irrigation development, and make you appreciate the enthusiasm that has been displayed at these meetings. The favourable attitude towards irrigation shown at all of the meetings has been a complete revelation to me, and it has seemed almost beyond belief that such a feeling can exist when I look back upon the feeling which existed just before the war.[125]

Peters strongly urged that if the enthusiasm continued to grow:

> we should be prepared to give the interested farmers every assistance possible in working this matter out, in the way of placing our advice at their disposal, and also by being prepared to make further surveys to determine more definitely the possibilities of irrigating certain small and local areas which have not yet been sufficiently developed by our general surveys of the project.[126]

Peters was of the view that this interest should be encouraged and all possible assistance provided to bring about a change of attitude on the part of the institution which he represented. Drake, however, was not impressed by the ground swell of support for irrigation developing in the West. He took exception to Peters' comments about people in the East being opposed to irrigation. He felt that it would be more accurate to say that they were indifferent to irrigation and that it was Peters' duty to:

> explain the existing situation, with its uncertainties and difficulties, as clearly as possible whenever [he was] called upon to address any gathering of people in the interests of irrigation. I do not think it would be wise for you to urge upon them the desirability of bringing pressure to bear upon the Dominion Government to cause surveys to be made or to have works constructed for the purpose of irrigating additional areas of land, unless at the same time, you clearly place before your hearers the difficulties in the way.[127]

The difficulties identified by Drake included the uncertain water supply, the high cost of reservoirs, uncertainty as to the willingness of settlers to pay a high construction cost when they know the facts and the difficulty of financing the construction as prescribed in the *Irrigation Districts Act* unless financial support was given by either the provincial or federal governments. Drake felt that there was no chance of this support. Drake, appraised of government attitudes realistically, and correctly accused Peters of making promises that neither the federal nor the provincial governments would keep. The disagreement between Peters and Drake was not made public in any way, thus preserving the impression that the federal government actually had a plan to deal with the water crisis in southern Alberta.

LETHBRIDGE IRRIGATION CONVENTION

The culmination of these mass meetings was a convention in Lethbridge on the 13 and 14 of March 1919. Its main purpose was to publicly demonstrate support for the construction of the Lethbridge South Eastern and the Lethbridge Northern Irrigation Projects. The President of the Lethbridge Board of Trade, George Marnoch, encouraged the participation of as many farmers as possible in the convention. As he indicated to the *Lethbridge Daily Herald,* prior to his departure for Edmonton to interview the provincial government, "the convention is no invitation affair, but is in the interest of every farmer who wants irrigation to be present and secure all the facts at first hand as well as providing the body of public opinion which will impress the governments of the Dominion and the province with the necessity of taking action."[128] The *Lethbridge Weekly Herald* on 12 March 1919 made an even more dramatic appeal for support:

> The salvation of the south country, and that means the salvation of the farmer, hinges on the great irrigation project of fifteen millions to be launched. The convention should be made a record one. Every effort should be made by those interested in irrigation to give the mighty project the mighty boost it deserves. A great agricultural revolution is on foot and the call is out for the farmers of the south country to man the guns.[129]

The convention began with a noon luncheon courtesy of the Lethbridge Board of Trade on 13 March 1919, at which time the plans for the following day were finalized. The luncheon, in the opinion of the *Lethbridge Herald,* was "the biggest gathering of any luncheon probably in the history of the Board of Trade."[130] George Marnoch called upon various people to make some remarks. Don Bark, who was a staff member at the CPR Experimental Farm at Brooks, commented on the contrast in the crops which he noticed on his trip to Lethbridge from Brooks. He mentioned the case of :

> two portions of land of 30 acres, on one sown to barley and the other to alfalfa, the former on dry and the latter irrigated land. The alfalfa was a complete success and the barley a total failure. In the case of the two areas, there was only the irrigation ditch between them so far as distance went. Knowing Alberta as he did, he felt sure that the irrigation projects under

> consideration would mean millions to the district. The soil is all right, the climate is all right and the water supply is there to make the prospects permanent.[131]

Lawrence Peterson, the MLA from Taber and a founding member of the TID indicated that the people of Taber were preoccupied with the final arrangements to extend the CPR irrigation system to the 17,000 acres located in the area but "the farmers of the district were anxious to help all their fellows to get the benefit of irrigation."[132] Mr. Floyd of New Dayton stated that on his farm there were 6,000 acres that needed irrigation and that the farmers in the area wanted all the land in the area irrigated. Mr. Bailey, who had organized the first meeting in Fort Macleod, spoke of his experience with an irrigated tract of land near Fort Macleod and commented upon the way irrigation had permitted Lethbridge to have beautiful parks and lawns for homes. "What can be done from a flower and lawn standpoint," he concluded, "can be equally done with regard to the farms from a crop point of view."[133] Mr. Beattie of Bow Island was not optimistic that irrigation would ever be extended to his area but he hoped that the farmers in the Burdett area would receive its benefits. John Powelson estimated that 90 per cent of the farmers in the Raymond area were "fully enthused" about irrigation. Christian Jensen of Magrath explained that the farmers of Magrath had been trying to get irrigation for the last 15 years and were:

> led to believe that it was impossible in that all the water in the St. Mary and the Belly River was used by the CPR. But the real situation had been explained by Mr. Peters and the CPR was ready to take hold of the matter in the way of bringing the irrigation needed. He had been long enough in the country to know that we had our good and bad seasons and that the profits of the good were spent in tiding over the bad. Irrigation would prevent this. With irrigation going there would be the establishment of canning industries, the benefit of keeping the money here instead of sending it out nor would there be the Russian thistle so much around.[134]

William Fairfield spoke on the potential for accelerated tree growth as a result of irrigation. Commissioner Peters noted that:

> since 1909 he had travelled every year in the south country and noticed a decided change of feeling among the settlers in favour of irrigation. Good land, good settlers, good water available, made a trinity which, aided by good spirit, will bring about the great development which was being advanced. The big irrigation project when completed will be the greatest ever developed in America and probably in the world and stands next to railway development in point of importance to southern Alberta. It needed all the cooperation of the people of the south country to be assured of the support of the government bodies.[135]

The luncheon meeting ended with the drafting of eight resolutions for the consideration of the convention the following day.

The convention began the following day at the Majestic Theatre at 10 am with an enthusiastic group of attendees estimated at over 400 including 30 from New Dayton, 20 from Nobleford, 25 from Coaldale, 50 from Monarch, 45 from Kipp, 15 from Milk River, 30 from Stirling, 25 from Bow Island, 75 from Taber, 2 each from Coutts and Foremost and 1 from Etzikom. Attendance was increased with the arrival of the Cardston train at 11.15.

Mayor Hardie's address of welcome indicated that the "interests of Lethbridge were tied up with the interests of the farmers, the prosperity of the farmers meant the prosperity of the whole country."[136] The Mayor's opening remarks were followed by presentations by Peters, Fairfield, and Marnoch. Peters used a map to show various projects under consideration. They included the Taber extension, the Lethbridge Northern and the Lethbridge South Eastern comprising in all a total of 500,000 acres. Peters encouraged the delegates by reporting that the surveys for the LSEP were not rough and incomplete but careful and in detail and that the water supply from the St. Mary River was probably adequate, but the Belly and the Waterton could be tapped if that was not the case. He estimated that the cost for the 314,000 acres to be included in the project would be about $31 per acre. Peters also "did away with the idea that the CPR was hogging the water in the southeast."[137] He said that the issue of water availability in this area had more to do with the interpretation of Article 6 of the International Waterways Treaty of 1909 by the International Joint Commission than with the CPR.

Marnoch then took the floor to report on his recent trips to Edmonton and Ottawa and to lead the discussion of the resolutions which had been prepared the previous day. Marnoch began by pointing out that the congress was "hoping today to lay the foundation of a development that will do more to stabilize our business and our homes in this new country than anything that we have hitherto attempted."[138] He paid tribute to William Pearce for having initiated the effort of the federal government to survey the water resources of the area. The main task for the irrigation industry, given its development to date, was to convince government of the value of the projects under consideration because:

> we know that we ourselves cannot, unaided, carry these works to completion. On the other hand, we have to recollect that we have to persuade the government that it is their duty on our behalf to help us to finance and to construct these works; we see that we can make them so profitable to ourselves that the repayment of the principal monies, and the payment of the interest charges and the costs of maintenance and operation will be easy for us to carry; but the work of persuading the governments to back us still lies before us.[139]

Among the eight resolutions considered, three dealt directly with irrigation. The first asked the federal government to complete all surveys and plans for the various irrigation projects that could utilize water from the Oldman, Milk, St. Mary, Belly and Waterton Rivers. As part of this same resolution, both the provincial and federal governments were asked to help in the formation of irrigation districts and in the financing and construction of the necessary works because the projects involved "the expenditure of such capital sums as would be difficult for us to raise on our own initiative."[140] The second resolution called for the creation of a committee to advise the Alberta Government on revisions that were necessary to the Alberta *Irrigation Districts Act*. The most urgent changes were to be made at the current session of the Alberta Legislature.

The final resolution dealing specifically with irrigation, called for the formation of an irrigation development association which would continue the important work initiated at the conference. The first of the three aims of the association was to "further the use of irrigation water from our mountain streams for the stabilization of our farming operations and for the steadying from year to year of our business operations."[141] The other aims were to assist in the formation of irrigation districts and to maintain the link between the farmers and the federal and provincial governments "to secure the assistance of these governments, financial and otherwise for the furtherance of the work of the association."[142] Membership was to be open to all interested farmers and businessmen for an annual fee of $2.00, of which $1.50 was to be provided to the association's central office in Lethbridge and 50 cents was to go to the local branch of the association. Local associations were to be established north, west, south and east of Lethbridge at Fort Macleod, Monarch, Barons, Carmangay, Coalhurst, Iron Springs, Turin, Grassy Lake, Bow Island, Purple Springs, Taber, Barnwell, Chin, Coaldale, Foremost, Etzikom, Nemiscam, Wrentham, Stirling, Raymond, Magrath, Cardston, Glenwood, Hillspring, Milk River, Warner, New Dayton and Masinasin. These towns included the territory of the proposed Lethbridge Northern and Lethbridge South Eastern Projects as well as the area proposed for inclusion in the United Irrigation Project. George Marnoch was elected chairman of this new organization.

The highlight of the afternoon session was the reading of telegrams from Arthur Meighen, the Minister of the Interior, and Premier of Alberta, Charles Stewart. Meighen indicated he was ready to continue the work of planning the expansion of the St. Mary Project based on a "clear understanding that responsibility from there on is provincial or local and not Dominion."[143] He was also willing to loan the province funds to help it purchase irrigation district bonds. The telegram from Premier Stewart also avoided making any commitments consistent with the objectives of the convention and simply indicated that instruction in irrigation farming would be an important part of the curriculum in the provincial schools of agriculture.

The convention ended with high hopes that the new Irrigation Development Association would be able to achieve its objectives. It also ended with the support for its objectives being obtained from outside the Lethbridge region. The *Calgary Herald* described the convention as:

> a splendid advertisement for irrigation and a remarkable testimony to the intelligence and enterprise of the south country agriculturalists. It is to be hoped the provincial government will

> consider the proposal favourably and that the work will be authorized at an early date. It will take some years to complete, and may cost something like twenty millions but will be worth much more than that.[144]

POST-CONFERENCE DISAPPOINTMENT

The Irrigation Development Association made immediate plans to carry out its objectives. At the first meeting of its executive, held on the 14 March 1919, a memorandum was prepared for presentation to Premier Stewart. The document argued that a fundamental change had to be made in the way irrigation development was undertaken in Alberta because the private sector could no longer be depended upon to take the information obtained from federal government surveys and use it to construct irrigation projects. Irrigation had entered a new phase in which government would have to go beyond its traditional role of simply undertaking surveys and assume greater responsibility for the construction of projects.

The executive of the Irrigation Development Association met Premier Stewart and his cabinet to present this document on 17 March 1919. Stewart made three points in his response. First, the Dominion Government had always maintained jurisdiction over irrigation and the Province of Alberta had not been officially informed of the federal government's desire that the province should accept this responsibility. His second point was that if the province accepted responsibility for irrigation it would be necessary for the province to review the projects already laid out by Dominion Government officials before any decision on funding could be made. His third point was that the province did not propose to borrow money from the Dominion Government and other financial arrangements would, if this responsibility was accepted by the Province, have to be made. Based on these considerations, Stewart concluded that it seemed to him that "the federal authorities should advise this government definitely as to their future position in connection with irrigation matters. If they decide to drop it altogether, then this government will endeavour to find a solution of the difficulty and protect the interests of those so vitally concerned."[145] In order to ensure that there was no delay in resolving the issue of jurisdiction, Stewart indicated that he would be opening this matter with the federal authorities at once. Resolving the issue of jurisdiction over irrigation was part of the larger issue of the ownership of the natural resources of Alberta which had been retained by the federal government when Alberta was created in 1905. This issue was not fully resolved until 1936.

This meeting finally revealed that support for expanding the St. Mary Project as a public work ended with Commissioner Peters and did not extend to more senior bureaucrats or politicians in Ottawa or Edmonton. Having decided not to build irrigation systems as public works, during the 1920s, both levels of government put the entire burden of expansion on the water users through the creation of irrigation districts. The federal government continued to provide technical assistance while the Province of Alberta provided bond issue guarantees for some districts. It was also made clear to Peters that he should not raise "hopes that will never be realized." The Irrigation Development Association quickly faded from view and there was no further coordinated attempt by farmers or the business community to persuade the government of the need for action until 1939.

Despite the fact that during the 1920s there was no chance that the federal or provincial governments would fund expansion of the St. Mary Project, farmers in the New Dayton, Warner and Masinasin areas initiated the creation of irrigation districts between 1919 and 1922. Local interest in the extension of the St. Mary Project east of Taber to Burdett continued into the 1930s with the creation of the Lethbridge South-East Irrigation Organization. Extension of the St. Mary Project east of Taber and east of the Milk River Ridge was beyond the means of the water users if they were expected to pay the entire cost. The only successful expansion of the St. Mary Project achieved in the 1920s was through the creation of irrigation districts which could obtain their water supply from the CPR. This approach to expansion used the existing headworks and main canal system and thus was an option available only to the communities of Taber, Magrath and Raymond.

ELEVEN

THE TABER IRRIGATION DISTRICT 1919 - 1946

Drought and the end of World War One released the pent-up demand for completing the task, begun in 1912, of putting irrigation east of Chin Coulee on a firm foundation. This began in 1919 with the negotiation of a construction contract with the CPR. Following construction, the TID implemented a number of very progressive ideas with respect to management, the most important of which was hiring a full-time manager and full-time ditch riders rather than depending on lateral associations organized by farmers to handle water delivery and maintenance. In the 1920s, operational problems such as surplus water, weeds and drifting soil were dealt with as effectively as possible through the expansion of the district and the use of chemicals. The TID also took an aggressive approach to the development of irrigation agriculture through such initiatives as the establishment of a cannery. The result was the creation of a mixed agricultural economy by 1929 in which sugar beets, alfalfa and barley along with wheat were the main crops.

The Depression and World War Two brought difficulties to the TID, but the organization had weathered adversity in the past. World War One delayed construction for at least five years. Drought and the resulting dust storms in 1920 filled the newly excavated canals with soil. The post-World War One depression in commodity prices made the shift from dryland to irrigation agriculture with its higher input costs and greater labour demands more difficult. The Depression delayed the payment of water rates, delayed payments by the TID on the bonds and the water rental to the CPR and limited some of the district's maintenance operations. The Depression ended with the Second World War which brought higher commodity prices and a more active economy, but also made it more difficult as staff were attracted by the higher wages paid by war-related industries.

The TID took the challenge of drought, depression and war in its stride and continued to pursue the development of irrigation agriculture. Having established a mixed farming economy by 1929, the TID shifted its emphasis in the 1930s to the development of industries based on the products of its irrigation farms. It was the second community, after Raymond, to successfully accomplish that goal. The TID's approach involved investing its own funds to build a cannery which attracted further investment. The farmers cooperated by growing crops needed by the canneries and other industries.

The contributing factors to this success included a stable water supply, courtesy of Chin Coulee Reservoir, which permitted the farmers to grow a wide variety of crops and not suffer water shortages at critical times in their growth. The advantages of stored water became increasingly obvious as the ability of the St. Mary River to provide a

Left to right seated: Wm.Valgardson, trustee; John W. Anderson, trustee; Carl C. Cook, trustee; B.R. McMullin trustee; Back row; F. P. Stevens, trustee; Wilbern S. Johnson, trustee; Ted Sundal, secretary- treasurer; E.N. Francis, trustee; Nephi J. Jensen, trustee; John G. Barton operating superintendent; Ken Anderson, Manager. (Irrigation Impact Museum, Taber)

steady water supply to the rest of the St. Mary Project declined to a new low in 1936. The terms of the contract with the CPR were also a benefit since payment of principal on the construction bonds was deferred for 10 years and did not include a charge for the construction of the AR&ICo. headworks. The TID was also able to control its level of debt and had extra income from the sale of land in the western extension and in the Barnwell Gardens subdivision. The services of Ted Sundal, as Secretary-Treasurer from 1919 to 1967, and Manager from 1920 to 1936, provided the TID with strong leadership and continuity during its formative years. Sundal took an interest in all aspects of the district's operations from the implementation of a centralized system of water distribution to the development of industries based on irrigation agriculture.

The TID was thus able to continue its operations with few changes from the way it began in 1921. Some adjustments were made to accommodate the Depression but overall the TID was able to build on the success achieved during the 1920s. By 1939, the TID had come to symbolize the potential of the St. Mary Project given the availability of stored water to all of its water users. John Evans, the Secretary-Treasurer of the Raymond Irrigation District, was particularly impressed with the results achieved. Following a tour in September 1939, he indicated in a letter to *The Lethbridge Herald* that for "those who had come from partially dried out districts, it seemed as if we had suddenly been transported to the Garden of Eden in all its pristine beauty and excellence. The fields were moist; the crops luxuriant and bountiful."[146]

CONSTRUCTING AND PRIMING THE SYSTEM

The immediate priority at the end of the First World War was to compel the CPR to build the system based on the 1917 contract or to negotiate a new one. The 1919 Board of Directors of the TID, Lawrence Peterson, Carl Cook and J.W. Anderson took advantage of the 11 February 1919 irrigation rally at the Rex Theatre in Taber to bring some pressure to bear on the CPR. Carl Cook advised the 500 delegates that:

> though the scheme has been worked out in detail and an agreement drawn up, the CPR were now holding back from carrying out the project on the grounds of high cost of material and labor. The Taber Irrigation District committee

> had done all they could to hasten the enterprise, and were now looking for a solution to the impasse. They were confident that it would pay to undertake the scheme even at an increased cost, and they would like to see it completed this year.[147]

Peterson followed Cook with a motion that:

> this convention petition the CPR to enter into contract immediately with the Board of the Taber Irrigation District with a view to having the work completed this year so that water can be turned into the system in 1920 thereby gladdening the hearts of the people and making the land to blossom as the rose.[148]

Despite the passage of the motion in the spring of 1919, the CPR continued to avoid completing negotiations for a construction contract. This delay prompted Peterson to request assistance from the Department of the Interior. In a 10 April 1919 letter to E. F. Drake, Director of the Reclamation Service, Peterson indicated that the citizens:

> would appreciate it very much if [he] could bring any influence on the CPR to get them to speed up in the matter, as the people have been disappointed so many times in the last five years on this project that I think there should be no unnecessary delay.[149]

The impasse was resolved in May 1919 when the CPR sent the TID a revised contract which increased the cost per acre from $11.50 to $16.00. It was unanimously approved at a meeting held at Barnwell School on 13 May 1919 with only minor changes requested with respect to the guaranteed amount of storage water in Chin Coulee Reservoir and a reduction in the duty of water. These issues were dealt with the following day in a telephone conversation between Peterson and Naismith, thus clearing the way for the drafting of a final contract which was signed on 12 July 1919.

The contract identified the work which was to be carried out and designated the CPR as the project manager. The CPR agreed to secure, at its expense, the rights of way for the main canal while the TID agreed to acquire, at its expense, the right of way required for the distribution system or to acquire easements in perpetuity over the land required. The TID agreed to obtain the approval of the Department of Pubic Works for all road allowance crossings.

Following the construction of the system, Chin Coulee Reservoir was to be filled from the existing works when the full capacity of the system was not required for supplying water to the CPR's own water users. The CPR was not to be liable for damages incurred by TID water users if the reservoir could not be filled for any reason beyond the company's control. The water obtained by the TID could only be used to irrigate land within the district and to fill ponds for watering livestock. Allowing water to run to waste was also forbidden. Refusal by the TID to take water and pay the rent for two consecutive years terminated the contract. The contract, therefore, had a number of similarities with the water contracts between the CNWICo. and their water users.

The TID agreed to pay $272,000.00 for the construction work and $8,475 annually for the rental of 34,000 acre-feet of water. This amount was to be paid even if less than this amount of water was diverted. The TID was to pay the CPR in debentures with an interest rate of 6%. Repayment of the principal was to be in 20 equal instalments of $13,600, beginning on 1 August 1929. Interest payments began on 1 August 1920 with the first 10 payments being $16,320.00. The interest payments dropped by $816.00 per year so the last interest payment, due in 1948, was $16.

In addition to the deferment of the payments on principal, the contract with the CPR had the advantage of not charging the TID for any of the costs of building the headworks or Chin Reservoir. The *Lethbridge Weekly Herald* pointed out that the TID had received the maximum advantage from the St. Mary Project system at minimum cost from its first year of operation. The Taber farmers, it said:

> instead of commencing at the river and building many miles of ditch before reaching their lands, have been able, by cooperation with the CPR Lethbridge irrigation system, to link up with the water within 25 miles of the land to be watered. The result is that they are able to get water delivered to them at a construction cost of about $16 per acre, as against $25 to $35 which they would have had to pay had they been required to build the headworks and to have brought the water all the way from the river. That is why the Taber and Barnwell farmers are patting themselves on the back.[150]

The advantage of stored water rather than depending directly on the flow of the St. Mary River was also pointed out by Sundal in his 1921 Annual Report, which noted:

> the convenience of reservoirs within a few miles of our district, is an outstanding asset that few perhaps fully realize. The feature of being able to run the canal at full capacity, whenever desired, is a matter of great importance to crops results. On the other hand if we were dependent upon a live river stream that often decreased in volume in mid-summer . . . it would be impossible to meet capacity requirements.[151]

The first sod was turned on 21 July 1919 at a site approximately two miles south of Taber on land owned by Bert Holman. The CPR awarded the general construction contract to Messrs. Grant, Smith and MacDonald with Herb Campbell serving as their senior representative. Scarcity of labour and dust storms delayed completion of the project. As late as June 1920, G. N. Houston, of the Department of the Interior, reported that while the main canal had been excavated for its entire length, the heavy spring winds had blown it full of sand in a number of places thus requiring re-excavation before water could run. A two mile stretch of the main canal had been filled to the top of the present banks thus requiring the handling of double the original yardage in order to clear the channel. About half of the laterals remained to be built and only a small force of about four teams was at this work. He estimated that one-third of those completed would have to be re-excavated. Almost no structures, with the exception of some highway bridges, had been built on the main canal. Bank gates were set in the main canal but almost none had been installed. The flume on the main canal southeast of Barnwell was about three-quarters finished but no men were working on it. About five teams were working on the south approach and the lateral taking off at that point. After pointing out other examples of the general lack of activity on the project, Houston came to the obvious conclusion that "on the whole it did not appear that any great effort was being made to hurry the work to completion."[152]

Construction was completed in October 1920, even later than Houston's prediction of late summer. Upon completion, the system included a main canal approximately 25 miles long, 21 laterals and 487 structures of all types with 280 drops and 100 branch gates. The major structure in the system was a box flume southeast of Barnwell. All structures were built of wood. A total of 662,000 bd ft of lumber was used in the construction. Waste water was returned to the Oldman River via the spillway at the end of Lateral Nine west of Taber. Waste water from the 12 laterals to the east of Lateral Nine tailed out into unoccupied land.

Priming the system began even before construction was completed. Water for this purposes was let into the TID main canal at 10.30 am on 10 September, 1920. Superintendent of the Lethbridge CPR Project, Sam Porter, and Ted Sundal turned on the first water to flow into the TID main canal. Sundal recalled that:

> it was a rare thrill to watch the first stream coursing down the canal and realize this event was the turning point for a people who so frequently had lost their crops through drought. The dried-out stock watering ponds and fodder shortages were acute problems at times but now the stream was actually on the way to solving the difficulty.[153]

The priming process moved systematically through the system while construction was being completed. A series of dirt dams were used to stop the water from interfering with the last of the construction work. The first dam was put in just beyond the junction of the main canal and Lateral One which could then be primed while the contractors removed dirt from the main canal to the east. The priming operations reached Lateral Nine by 22 October, 1920.

The priming process to this point revealed that the canal and the structures had many deficiencies. The dirt had not been properly backfilled and tamped thus leaving it loose and with hollow spaces. The wood used in the construction of some of the structures had not been properly nailed down with the result that planks floated away with the first little stream. In some cases, the structures were not complete thus the TID staff were "compelled to hunt around for a two by four or a two by six to fill in places that have been missed."[154] The banks on some of the laterals also needed to be raised. By early November 1920, water had been run through the entire system to a depth of at least three-quarters of the full supply. Sundal, with an emergency crew, went over the entire system while this full test of the system was in progress and identified additional problems. These included leaks in the Barnwell flume and inadequate approaches to a number of the highway bridges.

Staff from the Department of the Interior also inspected the system and found some of the same problems Sundal had identified. G. N. Houston's report indicated that two culverts were set eight inches too low. The banks of Laterals One and Three were too low while Lateral Three had not been excavated to a uniform grade. The Barnwell flume was also not satisfactory since it had not been built entirely according to the plans. The plans called for the use of No.18 Toncan metal, whereas, it was constructed of the best quality galvanized iron sheets of the same gauge. The manufacturers claimed that Toncan metal was non-corrective and therefore more durable than galvanized iron. The approaches to a number of the highway bridges were too abrupt and needed to be graded to an easier slope.

On 12 November 1920 inspection of the system continued when the entire board, along with officials of the Provincial Department of Public Works, including its Minister Archie McLean, inspected the bridges and culverts. McLean expressed the view that the bridge timber structures were satisfactory, but in nearly every case, the approaches required some dirt to meet the requirements of the department.

At a meeting with TID officials on 9 December 1920, the CPR agreed to pay the TID $200 to cover the additional work required to correct the deficiencies of the system identified in these various inspections. The TID formally accepted the system from the CPR on 17 February, 1921.

DISTRICT EXPENSES AND INCOME

Financing the TID began in July 1919 with the passage of its By-law 2 to provide for the creation of the $272,000.00 in debentures. The major expense in operating the TID through to 1945 was the payments to the CPR on the construction bonds and for the rental of water. The TID used bank loans to cover the operating costs pending the payment of the water rates. The operating loans were repaid each year with the payment of the rates. In 1932, the TID was forced to borrow more money because of the non-payment of the water rates and bank interest reached a high of $2,392.74. Besides the costs of debt, other operating costs included vehicle expenses which were $232.00 in 1920 and over a $1000 by 1937. Tools and equipment were generally less than $100 a year. Staff wages were $755.00 in 1920 but over $7000 by the 1940s. Weed cutting and lumber were two other regular expenses. The former reached a high of $1,562.45 in 1931 while the latter reached a high of $2,429.40 in 1941.

These expenses were covered by the water rate which was set in 1921 at $2.00 per irrigable acre. The water rights subject to assessment in 1921 were 16,824 acres thus giving the TID a total income of $33,648.00 for that year. This income provided the $16,320.00 required to pay the debentures debt and $8,475 for the water rental. The remaining $8,853.00 or 26% covered wages, materials, repairs and car operating expenses. In his 1921 report, Sundal expressed the hope that with careful management and the avoidance of serious breaks in the system, it would be possible to maintain an annual assessment of $2.00 per acre until the year 1929 when the first principal payment of 80 cents per acre was required on the bonds. Sundal was able to husband the district's resources well and there was no significant increase in rates until 1929 when they went from $2.00 to $2.60. The bonded debt and water rental payment made up $1.87 of the rate for 1929. The water rate dropped from $2.60 in 1929 to $2.00 in 1937 and remained at that level until 1945.

The TID also received income after 1928 from principal and interest payments on land purchased in the western extension. The total received by 1941 was $140,368.73 on a investment of $33,142.07. The district also earned income from rental of the dragline to farmers to dig ponds and to various government agencies for construction purposes.

RATE ENFORCEMENT

The Irrigation Districts Act, and the various revisions thereto, provided means by which the irrigation districts could deal with the nonpayment of the water rates. The Irrigation Districts Act indicated that all rates or arrears were considered as a debt to the board and that the assessment roll was the only evidence required to indicate the existence of this debt. The penalty for nonpayment of the rates by the 15 December in the year for which they were due was 5% of the amount owing after this date. In addition to assessing this 5% penalty, the district was authorised to seize property in the same way a landlord might recover rent in arrears.

The ultimate weapon which could be used by the district to recover rates was the confiscation of land. The procedures followed in this case required the preparation of a rate enforcement return which listed all the rate payers who had not

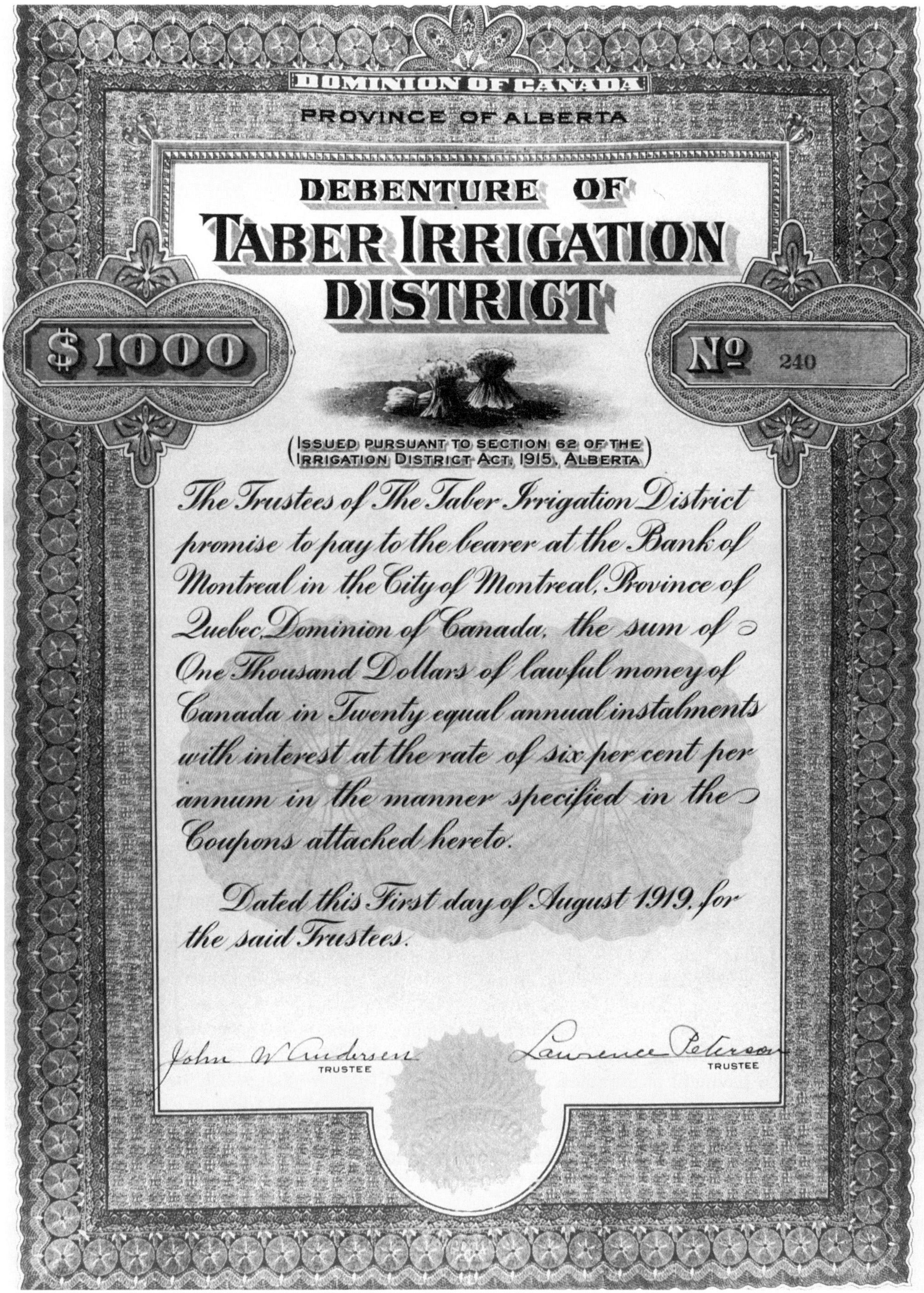

DOMINION OF CANADA

PROVINCE OF ALBERTA

DEBENTURE OF

TABER IRRIGATION DISTRICT

$1000

№ 240

(ISSUED PURSUANT TO SECTION 62 OF THE IRRIGATION DISTRICT ACT, 1915, ALBERTA)

The Trustees of The Taber Irrigation District promise to pay to the bearer at the Bank of Montreal in the City of Montreal, Province of Quebec, Dominion of Canada, the sum of One Thousand Dollars of lawful money of Canada in Twenty equal annual instalments with interest at the rate of six per cent per annum in the manner specified in the Coupons attached hereto.

Dated this First day of August 1919, for the said Trustees.

John W Andersen — TRUSTEE

Lawrence Peterson — TRUSTEE

(Taber Irrigation District)

paid their rates for at least two years. This list was then certified correct in a Court of Confirmation conducted by a Judge of the District Court of Alberta. The district then submitted this document to the Land Titles Office which was authorised to transfer title to the land from the farmer to the district. The land could then be sold to recover the amount owing. Failure to follow these rules on the part of district officials was punishable by a fine of $1,000.

Despite the economic conditions of the early 1920s, the non-payment of rates was not a problem for the TID during its early years of operation.. In fact, by 1923, the TID generated a financial surplus in its operations prompting it to consider purchasing some of its own bonds back from the CPR to create a sinking fund. In 1923 each water user was also given a rebate of 25 cents per acre of water right.

In the early 1920s the Board was mainly concerned that rate enforcement not be a hardship on the farmers. In July 1922 it requested a 90 day extension to all requirements under the Rate Enforcement Return, so that a delinquent would appear before a judge on or before 15 November instead of 15 August as required in the Irrigations Districts Act. The Irrigation Council revised the date to 26 October. On 4 November 1922 the Board requested the Irrigation Council to make a further 15 day extension since a number of the ratepayers had not finished threshing. The TID first acquired the ownership of land through rate enforcement in 1929 when seven parcels of land came into its possession by this method.

The Depression was not reflected by an immediate increase in the arrears on the payment of the water rates. From 1927 to1932, the amount of money owed the TID in water rate arrears averaged about $40,000.00 per year. In 1933 the amount of the arrears increased to over $100,000 and remained at that level until 1936. In 1936, 226 parcels or 72% of the parcels in the district were in arrears. In 1936 a total of 49 parcels of land had been acquired by the district by rate enforcement. On 16 March 1934 the TID held an irrigation rate land sale to recover unpaid irrigation rates for 1929 and 1930. Land not sold by the TID was generally rented on a share crop basis. The board also dealt with the nonpayment of rates by making limited use of its right, under 1933 revisions to the Irrigation Districts Act, to claim the income from the sale of wheat and beets.

The board was opposed to any concessions on the payment of water rates since it felt that it was important to protect the credit of the district. The board's strategy was to keep the rates low but not to offer any concessions on its payment. As the 1934 annual report indicated, the TID was a "full co-operative unit and [did] not make assessments for profit purposes, and in the event any form of discount or arrear reduction is offered, it simply means, the water users paying current rates, are asked to share an additional burden by paying a portion of the adjusted rates."[155]

The Depression also affected the payment of principal and interest on the land contracts in the western extension. These land holders felt the immediate impact of the Depression and a majority of the land contracts were in arrears in 1929 and 1930 as a result of the falling commodity prices. The interest rate on these contracts went up to 8% for failure to make payments. Because this was considered an excessive rate of interest in 1933 the board offered a reduction in the rate of interest as an inducement to encourage payment. This reduction had the desired effect since close to two-thirds of the contract holders had paid out the interest in full by 1934.

The drop in farm income during the early 1930s required that the TID make a temporary adjustment in the way it met its financial obligations to the CPR. The TID suspended payment of principal and paid only the interest on the CPR bonds between 1931 and 1935. In 1936, the situation improved to the point where both principal and interest could be paid. Payments made up to 1945 reduced the bonded debt to $95,000. The TID had intended to raise its rates in 1930 to permit it to pay the annual water rental and reduce its debt to the CPR. The Depression made this impossible. As Sundal explained to H. A. Harvey, Superintendent of Inspections & Collections in March 1934, the policy:

> we have lined up for the next two or three years contemplates carrying the water rental account one year in arrears as at the 31 December of each year. When the next spring arrives we realize the current rental is due the first of May, however, we cannot realize financially on the service and production until the end of the season.
>
> We are anxious to get away from the 6% interest burdens, however we can only administrate in accordance with the ability of the farmer to

> pay his assessment accounts. Some commodity prices are slowly recovering, however, as stated above I am sure it will at least take two or three years before we can hope to go back on a current year payment basis.[160]

This policy was consistently maintained up to 1944 despite frequent letters during the early 1930s from the CPR expressing their "disappointment" at not receiving full payment on time. The consistency with which the TID implemented this policy convinced the CPR of their willingness and ability to pay, thus it made no effort to force a change in this policy. In 1945 the TID was able to clear its account early when it paid the water rental charge on 28 December of that year. The CPR waived the interest charge on this amount for 1945.

ADMINISTRATION

The TID wanted to establish an equitable water distribution system for all farmers regardless of their location whether on the main canal or at the tail end of a lateral. Unlike the CPR, the TID decided to institute a ditch rider system from the very beginning of its operations rather than depend on lateral associations. This was a revolutionary step in the history of irrigation in Alberta since not even the CPR was prepared to assume full responsibility for the operation of the system. Having a full-time staff was considered the most satisfactory because of "the newness of the system where daily attendance would be required at all structures handling water."[161]

The board first considered the selection of a water master or office manager to oversee the project for the year at their 27 January 1921 meeting. Prior to the meeting, some of the water users in the eastern section of the project had indicated to the board the type of individual who should be hired. These views were conveyed to the board by Trustee Willard who indicated that the farmers "favoured securing a practical and qualified engineer who might demonstrate methods of handling irrigation water and be of general assistance to farmers who have had no experience in irrigation."[162] The board decided to accept the offer of Secretary-Treasurer Sundal to handle the office work of the TID, furnish it with an office and telephone and assume charge of all operations and maintenance of the system together with three ditch riders for $125.00 per month for the year. The contract with Sundal was renewed on an annual basis. During his 47 years with the TID, Sundal used the Annual Report to outline his ideas on such topics as new crops which could be grown by Taber area farmers and the industrial growth that could be based on these crops. His interest in the development of the TID as a area of specialty crops led him to conduct his own crop experiments and to become involved in a number of related organizations such as the Taber Sugar Beet Growers Association.

In March 1921 the board considered the applications of Parley Palmer, Chester Asplund, S. Stubbert and F.O. Pickett for the positions of ditch rider. After a full discussion, the board accepted the applications of Parley Palmer and Chester Asplund at the rate of $135 per month during the irrigation season. The district's solicitor was authorized to prepare an agreement for hire to be signed by each ditch rider. The agreement included the requirement that the ditch rider was to furnish his own horse without expense to the district for canal and lateral patrol purposes. When the terms of the contract were later reviewed by the board, the horse provision was changed to require the ditch rider supply a dual purpose horse which could also be used for maintenance work E.L. Fulmer was hired as the as the third ditch rider for the 1921 irrigation season with a similar contract.

In 1923 the board continued with the approach taken in 1921 and 1922 to provide for the day to day management of the project with the exception that Sundal added the duties of a ditch rider to his other activities. Sundal was the only applicant for the position of Secretary-Manager while eight individuals applied for the two positions as ditch riders. Sundal's offer to carry out the same duties at the rate of $150.00 per month was accepted. Parley Palmer was hired to operate the western division at the reduced salary of $110 per month while George Ellison was hired to operate the eastern division at a salary of $95 per month. In 1924 and 1925 the Board once again advertised for bids but only received responses from the three incumbents. Sundal was hired at the same salary while salaries for Ellison and Palmer were raised to $105 and $110, respectively. In this year, the ditch riders were required to supply a cart along with a horse for maintenance purposes. For 1925 Palmer and Ellison were requested by the board to be "a little more attentive to repairing and water distribution during the coming season."[163] In 1926 the board employed John Barton as a third ditch rider to take over Sundal's duties in the central section.

Seasonal staff was hired to provide extra hands for such tasks as weed cutting. The TID paid day labour at the rate of 50 cents per hour with an additional 25 cents per hour for the use of a team. In 1930 two ditch riders' homes were established, one in the Jamieson division with 14 acres of land and the other in the Fincastle division on nine acres of land. The houses, which were originally located at the Regal Mine, were purchased for $300.00 each and moved to the respective locations. The ditch riders occupying these places were on a lower salary schedule and thus the District was able to recover the $1,972,72 paid in the development of both sites.

In 1936 Sundal's dual role as Secretary- Manager came to an end at the insistence of the water users. At the annual meeting in March of that year, a motion was passed indicating that it would be in the best interests of the TID that the position be divided "because the duties of each [were] so far removed from each other that it [was] very difficult to find a person capable of doing justice to both."[164] The board was thus instructed to advertise these positions for tender each and every year and that these positions were not to be considered permanent. Sundal continued as Secretary-Treasurer after 1936 while John Barton became the new manager of operations.

The separation of the offices of manager and secretary-treasurer was only one example of the role played by the water users in the operations of the TID during the 1930s. The Board of Directors received a number of other suggestions as to how it should operate the district from the water users who took a greater interest in the affairs of the district because of the Depression. Instructions to the board came out of a number of special meetings which were held along with the regular annual general meetings. In addition to passing motions at these meetings, the water users created special water users committees to work with the board on specific issues. These committees were provided for by The Irrigation Districts Act. One of the first water users committees was appointed on 17 March 1934 to deal with the problems caused by the Depression for the purchasers of land in the western extension. In March 1936, a water users committee was created to negotiate a reduction or even an elimination of the interest on the CPR debentures. This committee met with Sam Porter of the CPR and suggested a reduction in the interest rate to 3%. Porter suggested that the CPR would be willing to discount the bonds if they could be sold so that the company could get its money out. The committee also interviewed the provincial government, at which time they were advised of the possible creation of a commission to investigate the issue of the financial situation of the irrigation districts and irrigation farmers.

FIRST IRRIGATION SEASON

The first full season of operation of the TID began on 11 April 1921 when the various deficiencies identified during the priming process in the fall of 1920 began to be remedied. These included backfilling dirt around drops and delivery gate structures as well as the raising of banks on a number of small levees that had been found too low to carry the full stream. Crews of 2 men on laterals with north-south winds and 4 men crews on laterals with east-west winds burned weeds and prevented wooden structures from burning. Repairs were also required to structures damaged by weeds which had entered the system during the priming process. The extent of the damage prompted Sundal to conclude that:

> the policy of running the system into late fall until it is either frozen up or so plugged with weeds that it is impossible to make regular deliveries of water, is a poor and unsound policy and may prove costly to the district. The heavy weed jams often push structures out of place, and it is difficult to set them to proper grade again and make a first class job of it. I would therefore in the future, recommend, that the district attempt to make regular deliveries to the middle of October, and any delivery beyond that date will simply be from day to day while weather conditions permit.[165]

Water was turned into the system to flush out soil and the remaining weeds on 17 May 1921. Damp, partly burned weeds northeast of Chin caused a temporary closure of the system on the 27 May 1921. The water was first turned into the system for irrigation on 4 June with full capacity reached on 16 June. Water flow was maintained at full capacity until the end of July when it was reduced by two-thirds. The staff took advantage of the low flows in August to continue repair work. The rate of flow was increased in September to provide for fall irrigation. The total stream was used in the western and central divisions until the 26 September when the first request for water

Main Canal, Taber Irrigation District showing measuring flume, 1920 (Sir Alexander Galt Museum & Archives, Lethbridge)

was made in the eastern division. The first irrigation season ended in late October, as Sundal had recommended.

The water distribution policy, as explained by Sundal, was to deliver water as requested by the farmers until such time as the demand became general through out the district. When the demand for water became general, Sundal made up a schedule on the basis of thirty minutes to the irrigable acre. The schedule was mailed to the farmers. During the schedule period the district:

> made an effort to deliver a stream of four cfs to the individual users, a total of thirty-one streams in the district. Prior to this we often [delivered] streams of six, eight, and ten second feet when requested. As near as we could balance it according to the capacity of a lateral as well as the irrigable acreage on the lateral, we figured that one stream should serve a little over five hundred acres in approximately two hundred and fifty hours, so that a user with a hundred acres of water right took the stream for fifty hours, and would secure it again at the expiration of two hundred hours.[166]

The information on the schedule cards provided to the water users was also provided to the ditch riders who then knew exactly when and where each stream was to go. Sundal anticipated that once the system was thoroughly saturated and the land had been well watered, the TID would be able to increase the scheduled streams to about 40 therefore serving considerably lower acreage with each stream. This approach would give each farmer more water during critical periods when water availability has a dramatic effect on crop yields.

Following the end of the irrigation season, the system was once again inspected by the Department of the Interior and its operation reviewed with Sundal. Most of the problems identified during the priming process had been corrected. The structures on the main canal were in good working order and had worked satisfactorily throughout the season. The problems that remained included erosion on the downstream side of both wings on the headworks and some other structures. This occurred because the riprap had been placed without a gravel backing. Sundal's solution to this problem at the headgates was to backfill the wings with brush and boulders held in place by means of two by fours and chicken wire. A modified version of this method of preventing erosion was used by Sundal at other structures. Brush, straw or tumbleweed were placed in the cavity and held in position with wire and stakes. A heavy silt

Dam, headgates and wastegate for the Taber Irrigation District in Chin Coulee, 1920 (Sir Alexander Galt Museum & Archives, Lethbridge)

layer rapidly formed and bound the material together.

The flume southeast of Barnwell had leaked throughout the irrigation season and created a large pond of water in the depression immediately below. At the fall inspection, the problem was traced to the complete lack of bevel strips which should have been installed on the inside of each joint. These items were provided by the manufacturer and installed after the system was closed down in the fall of 1921.

The third deficiency revealed in the first year of operations was the excessive loss of water from seepage and evaporation. The portion of Chin Coulee between the Stafford Reservoir and the TID headgate was flat and the water spread out over it to a very shallow depth. Sundal wanted this eliminated by the construction of a canal along the east side of the coulee. The cost of such a canal prevented any immediate action on his request. P.J. Jennings of the Department of the Interior ended his inspection report by noting that "there has been no complaint whatsoever from any of the water users in this district during the last year. Considering this is the first year that operation was undertaken in earnest, I consider this a very creditable record."[167]

MAINTENANCE AND OPERATIONS

The 1921 irrigation season established a routine which was followed through to the 1940s. The spring construction period began in early April and lasted until the water was turned into the system in early May. Fall irrigation depended on the weather. In 1935 the system operated until 26 December for pond filling purposes. Interruptions in service only occurred in 1936 because of the record low levels of flow in the St. Mary River and in 1939 when someone damaged the headgate while fishing with dynamite. The construction of the North East Reservoir north of Coaldale in 1937 increased the water supply available to Taber. The water from this reservoir, however, carried a large percentage of yellow clay silt which was deposited in the main canal and laterals, increasing ditch cleaning costs.

Before the irrigation season began old timber structures including bridges and rock drops were repaired and replaced, soil which had drifted into the canal was removed and low banks on levees were built up. The Barnwell flume received many coats of Barrett's Roofing Plastic Cement to prevent it from rusting. The mixture was thinned down with a small quantity of gasoline and applied with long handled brushes.

Horse and cart used by George Allison a ditch rider in the Fincastle area (Irrigation Impact Museum, Taber)

In 1926 the TID introduced the use of stair-way rock drops as a substitute for the traditional wooden drop. They were first tested on the larger laterals in 1926 and built at other locations following the success of these tests. Over 40 stair-way drops had been built by 1929. The rock drop had the advantages of being permanent and less susceptible to damage from muskrats. The rock structure was 10 to 12 feet long with a maximum drop of two feet. A quantity of old straw, manure or flax straw was used as a bed for the rocks and as a binder between them. Rocks varying in size from 25 to 300 lbs were carefully placed so no sharp corners in an upright position to catch floating weeds. The extension of Lateral Three was built in 1936-1937.

Muskrats became a problem during the 1920s. The TID prepared a muskrat destruction program to avoid the possibility of a serious washout when the canals were at full capacity. It was proposed in 1932 to put in deeper and double cut-off walls under a number of the bankgates for greater safety. The measure, however, was not expected to stop the increasing populations of muskrats from burrowing underneath the structures

EQUIPMENT

The TID's first equipment acquisition was a model T Ford purchased from J.W. Anderson in 1921 for $250.00 in cash. Other equipment purchases included a fresno in 1922 and six scythes in July 1923. In 1926 the TID purchased a Sawyer-Massey No.5 wheel ditcher for $275.00 to clean the smaller laterals of three to four feet wide. The wheel ditcher was replaced in 1931 with a Chattin ditcher for silt cleaning work. The first Chattin ditcher was brought into the district for experimental purposes in 1930. No one locally, however, understood the special hitching required to ensure its proper operation thus its operation was unsatisfactory. In the spring of 1931 the inventor, Mr. Chattin, demonstrated the way to hitch the machine for various types and depths of ditches. The improved performance of the machine reversed the board's negative view and one was purchased. Sundal claimed that this was the first small and inexpensive machine which could adequately clean silt from small laterals. Prior to the development of the Chattin ditcher, the only machine available for small ditch cleaning cost $10,000 which the board considered excessive for a small project. The Chattin dtcher also proved less expensive to operate than the DNR excavator rented from the CPR.

A dragline was purchased from Frank O'Sullivan in 1940 and was immediately put into service in the construction of a new spillway ditch on Lateral 15 on April 23. The machine was rented by various government agencies for construction projects and custom work was undertaken for farmers who required the excavation of stock water ponds. The first garage and shop to house and repair the district's equipment was built in 1937.

First vehicle.((Irrigation Impact Museum, Taber)

The TID also made use of equipment from the CPR. In 1928 it rented a DNR excavator and an Austin dragline. The DNR excavator was needed because, after eight years of water service, a number of district laterals particularly on flat grades had accumulated considerable silt deposits and weed growth. The silt and weeds retarded the movement of water and lowered the carrying capacity of the ditch. The district grader had frequently been used in an attempt to clean out three and four foot bottom ditches but the straight blade machine did not make a satisfactory job. The DNR machine was rented for $280.00 per month, with the TID providing necessary repairs. The TID also had to provide accommodation for the crew, a vehicle for their use and pay them at the rate of $5.50 per day for the operator and $4.00 per day for his helper. Machine cleaning operations began on the 16 July on Lateral One and finished on the 30 October on Lateral Three with 18.4 miles of canal cleaned at a cost of $2,960.00.

DRIFTING SOIL

Drifting soil, weeds and surplus water which were encountered in the operation of the system in 1920 and 1921 had to be dealt with on an ongoing basis. Soil blown into the main canal and laterals created drift bars and sometimes filled the canal entirely, as happened in June 1920. In the spring of 1924, the main canal northeast of Chin was practically full from bank to bank with drift soil. As an experiment, it was decided to try and flush out the drift with a 50 cfs stream of water. It was hoped the water would carry the soil 15 miles to the spillway at the end of Lateral Nine. The experiment was a success and the drift was removed from the canal section in less than thee hours with none left as a silt problem.

The cost of removing soil drifts and the delays in service prompted the Board of Directors in 1927 to request that the province enact legislation covering careless handling of drifting soils in irrigation districts and bordering two miles along any system of canals. As Sundal explained in a letter to L.C. Charlesworth, Chairman of the Irrigation Council:

> a drift bar in a ditch is usually mixed with weeds that have also lodged there from high winds, making it a rather difficult and expensive job to remove with slips or fresnos. A drift bar may be fully removed with a team outfit and the following day another strong wind comes along to undo all previous work and further delay water delivery. [168]

The board suggested that irrigation district officials should have the right to enter land held by residents and non-residents "with suitable implements for the purpose of cultivating or ridging in an effort to check the drift. If carelessness or indifference can be proven, the cost of the work should be charged up against the land as taxes."[169] Sundal pointed out that the control of soil drifting was also a advantage in the promotion of settlement since "any country or district subject to extensive drifting of soils every spring must be looked upon with a certain amount of suspicion by prospective settlers. It is therefore a safety measure for the country at large to . . . limit or remove this evil."[170] The Prairie Fires Act was cited as a model for this legislation since it provided for fines for anyone who permitted a fire to pass from his own land to that of another. Drifting soil, like fire, constituted a public menace to farm land:

> where in carelessness in handling a fibre-less soil may be the means of injuring or destroying valuable crops on adjoining farm. It should be equally necessary, therefore, to provide for protection against it. In this connection it should be noted that drifting soil rarely occurs, even in the heaviest winds, except on lands which have been very finely worked up without proper precautions being taken against drifting.
>
> Therefore, the man who works up his soil to a condition where it is liable to drift, thus becoming a menace to irrigation ditches and his neighbour's crops, should be compelled to take suitable precautions against drifting or in the alternative to be liable for the damage caused.[171]

The legislation requested by the TID and other districts was ultimately passed in 1935. The development of a program to deal with this problem was initiated in 1935 by the PFRA. Dr. Asael Palmer of the Dominion Experimental Farm in Lethbridge played an important role in this program.

WEEDS

Weeds were the second ongoing problem for the TID which were initially combatted by cutting them with scythes or burning them, but at the risk of destroying wooden structures on the canal or farm buildings adjacent to the canal. In 1922, the board paid a landowner $200 as compensation for the alleged destruction of a barn. In August 1922, the TID used chemicals for the first time. Seeding the banks of the canals with grass was a third approach to weed control introduced in the 1920s.

In 1930 Russian thistles were replaced by sow thistle as the most prevalent weed which needed to be controlled by the TID. Two years of good moisture reduced Russian thistle but allowed small patches of sow thistle to develop on nearly every farm and irrigation ditch over the district. The chemical Atlacide was found to be 75% effective against sow thistle. Some patches were fully eliminated by the first application while others required follow up with a second and third spraying to eliminate the new shoots from underground runners.

In 1931, the sow thistle were attacked with mowers and scythes commencing on 2 July and Atlacide on the 6 July. The Atlacide spraying program, while effective, was also expensive. The 1931 Annual Report noted that the price of 11 to 12 cents a pound for the chemicals was too high for the average landowner. It was also noted in this same report that even if the TID supplied the chemical, a considerable portion would never be used because of the lack of spraying equipment in the neighbourhood, some farmers were not prepared to try and control weeds. The 931 anti-weed campaign was also frustrated by farmers in the Coaldale area who simply pulled or cut sow thistles, permitting the seeds to mature and be carried by the west wind into the TID water supply.

The TID 1931 Annual Report made a special appeal for cooperation "since a spirit of neighbourly cooperation will accomplish effective weed work without a great deal of extra expense, and when conditions return to normal the more drastic measures can come up for consideration."[172]

SURPLUS WATER

During the first year of operation surplus water created problems for area residents, individual farmers and the entire district. As Sundal pointed out in his 1921 annual report:

> there was sufficient evidence this fall in the flooding of so many roads to indicate that each farmer will soon have a problem of drainage to work out to handle the waste water. Once the subsoil has been thoroughly saturated the top soil will absorb but a limited quantity, that will necessitate frequent changes and a more constant attention to spreading, otherwise damage from over-irrigating will injure the standing crops and the water seeking lower levels will soon flood the roads. The public have been very considerate in their traffic through the flooded roads the past season but there is a limit to patience and unless the practice is reduced to a minium there will undoubtedly be court penalties to pay.[173]

The problems of the individual farmers was made worse by the unauthorised release of water. As Sundal noted in this same report:

> the practice of releasing a full stream down a lateral and out of turn, has frequently worked a hardship on your neighbour. It may be necessary to pass a by-law to stop this evil, when water shall not be released without proper notification to the ditch rider, or an arrangement made with a neighbour to handle the stream.[174]

The surplus water created two problems for the TID in operating the system. The first was in the eastern division which did not have a means of collecting waste water and returning it to the Oldman River. The excess water in the eastern division tailed out into unoccupied Crown land. The inspection of the system in 1922 by staff from the Department of the Interior indicated that a large area was becoming gradually submerged and unless steps were taken to dispose of this tail water, the land would ultimately be damaged.

The second problem was created by the return of water to the Oldman River via the spillway at the end of Lateral Nine a mile and a half west of Taber. This area was underlain with a seam of coal, portions of which were being worked by the Domestic Mine and Canada West Mine. Surface spillway waters increased four fold the amount of water pumped from certain areas of the Canada West mine.

By 1922, the TID was receiving complaints from the residents who received their domestic water via the Canada West pumping plant which was 300 yards downstream from where the waste water from the TID emptied into the Oldman River channel. Residents claimed the "irrigation water was dirty and decidedly contaminated."[175] Water samples from the Oldman River and the irrigation overflow to the river tested by the Provincial Laboratory "indicated a marked difference of the two samples with the largest percentage of contamination in the irrigation water."[176] Sundal concluded that the TID's legal liable for the damage to the drinking water and the mine might force it to abandon the Lateral Nine spillway.

The TID dealt with the surplus water problem in a variety of ways, most obviously reducing the amount of water diverted from Chin Coulee Reservoir. Starting in 1922, only 23,894 acre-feet were used compared to 34,998 acre-feet the previous year. The high rainfall in 1927 resulted in the smallest diversions when only 6,278 acre-feet were used. The unused surplus by 1929 was 102,402 acre-feet. This was not the best solution since the TID paid for 34,000 acre-feet whether it used it or not.

The TID tried to reduce waste in the eastern division by providing an amount of water as near as possible to meet the actual demand. It also sought the cooperation of the farmers at the end of the system. W.W. Russell was given free water when he agreed to construct a two-way gate on Lateral 21 and a culvert across the road to divert waste water to his land. The Russell gate took a portion of the surplus water but was not adequate to meet emergencies. A second eastern division area farmer agreed to build a two-way gate, a culvert across the road and a ditch to put additional waste water.

Converting surplus water from a liability into an asset by selling it to farmers outside the district was first suggested at the end of the second irrigation season in October 1922 when the surplus was over 11,000 acre-feet. In a letter to the Superintendent of Irrigation, Frederick Drake, Sundal noted that as of 1 October, when the irrigation season was practically over, the TID had only used 17,235 acre-feet of its 34,000 acre-feet allocation. The surplus was available despite the fact that year had been dry and there was a general demand for water. Sundal concluded that the TID had spilled sufficient water to irrigate another 1,000 or 1,500 acres. He, therefore, asked that the board be empowered to make annual contracts with outside landowners who would agree to take any waste water in the system and at the end of the season pay an amount according to the acreage benefited. The suggested rate for this water was $1.00 per acre. The waste water would then benefit more crops in the district and benefit the irrigation district financially. The TID was prepared to make application for amendments to the legislation at the coming session if this power could not be granted under the present Dominion irrigation act. The amendments would ensure that the board had the authority to collect a water service charge under the annual contract. This contract would in no way establish a regular water-right or oblige the district to deliver water. It was possible that in one season there might be sufficient to cover 30 or 40 acres of a farm while the next year it might be double this quantity. Dryland farmers, Sundal concluded "have observed our waste water going to the river and we have received many requests from them to buy this water according to the benefits derived and it is now time to make the adjustment."[177]

No further action was taken on surplus water until the end of the 1923 irrigation season when the surplus was over 19,000 acre-feet. The idea had greater appeal in 1923 because farmers in the district due to low commodity prices faced an ongoing economic crisis which was making it difficult to make the conversion to irrigation farming. The first step taken was a letter dated 23 November to Peter Naismith, the Manager of the CPR's Department of Natural Resources requesting a meeting. The TID's contract with the CPR specifically forbade the irrigation of land outside the district with water from Chin Coulee Reservoir. In the letter requesting a meeting on the water surplus situation, Sundal provided Naismith with background information on the economic situation in the district, the crops that were grown in the region and the reasons for the water surplus. The Taber area, according to Sundal, was facing a serious depression because crop returns did not balance farm expenditures. The economic situation of the irrigation farmer was made worse by the disappearance of the market for hay which had been purchased by the dryland farmers. At the beginning of the decade the dryland farmer paid $30, $40 and $50 a ton for hay to save livestock. The present financial condition of the dryland farmer was, however, forcing him to dispose of his cattle rather than purchase hay.

Given the lack of capital and credit, Sundal said, the farmers did not have the means to buy or retain livestock, erect suitable buildings to house them, provide an adequate water supply, put up better fences or acquire implements for mixed farming operations. Lacking the means to make the shift to mixed farming, Sundal concluded that the farmers in the Taber area would continue to grow wheat and limited quantities of Grimm alfalfa, corn and hay. Alfalfa hay was the only crop that required the full duty of water. Grimm alfalfa and corn had a very light water demand. Sundal, therefore, wanted to sell the surplus water not required by these crops. He estimated that 8500 acre-feet could be temporarily loaned to adjoining areas and, deducting the usual conveyance loss, would serve an additional 6000 acres of land.

To implement this plan the board wanted to introduce a system whereby the extra water rights could be loaned on a five year contract basis. At the end of the period, if the general conditions had not improved, a further extension would be granted or the water-rights restored to the original unit. The water-rights were only loaned for the period and were subject to recall when the contract expired. Sundal argued the proposed adjoining area could be served with a minium construction cost, requiring in most cases only bankgate construction on the main canal with the ordinary capacity lateral ditches. The effect of spreading the present district water-rights over a larger area would reduce the TID's annual assessment to approximately $1.25 per acre with a similar annual charge to contract holders.

Sundal and Naismith met on 30 November and according to Sundal, Naismith was in agreement with the proposal so long as the district did not request more than 34,000 acre-feet of water annually. Given what Sundal regarded as approval from the CPR, he requested approval from the Department of the Interior and the Irrigation Council. He sent the information provided to Naismith to Victor Meek and L.C. Charlesworth. He also added that many farmers had water rights far in excess of their actual requirements. They had attempted to farm almost as large an area under irrigation as they would have done under dryland farming conditions, with the result that their lands were not properly prepared and irrigation had been inadequate.

Meek agreed to send the proposal to Ottawa for review. The review he sent to Ottawa, however, was very critical of the idea. He indicated that both Sundal and the farmers in this district needed education since " they evidently cannot break away from the old dryland methods of farming and do not yet fully appreciate the value of a water grant."[178] Meek also disagreed with Sundal on the value of European immigrants as irrigation farmers. He felt that there were many:

> instances of settlers from various parts of the old country who [had] made a success of farming not only under dry conditions but also under the peculiar conditions which irrigation farming in Alberta introduces. Of course new settlers taking up irrigation lands should undoubtedly be given some instruction and advice as to the methods to be allowed to insure success and I am of the opinion that these new people will successfully farm under irrigation conditions and adopt the methods necessary to acquire success much sooner than will the farmers of the old pioneer type whose views and opinions are usually hard to change.[179]

Meek also pointed out the terms of the contract with the CPR, which precluded using water to irrigate land outside the district. Meek was not convinced that the CPR had agreed to the plan. According to Meek, Naismith indicated to him that he would not consider any change to the existing contract. The federal government interpreted this to mean that the CPR would enforce the precise terms of the contract. Meek considered the problem as one of surplus land not surplus water. The solution, he said, was for the trustees to determine the areas in each parcel which were in excess of the owner's requirements or capabilities and endeavour to lease them to outsiders who had no water rights. If a rate per acre were changed to take care of the irrigation assessment and the overhead, the owner should be satisfied and such charges should be an inducement to neighbouring dryland farmers to procure water rights. Meek suggested that some of the present owners might consider selling on easy terms portions of their present holdings if a selling campaign was organised. The solution of their surplus water problem in Meek's view was "undoubtedly more intensive development and smaller holdings with an improvement in farm methods and the application of water."[180]

Despite Meek's comments, Sundal's proposal was approved by Ottawa because it did not want any water to go to waste. Sundal was requested to provide a list of the lands and the owners who wished to avail themselves of the temporary water supply. Before Meek took action, however, he contacted L.C.. Charlesworth, Chairman of the Irrigation Council for a provincial perspective on the issue. Charlesworth, like Meek, was very critical of temporary water rights, pointing out that he did not believe that the farmers of the TID were in such financial conditions that they could not help themselves by increasing their livestock. He suggested acquiring another milk cow or two, or a brood sow or two, or a few sheep would give them a new beginning. He also agreed with Meek that it would retard the very type of development the TID was trying to promote:

> If it is their intention, as all agricultural authorities agree it should be, to drop grain growing except where it has a proper fit in a crop rotation, to diversify more and more each year and to confine operations to an area that can be properly worked, along these lines disposing of their surplus land, it would not be long before they had water running to waste. If, on the other hand they dispose of the water they should be arranging to use, they are likely to continue in the way they have been going and stick to old methods and grain growing. The probable effect will be that they will soon be back to the same old conditions which brought about the burden of debt Mr. Sundal speaks about.[181]

Charlesworth further suggested that the program would be difficult to administer since no guarantees of water delivery were to be given. The plan was in his view a source of litigation.

ADDING ACRES TO THE DISTRICT

The ultimate solution to the surplus water problem was the addition of acres to the district on a permanent basis. Expansion of irrigation in the area east of Chin Coulee was under consideration even before the TID system had been built. The area from Chin Coulee to Burdett and north from the 2,700 foot contour to the Oldman River, designated as Tract 5A in the 1918 plan for the LSEP, had 116,000 acres of irrigable land in addition to the 17,000 acres included in the TID. To ensure that Taber would be part of the development of this tract, its residents participated in the irrigation district movement which developed as a result of the 1917 to 1926 drought. In an effort to protect its future expansion options, the TID also made its views on expansion known to the provincial government. In a 15 July 1919 letter to the Deputy Minister of Public Works, Sundal noted that:

> undoubtedly in the very near future there will be attempts to organize another district to the east of Taber... At present the Taber Irrigation District comprises only a small portion of the lands in the district which will come under the larger project, and in view of this, perhaps the territory lying south of Taber to Chin Coulee, then the district northeast to the river running east to the western boundary of Range 14 (Purple Springs) in Townships 8, 9 and part of 10 should be included in the Taber Irrigation District.
>
> Or, as the headgates are west of Taber with the source of water from Chin Reservoir, would the entire district to the east as far as Burdett be comprised in our district and administered from here.[182]

Taber's effort to promote the development of Tract 5A was continued in the early 1920s by the Taber Irrigation Development Association, which had as its objective the irrigation of 150,000 acres. It was a separate organization from the TID but was linked to the TID since Sundal served as its secretary. This organization promoted expansion while leaving the TID board of directors to concentrate on the immediate tasks of construction and operation of its system.

Drought and dust storms, which were particularly intense in 1920, were a further reason for expansion. As Sundal noted in a 9 June 1920 letter to Irrigation Commissioner Peters:

> we experienced a very bad dust storm again yesterday and quite a number of farmers have suffered heavy losses from blown out crops. I tell you they are up against a tough proposition and we cannot bring water on the land any too soon. We require a good general rain storm before too long otherwise we can look for a short crop this fall.[183]

In addition to the movement to bring about the construction of the LSEP by the government, owners of land in the area of Chin Coulee lobbied for inclusion in the district. This land had been included in the TID in 1915 but removed in 1917 because of its status as school lands. The first request came in 1918 from the Manager of the Rock Springs Coal and Brick Co. Ltd., B. K. Bullock. He was advised by Sundal that under the present circumstances it was unwise to ask for any alterations in the boundaries of the district since it would to delay construction. What we are all interested in, Sundal said, was "to see the dirt moving as soon as possible."[184] As soon as the system was in operation the idea of expansion could be reconsidered. The second request came from Lethbridge resident John Perry, the owner of 960 acres two miles west of Cranford. The TID was advised on 5 July 1919 by Perry's lawyer that the main ditch from Chin Coulee Reservoir to the TID would cross his land making it convenient to irrigate 350 acres. Perry was encouraged to make his request for inclusion in the district because the TID intended to distribute1½ acre-feet per assessed acre rather than 2 acre-feet leaving the TID sufficient water to irrigate an additional 5,266 acres. Perry was advised that the government had granted permission to irrigate 17,000 acres, which was the storage capacity of Chin Reservoir, and it was impossible for the district to grant further water-rights until a larger water supply was available.

Given that the critical issue was water supply, Perry and Bullock, along with other owners, organized a meeting with the CPR and various government officials to discuss the issue. At this meeting held on 1 December 1921, Sam Porter of the CPR indicated that additional water would only be available if a system of reservoirs was built on the St. Mary River and that it was not the intention of his company to undertake their construction.

THE WESTERN EXTENSION

In 1925 the TID decided solve the water surplus problem and accommodate the requests for the addition of land to the district by expanding to the west. In addition to adding privately owned land, the TID also purchased the remaining school lands in the area which had not been sold by the federal government. Most of this land had originally been included in the TID when it was first created in 1915 but was removed at the insistence of the CPR. Sundal went to Ottawa in May 1926 and arranged with officials of the School Lands Branch of the Department of the Interior to buy 5440 acres for $1.00 per acre. The School Lands Branch agreed to the purchase provided it was approved by the Premier of Alberta, John Brownlee. His approval was obtained in August 1926.

The Department of the Interior provided a plan for the irrigation of 4661.5 acres which it estimated could be built for $25,182.85. In 1927 an agreement was executed between the School Lands Branch, Dept of the Interior and the TID for the purchase of thirty-two quarter sections (5120 acres) with 3008.3 acres classified as irrigable. The board eventually received title to 5149 acres of land. The water users ratified a by-law to borrow of $35,000.00 to build the extension on 28 April 1928 by a vote of 51 to 13.

In June the TID called for tenders to excavate the new lateral. W.S. Johnson & Sons of Barnwell was the only company to respond to the call for bids. Since their bid was considered fair, the company was awarded a portion of the ditch work at the rate of 23 cents per yard. The first sod was turned on 7th July 1928. Camps were established by W.S. Johnson & Sons and the TID which employed men and teams on day labour. Both camps disbanded the middle of August for the harvest and threshing season. Further ditch work was accomplished during November after the completion of harvest. The ditch work finished in the fall of 1928 was approximately 90% of the total construction required. The remaining construction was completed in the spring of 1929.

The TID placed all the timber structures used in the extension using day labourers. A total of 42,978 feet of lumber was used in the construction of eight bank gates on the main canal, 13 highway culverts, two private farmer's culverts and 50 single delivery units. The extension was completed in the spring of 1929 at a final cost of $33,142.07.

The TID had sufficient cash on hand to cover the cost and did not have to issue the construction bonds.

The land was sold at a public auction conducted by E.O. Wright on 23 February 1929 at the Social Hall in Taber. The 57 parcels of land offered for sale were at four locations: 19 were northwest of Jamieson (now called Cranford), 12 were east of Jamieson, 4 were a mile south of Taber and 22 were immediately northwest of Barnwell. The parcels of land in the Jamieson area and south of Taber were 80 and 160 acres and had prices of $15 to $50 per acre. The improvements on the land varied from parcel to parcel as did the acres of water right. In at least one case, the land included a mile of fence plus a house, barn and coal shed. The land in the Barnwell area consisted of parcels which had been created by the subdivision of one section of land. The original townsite of Barnwell had been created in 1909 during the urban land boom in western Canada. Only a small number of lots had actually been developed before most of the townsite plan was cancelled. The TID subdivision was located on the section northwest of this original townsite. The plan, called Barnwell Gardens, created 23 parcels of 9.5 acres to 77 acres.

The terms and conditions of sale required a one-tenth payment in cash at the time of the sale, and the balance in nine equal instalments payable on the 31 December each year thereafter. Interest was to be charged at 6% if the payment was made on or before the due date of 31 December and then increased to 8% as a penalty for late payment. For land which was sold at below the upset price, the purchaser had to immediately pay the first instalment, otherwise the land would be put up for sale again. All land was subject to an irrigation and drainage easement in favour of the TID. The land was also sold subject to further reclassification by Dominion Reclamation Engineers to increase or decrease the water rights. Any tenant on the land was allowed 60 days to remove buildings or fences. No purchaser was permitted to acquire more than one parcel in the farm land and one parcel in the Barnwell Gardens subdivision. The land was sold subject to an irrigation assessment for 1929. The purchasers agreed to plant at least one thousand trees on their parcel within three years. Thirty parcels were sold during the auction.

The remaining twenty-eight parcels were sold later by sealed tender. The total value of the land sold was $143,538.22 with $17,661.01 received in cash. Following the sale, the TID also subdivided a small area of land between the Barnwell Consolidated School and the CPR station into 34 lots which were 50 by 130 feet in size. These lots were offered to the public at prices ranging from $50.00 to $100.00 depending on their location.

A fourth ditch rider district was added to the TID with the completion of the western extension, originally called the Jamieson Division. The addition of the western extension increased the TID's annual water rental to the CPR. The additional charge was $2,330.75, with the first payment due on 1 May 1929. The TID did not receive any additional water but was permitted to use the 34,000 acre-feet provided under the 1919 agreement on land not in the district in 1919.

MIXED FARMING AND SPECIALTY CROPS

The promotion of mixed farming, the growing of specialty crops to meet specific markets and the general problems of the transition from dryland to irrigation agriculture were concerns of the TID administrators as soon as the system went into operation. It also worked with other community groups to promote the over-all welfare of the agricultural community.

In the TID's first annual report Sundal noted that exclusive grain farming under irrigation did not pay and that dairying, beef, pork and poultry must take their places on the farm. In this same report, Sundal discussed the challenges for farmers in the Taber area in making this transition. The first difficulty in the transformation of a dryland farm to an up to date irrigated farm, Sundal noted, was that it required a great deal of capital and labour. The farmers in the Taber area, because of the poor years before the introduction of irrigation, did not have the extra funds to purchase cattle despite the low prices for livestock. Sundal suggested the passage of emergency credit legislation for the purpose of financing stock purchases in the various irrigation districts. He felt that this was appropriate since it was as important for the government to strengthen the credit of individual farmers as it was to guarantee irrigation bonds. Sundal also suggested a committee of farmers meet a committee of Taber bankers to discuss the financial situation and if possible, devise ways and means of financing livestock purchases.

The distance from Taber to large urban centres which could serve as markets was another problem. As Sundal pointed out "large cities consume unlimited quantities of food and garden products, but the long distance to these centres and the excessive cost of transportation makes it impossible to reach them and realize a profit for your labour."[185] Within a radius of four or five hundred miles for root crops was:

> very limited and uncertain, as a favourable year with overproduction quickly gluts the market, and you get little or nothing for your season's work. True enough, there are considerable quantities of vegetables shipped into the district by Taber merchants and it would be advisable for individual farmers to interview these people and endeavour to grow the various varieties to meet this demand.[186]

Given this situation, Sundal suggested that the effort of the Taber farmers should be toward:

> growing the standard food crops such as wheat, beef, pork, butter and alfalfa seed that are transported and in constant demand on the world's markets, and have a daily quotation of their value". Any particular one of these may have a slump in price for a time, but recovery is certain as the world must have food. The slump feature suggests the danger of engaging exclusively in any one of these branches as you may easily fall a victim to unfavourable weather and pest conditions, that may leave you stranded for an income to meet farm and living expenses. Our products should be grown in a balanced quantity, so that if one fails the others will maintain a normal income. Keep in mind that the cost of production in irrigated farming, is high, and therefore it is necessary to keep every acre busy.[187]

The most important step towards diversification of the crops was taken in 1923 when Taber area farmers participated in the beet test plot program organized by the Raymond Board of Trade as part of its campaign to bring a sugar beet factory back to southern Alberta. Alfred Anderson won first prize with a yield of 20.27 tons per acre and N. J. Anderson came a close second with 18.96 tons per acre. Sugar beets provided a cash crop in the fall with succeeding payments arriving in early spring to finance general seeding operations and helped to control weeds. The overall success of the effort to diversify is evident in the 1929 crop production statistics which indicate that wheat was grown on 7,349 acres or 45% of cropped land. Sugar beets, alfalfa, and barley were the other crops grown in significant amounts by 1929.

SPECIALTY CROPS AND THE CANNING INDUSTRY

In 1928, the TID, in cooperation with the town of Taber, began an effort to establish a cannery to process the vegetables grown in the area. This effort was prompted by the construction in 1927 of a cannery in Edmonton by the Vancouver-based Broder Canning Company. In March 1928 Ted Sundal met with a company representative and later wrote that:

> our district, composed largely of medium to light sandy soils, is particularly adapted for the production of high quality vegetables and fruit products. To my knowledge we have been growing field and sweet corn for the past seventeen years, and during that period I do not remember a failure in growing Golden Bantam Sweet Corn. Cobs fill perfectly from butt to tip, an indication of suitable soil and climatic conditions.[188]

The most successful fruit crop, according to Sundal, was strawberries which in 1927 "assumed a commercial production, supplying a portion of the Taber and Lethbridge markets."[189] The area held potential for the production of other vegetables such as tomatoes, peas, beans and asparagus. All that was required, in Sundal's opinion, was a paying market for these products. Sundal offered to organize a meeting of interested growers for a discussion of growing crops for the cannery.

This first effort to sell the Taber area as source of fruit and vegetables was only partially successful. Taber, in the opinion of the company, was too far away to supply the Edmonton packing plant with fruit or corn. It was possibly close enough, however, to produce tomatoes for transport to Edmonton. The company suggested that "if your locality can produce a good reliable crop, we see nothing that would prevent us from drawing quite a considerable volume in that variety."[190] It also suggest that the TID organize the growing of test plots of tomatoes during the present season.

Later in 1928, the Broder Canning Company built a cannery in Medicine Hat. In addition to the construction of the plant, the company also planted 100 acres of tomatoes on the river flats in Medicine Hat which could be irrigated by pumping water

Taber Cannery Building, 1934.(Irrigation Impact Museum, Taber)

from the South Saskatchewan River. In an effort to increase supply, Mr. R Hawthorne, the field manager for the company, canvassed Taber area farmers in January 1929 to determine their interest in growing seed peas. He gave a presentation at the annual meeting of the TID water users on 26 January 1929. At this meeting, Hawthorne explained that the company wanted to test the potential of the Taber area for the production of seed peas. Canadian canneries, he explained, obtained their seed peas from Montana where they were grown under similar climatic conditions as southern Alberta. The company was interested in finding a suitable seed growing area in Canada which could replace the imported seed.

A number of sugar beet growers in the Taber area responded to this invitation to supply this new potential market. In 1929, they grew 300 acres of peas for seed test purposes. The TID Annual Report for 1929 indicated that it was an exceptionally good quality seed with a high germination test and has proven the possibilities of the district in growing seed for the canning pea industry in Canada. Tests of varieties and methods of seeding showed that broadcast seeding and high seeding rates gave the best results.

The tests were so successful that the Broder company asked the area farmers to continue to grow seed peas and to expand the range of their crops. At the 23 December 1929 meeting of the Taber-Barnwell Sugar Beet Growers Association, Mr. R. Hawthorne indicated that the company wished to test the district another year. It required 200 acres of seed peas, about 150 acres of string beans and a limited acreage of sweet corn, spinach and pumpkins. Hawthorne indicated produce would be canned at the Medicine Hat plant one more year while the company evaluated Taber for the possible construction of a canning factory. In 1930 Taber farmers grew 80 acres of string beans, 14 acres of pumpkins and 125 acres of seed peas which were trucked to the plant at Medicine Hat. Taber growers continued trucking vegetables to the Medicine Hat plant in 1931. The plant closed in 1932 because acreage in the river flat pumping area was too small for a general cannery production and the distance from Taber made the delivery costs excessive .

With the failure of the Broder Canning Company to follow through with the construction of a plant in Taber, the TID, the Board of Trade and the Town of Taber joined forces to build a plant using their own resources. In December 1933 a special committee was established which worked out a plan to build a plant and then lease it to the Broder Canning Company.

The TID was authorised by an Act of the Provincial Legislature to borrow up to $7,500.00 upon the security of all moneys owing on the land in the western extension. These funds were used to cover the cost of labour, extra materials and construction of a railway track into the plant. Some of the additional expenditures were carried for a time as interest bearing accounts and paid from rental income. The total cost of construction was $18,000.00 with $11,233.72 supplied by the TID and the remainder by the Town of Taber.

The construction of the canning factory began in 1934 and finished in the spring of 1935. Some of the construction materials such as lumber, brick, windows, doors and heating equipment was provided by the demolition of the Rhonalda Hotel which the town acquired because of unpaid taxes. Relief labor was used in dismantling operations and in the delivery of all usable material for factory construction purposes.

The building was leased for a period of five years in April 1934 with an option to purchase to Robert Broder of New Westminister, B.C. and Elwood L. Cross of Kelowna, B.C. at a rental of $133.00 per month. The construction program was sufficiently advanced by mid-August of 1934 to permit a limited pack of corn and pumpkins. Robert Broder subsequently purchased the partnership interest of E.L. Cross and organized the Taber Canning Company Limited. In August 1941, seven years after construction, a settlement in full with interest was received by the Town of Taber and the TID. In February 1948, the Taber Canning Company Limited sold the canning factory to the Cornwall Canning Company, a subsidiary of the Safeway organization.

OTHER INDUSTRIAL DEVELOPMENT

Taber corn was also the basis of a broom manufacturing company. In 1933 Sam Russ, a local beet worker, and W.S. Johnson of Barnwell grew experimental plots of broom corn. Because of the success of these test plots, Russ was able to interest Thomas Maxim, manager of the Queen City Broom Works in Regina, to test the production possibilities on a larger scale. In 1934 30 growers agreed to plant 375 acres of broom corn as part of this test. The seed was planted in rows 22 inches apart and at the rate of one bushel to seven acres. Because of the extremely dry weather in May, germination of the seed sown in April was slow and irregular thus raising the fear that the crop would be a total failure. Rain in June finally assisted seed germination but the moisture came to late for the crop to grow to maturity before the first frost on 25 September. Other problems with this first crop, cited in the 1934 Annual Report, were the many variations in thinning, suckering and cultivation methods followed on the various farms. Despite the various difficulties, 30 tons were harvested, baled and shipped to Regina in three rail cars. These results were sufficient to encourage Sundal in his hopes for another industry. He expected that a variety of seed would be developed which would allow the Taber area to eventually supply all the broom corn required in Canada.

In 1935 a broom corn growers organization was established and 47.5 bushels of seed was shipped in from the USA. The seed was used by 29 growers to seed approximately 350 acres. Weather conditions in 1936 were again unfavourable for the production of broom corn. A cold, and late spring made seeding operations two to three weeks later than in average years. Seed germination was retarded and irregular and the summer was cool and extremely dry. From a broom corn growing standpoint it was perhaps one of the most unfavourable years experienced for 25 years. These adverse conditions, however, did not prevent the crop from reaching maturity by the time of the first killing frost on 26th September. The total production was approximately 100 tons or 700 lbs per acre.

The local effort to grow broom corn in 1934 and 1935 was based on the expectation that a small broom factory would be built in Taber in 1935 with financing provided by Regina interests. When the Regina businessmen failed to undertake the project, a broom manufacturing company was established by a group of Taber businessmen. Given its traditional optimism, the TID anticipated success because an "Alberta market is available for a substantial volume of medium grade brooms and by manufacturing an article that will be competitive in both quality and price we can look forward to favourable support from the distributors and consumers of the province."[191] The plant went into operation on 6 July 1936 with a daily production of from seven to nine dozen. It manufactured four lines of house brooms ranging in a price from 35 to 95 cents, two grades of curling brooms and one grade of warehouse broom. The new plant used four of the six carloads of the broom corn produced in 1935. Because the local product was too short for the outside covering of the broom, a carload of Oklahoma brush was imported.

Sundal continued to explore the use of corn as the raw material for a variety of products in his never ending search for potential agriculture-based industries for Taber. In the 1943 Annual Report he outlined the successful efforts of the Iowa State College to use corn cobs and corn stocks to make plastics. Sundal anticipated that the local farms could earn $5 to $8 a ton for the corn stocks after the cobs had been sent to the cannery. He even considered the possibility that the entire plant would be used for plastic, permitting the crop to be harvested in one fully mechanized operation. He advised the water users to prepare now for the expansion of corn production since "corn is undoubtedly destined to play an important role in the gigantic plastic age coming up after the war . . .."[192]

The pickle manufacturing company, Dyson Limited, of Winnipeg, established the third food processing plant in the Taber area during this period. In 1942 it contracted with farmers in the Taber, Barnwell, and Cranford areas to grow 47 acres of cucumbers. The 117 tons produced were shipped to Winnipeg by rail and truck but arrived in poor condition. In 1943 a plant for the processing of the crop on site was built.

FURTHER SUGGESTIONS FROM TED SUNDAL

During the Second World War Sundal continued his efforts to test new crops. In 1941, the Canadian Sugar Factories Limited supplied Kobott soybean seeds. In 1944, Sundal experimented with sweet potatoes and tomatoes. In 1944, he planted 8,000 tomato plants of 32 varieties. An extremely dry spring made it necessary to irrigate the field before planting, and the majority of the plants were set out between the 3rd and 10th of June, 10 days too late for the best results. To plant this many plants, Sundal used the regions's first Cockshutt planting machine. This type of machine was used to plant tobacco in the East. Two men on a planter pulled with a tractor could plant about 1,000 plants per hour. The planter cost was approximately $125.00. From the 32 varieties the two outstanding early and heavy yielding were the Bounty and NDAC No. 38. Sundal believed that a program of early seed selection would allow Taber to supply the juice and canning market of the prairies.

EDUCATION AND RESEARCH

Education and research programs by government agencies were also encouraged by the TID in their effort to ease the transition to irrigation agriculture and develop specialty crops. This cooperation started in February 1923 when staff from the Department of the Interior gave a presentation to 40 farmers. The presentation stressed the importance of the initial levelling of the land with a fresno scraper and float. Proper land levelling resulted in the maximum smoothness of the surface which ensured the uniform application of water and consequently heavy stands of grain and high yields. Emphasis was also placed by the Department of the Interior on the necessity of locating the farm laterals accurately. The timing of irrigation and the amount of water to be used to ensure maximum crop growth was explained. Light and frequent applications of from 4 ½ to 6 inches were recommended.

It continued through to1944 when a small substation was established south of Taber by the Lethbridge Experimental Farm. The objective of the sub-station was to carry out special tests on light sandy soils which were characteristic of the TID. The experiments include corn, peas, and pumpkins for the canning industry.

TWELVE
MAGRATH IRRIGATION DISTRICT
1922-1946

Like Taber, the citizens of Raymond and Magrath were very much involved in the effort to obtain a government commitment to build the LSEP during the irrigation districts movement after World War One. The Southern Irrigation District was a joint effort of the two communities to take advantage of the expansion of the St. Mary Project based on this plan. Both communities, however, had the option to create irrigation districts which would receive their water supply via the existing system through an agreement with the CPR. The failure of the Irrigation Development Association to convince the federal and provincial governments to invest in irrigation doomed the Southern Irrigation District. The creation of two separate districts, while a more modest proposal, met the immediate needs of the two communities.

SOUTHERN IRRIGATION DISTRICT

Organization of the Southern Irrigation District began on 15 February 1919 at a meeting organized by the Raymond Board of Trade. The land of the proposed district was in Tracts 1A, 1B, and 4 of the LSEP. Commissioner of Irrigation Peters, however, explained that the Department of the Interior had done little detailed survey work in the area north the Milk River Ridge and that the irrigable land estimates were based on old contour maps and surveys borrowed from the AR&ICo. Given this information, the organizing committee decided to wait for the department to complete its detailed surveys of the area and the design of the canals. Peters was impressed with the degree of commitment shown by the delegates and recommended that funds be put in the Department of Interior's budget to cover the necessary surveys.

The Southern Irrigation District organizing committee, however, only delayed action for a matter of weeks. In March 1919, petitions signed by 258 land owners was sent to the provincial government. The individuals who signed indicated their willingness to bond their land to pay for the project. The petitions were approved and the notice of the intention to create the Southern Irrigation District was published in the 15 January 1920 *Alberta Gazette*. The creation of the district was approved by a vote of 350 to1 at a vote held on 1 February 1920. The district covered 190,000 acres, almost 100,000 acres of which were considered irrigable. The meeting also elected Herber S. Allen of Raymond, A.E. Fawns of Stirling and Christian Jensen of Magrath as the first trustees.The ministerial order establishing the Southern Irrigation District was published in the 31 January 1920 issue of the *Raymond Recorder* and the 31 March 1920 issue of the *Alberta Gazette*.

The plan proposed by the federal government to supply water to the Southern Irrigation District was based on the enlargement of the AR&ICo. canal to 1,200 cfs at a cost of $50,000; construction of a new canal to take water from the AR&ICo. canal east of Spring Coulee and south of Bradshaw to the proposed Ridge Reservoir at a cost of $386,000; and the development of the Ridge Reservoir to a capacity of 79,000 acre feet at a cost of $391,125. Other costs brought the total to $48.25 per acre based on the irrigation of 35,000 acres. This cost was prohibitive.

MAGRATH ORGANIZING COMMITTEE

Magrath was the first to abandon the Southern Irrigation District and concentrate on expanding the irrigable acre in the immediate vicinity of their own town. This change was led by the executive committee of the Magrath local of the United Farmers of Alberta. At their 14 January 1922 meeting, it discussed the idea of irrigating 18,000 additional acres north of the existing Magrath Lateral and west of Pothole Coulee. Earnest Bennion and Edward Hodges were given the responsibility to contact Sam Porter, CPR Chief Engineer, to see what could be done toward obtaining water for this area through the existing CPR system.

Christian Jensen, Chairman 1928 - 1947
(Magrath Irrigation District)

When contact was made with Porter, an invitation was extended to not only meet with him but also with Peter Naismith, Manager of the CPR Department of Natural Resources. The CPR offered to deliver up to 100 cfs of water in addition to the current diversion at a cost of $4,500 per acre-foot. The CPR expected that the distribution of this new water would be handled by the farmers themselves. This Magrath initiative was welcomed by the CPR because it was another opportunity to increase water users administration of the lateral system. The CPR had also undertaken some engineering work on the possible extension of the Magrath lateral.

Victor Meek of the Reclamation Branch of the Department of the Interior also showed interest in the idea and advised the organizers to request its evaluation with respect to water supply and cost. This request was to include the amount of water which each person was willing to accept and the legal description of their land. Victor Meek also advised that a petition would have to be made to the trustees of the Southern Irrigation District for the removal of land from that district. The results of these contacts were presented to the members of the UFA local who voted unanimously to accept the CPR offer.

The membership also created a permanent organizing committee made up Earnest Bennion, Fred Bradshaw, George Baily, Christian Jensen and Isaac Sabey to further irrigation development. Bennion served as Chairman and Bradshaw as Secretary. The new committee expanded the area under investigation to include land on the west side of the ridge between Magrath and the St. Mary River and southwest Magrath in the area of Bradshaw. Cooperation with Raymond and Stirling with respect to bargaining over the delivery charge for the water was discussed at the new committee first meeting on 20 April 1922. The decision was made to keep the cooperation at "an informal and unofficial level without in any way losing the identity of the representatives of each town."[193] It was recognized, the minutes continued, that "our scheme would have to be developed separately and apart from theirs consequently [Magrath and Raymond] could only work together in the obtaining of a reasonable price."[194]

The new committee also confirmed support from the CPR and the Department of the Interior for this approach to the expansion of irrigation in the Magrath area by contacting its Member of Parliament L.H. Jelliff. In a letter to Jelliff, Charles Stewart, Minister of the Interior, indicated that the land north of Magrath could be irrigated by an extension and enlargement of the company's Magrath lateral and that the CPR, had given the:

> settlers to understand that the company will, subject to the approval of [the Department of the Interior] be willing to enlarge its main canal system and to supply an adequate quantity of water in addition to that now supplied at the headgate of the Magrath lateral, subject to certain conditions.[195]

The petitions required to initiate the Department of the Interior's review of irrigation expansion in Magrath were prepared and sent to the federal government in April 1922. Separate petitions were prepared for each of the three areas since the intention was to create three separate irrigation districts referred to as North, West and South Magrath.

In June 1922 M. H. Marshall of the Department of the Interior was sent to evaluate the Jensen Plan which was the name originally given to the High Line Lateral. This plan called for a diversion from Pothole Creek north of Spring Coulee. The canal would then run northwest along the ridge separating the St. Mary River from Pothole Creek. Marshall concluded that the plan was impractical because the land sloped in the wrong direction and was unsuitable for canal construction. Marshall was followed in April 1922 by R.S.B. Lillico who conducted a second review of the Jensen Plan. He found that the scheme was feasible to the extent of irrigating several thousand acres by adjusting the water level at the diversion point through a check structure. Lillico also suggested that the construction of a diversion structure south of Spring Coulee on the AR&ICo. main canal would permit the irrigation of the land in the Bradshaw area.

Following Lillico's investigation, the Department of the Interior designed a distribution system which included two diversion structures on opposite banks of the main AR&ICo. canal at a point just south of Spring Coulee. The lateral on the west side would serve what was now called the West Magrath areas while the lateral on the south bank would serve the Bradshaw area referred to as the South Magrath district. A final survey by Ben Russel in June of 1923 resulted in the extension of the Bradshaw lateral and the addition of 2,442 acres

The organizing committee also had to counter the declining the level of support for irrigation which was evident in the fall of 1922 when it was discovered that the area farmers had not signed up for the water to the extent expected. This situation prompted the calling of a public meeting on 5 November to "get the true facts and circumstances of [the] proposition before the people concerned."[196] The Committee outlined the terms of the offer made by the CPR to provide water and reported on the extent of the support received from the farmers in the Magrath area which had potential for irrigation development. The committee indicated that only 6,200 acres had been signed for out of the total of 20,500 acres which could be irrigated, confirming the committee's conclusion that many of the resident land owners were apathetic and indifferent.

The meeting featured William Fairfield, Director of the Dominion Experimental Farm as the guest speaker After a few questions were put to the committee, Dr. Fairfield gave his presentation, which dealt with the question of the return on the investment if a farmer went to the expense and work necessary to put water on the land. Fairfield used the results achieved at the Experimental Farm. The most important benefit of irrigation according to Fairfield, was that the farmers would always be assured of their hay, which was more than they could expect from a dryland farm. He concluded his presentation by predicting that the time would come when those who took a full water right would be glad that they had displayed such good judgement. Despite the lack of an overwhelming endorsement of irrigation, the organizing committee at the 14 December meeting decided to move from the planning stage to the steps required to creating a district. They were encouraged by the results of the federal government surveys to take this step. The organizing committee also decided to create one district rather than three. The committee was also joined by John Steele, the Secretary-Treasurer of the Town of Magrath who became its new secretary.

CREATION OF THE DISTRICT

The first step was the preparation of a petition requesting that certain lands which were then part of the Southern Irrigation District be removed from that district. A second petition was then required requesting that these and other lands be established as a separate irrigation district. The petition requesting the removal of land was sent to the Irrigation Council on 21 July 1923. Steele was advised that as soon as the Magrath portion of the debt of the Southern Irrigation District was paid and the Chairman of the Southern Irrigation District gave permission, the petition for exclusion would be considered. One of these requirements was met on 19 June when the trustees of the Southern Irrigation District approved a resolution excluding the lands to be included in the Magrath districts. The liquidation of the Magrath portion of the Southern Irrigation District debt was dealt with by the end of July.

The petition requesting the creation of the Magrath Irrigation District (MID) was sent in immediately. The review of this petition by the Irrigation Council revealed it did not have a sufficient number of signatures of registered land owners. The Irrigation Districts Act required that the petition for the creation of an irrigation district had to be signed by the owners of at least one-half of the total land to be included in the proposed district.

In addition to advising Steele of the deficiencies of the petitions, L.C. Charlesworth of the Irrigation Council tried to persuade Earnest Bennion not to create a separate district but remain in the Southern Irrigation District. Another one of his concerns was the proposed methods of assessment of the new district which was to let the land owner decide the number of acres to be irrigated rather than determining the amount by a land classification survey. Charlesworth dropped both of these objections when he was advised by the CPR that it would not extend irrigation in the Magrath area by entering into individual contracts and when the federal government indicated that it was not opposed to the method of assessment proposed for the MID.

John T. Steel Secretary, 1925 - 1933
(Magrath Irrigation District)

The addition to the petition of the required number of signatures of land owners was made by the fall of 1923 and approval was given to proceed with the publication of the notice of the intention to create an irrigation district in mid-December 1923.Publication of the notice of intention to create the Magrath Irrigation District was placed in the 26 January 1924 issue of the *Alberta Gazette* and in the 2, 9 and 16 February 1924 issues of the *Lethbridge Herald*. The notice listed the legal descriptions of the 31,437.9 acres to be included in the district and advised anyone who objected to the inclusion of their land in the district to make their views known by 29 February 1924.

The publication of the notice resulted in many requests for the removal of land from the district. These petitions covered land of 17 area farmers located along the proposed Bradshaw lateral. Some of these land owners, such as the Town of Magrath and Levi Harker, already had sufficient water because of their current contracts with the CPR. The strongest protest came from Edmond Balderson who not only wanted to withdraw his land but objected to the entire project because:

> the time [was] inopportune as the cost of same would be unbearable at this time, owing to the low price of farm products also the farmers in general are burdened with financial obligations already brought on by the late war and careless financial ventures and to undertake at this time the financing of an irrigation scheme in the district would mean individually a financial failure for most of the people in the district.[197]

Creating an irrigation district under these conditions he suggested, would be like "a man running his head into a loose halter and hanging himself."[198]

The combined effect of the various additions and deletions reduced the size of the district to 28,101.18 acres of which 14,850.84 acres had been committed by its owners for irrigation development. As irrigation was requested for more than 50% of the irrigable land in the area, the district met the terms of the Irrigation Districts Act. Having met the size requirements, George Card was appointed returning officer for the vote on 17 May 1924 of the petitioners which gave their final approval to the creation of the district. On 2 June1924, a ministerial order was issued creating the Magrath Irrigation District. Earnest Bennion, Leslie Pharis and Isaac Saby were appointed the first trustees of the district.

CONTENT ADJUSTMENTS

The first meeting of the new Board of Directors was held on 16 June 1924 at which time Earnest Bennion was elected Chairman for 1924 and John T. Steele was appointed Secretary. Bennion remained as Chairman through to 1928. The first task of the Board of Directors was to determine the final size of the district as many petitions were received for inclusion and exclusion of land. Numerous requests were also received for changes to the water rights for the land. The first request came in December 1924 when a petition was received from the residents living south of Bradshaw asking to be admitted into the district. The number of petitions increased dramatically in April and May of 1925. These requests were handled by the Board of Directors seating as a court of revision.

AGREEMENTS WITH THE CPR

The Board of Directors had to negotiate agreements with the CPR for the construction of the high line lateral and operation of the new district. On the advice of Peter Naismith, the Board hired P. H. Morgan to prepare a general plan and estimate construction costs. This information was submitted to the officials of the CPR for use in future negotiations. His May 1925 report included a statement showing the amount of water to be taken by each individual water user in the district together with description of the lands to be irrigated.

An agreement was signed 18 June 1926. The CPR was to construct the high line lateral at cost and take $80,000 worth of 6% district debentures at par as payment. The district was to obtain all rights of way and permission for road crossings. The CPR was to construct and maintain the headgates in the main canal and the measuring flumes. The CPR had the right to shut off the water for repairs upon reasonable notice to the Board and to prorate water during drought. The company agreed to carry 30 cfs of water through its system to the headgates of the district when not prevented by causes beyond its control at a cost of $120,000 payable in district bonds at par. This charge represented a portion of the cost of constructing the original AR&I canal system at the turn of the century. The MID was thus paying a total water right of $200,000. The MID agreed to take over the Magrath Lateral and to maintain and operate all the works necessary to supply the district except the headgates on the main canal and the rating flumes and to indemnify the CPR against all liability arising therefrom. The MID agreed to pay an annual carrying charge or service charge of $3,375 or 67 ½ cents per acre. Water was to be used only for irrigation and filling stock ponds and not transferred to other lands or wasted. The District agreed to act as the CPR's agent and deliver water to lands covered by outstanding water agreements on the Magrath Lateral and collect the water rentals at the rate of $112.50 per cfs and turn them over to the CPR. The CPR agreed to cancel any old water agreements upon request of the District if they were in arrears. The MID contained 5,000 acres of irrigable land under the new water agreements and 2,006 acres under the old water agreements.

The contact between the CPR and the MID was similar in some respects to the contract between the TID and the CPR. It was, however, more complicated because the MID took over an existing system as well as adding new water users. The old water users were covered under water agreement number 1374 which required that the MID pay the CPR $1,477.70 every May 15. Water agreement number 1375 for the delivery of water to the new water users in the MID required an annual payment of $3,375.00. These two contracts gave rise to the terms old and new water.

OFFICIAL OPENING

The MID was officially opened on 23 June 1927. This event was covered by an unidentified staff reporter from the *Lethbridge Herald* whose enthusiasm led to an overestimate in the size of the new district by 13,000 acres but also produced the most poetic description of any official opening in the entire history of the St. Mary Project. The ceremony began when a cavalcade of 30 cars went from the

Number 117

Dollars $1,000.00

Magrath Irrigation District

Head Office: Magrath, Alberta, Canada

Irrigation District Debenture Instalment Plan

The Board of Trustees of the Magrath Irrigation District hereby promises to pay to the bearer the sum of One Thousand ($1000.00) Dollars of lawful money of Canada at the office of the Bank of Montreal in the City of Montreal, in the Province of Quebec, in twenty-eight (28) equal annual consecutive instalments on the 1st day of December in each year commencing with the year 1929, and interest at the rate of six per cent. (6%) per annum yearly on the 1st day of December in each year commencing with the year 1927, the annexed coupons for such payment of principal and interest, or interest alone to become payable as they become due upon presentation thereof at the said office.

Dated at Magrath in the Province of Alberta, this 1st day of June, A.D. 1926.

In testimony whereof and under the authority of By-law No. 3 of the Board of Trustees of the Magrath Irrigation District, duly passed on the fifth day of May, 1926, this debenture is sealed with the seal of the said Board and signed by the Chairman of the said Board and one of its Trustees.

Chairman

Trustee

(Magrath Irrigation District)

extreme end of the project to the new intake for the High Line Lateral, north of Spring Coulee. As the time to open the gates approached, the reported noted that even the "flow from the main CPR canal seemed to be pressing restlessly against the iron barrier. Did those water yearn to spread out over the fields and meadows, groves and gardens, to do their bit in filling the South's Horn of Plenty? One standing there almost felt that they did."[199] The gates were opened shortly after one o'clock by Earnest Bennion who declared as he reached for the wheel that controlled the intake that "this event makes another milestone in the agricultural development of southern Alberta, for it is just another factor in removing the hazard from our operations and insuring production of crops year after year."[200] The flow of the water into the new canal brought forth "from more than a hundred throats, lusty cheers [which] broke the tranquillity of the countryside."[201] The opening of the gate was followed by what the reporter called the historic part:

> Again those Southern Alberta pioneers standing on that fresh embankment recalled a similar ceremony back in 1901 when the first irrigation water in this region was turned into the canal, an assurance to the new settlers that Alberta offered them a permanent home. Still farther back the memory moved, back to the early days in Utah - the old home of these people - when with the sagacity of the true empire builder, Brigham Young and his band, in 1847 turned the waters of the mountain streams onto the sterile wastes of Salt Lake Valley opening the door to the development of the great intermountain commonwealth, today the home of millions.[202]

The cavalcade then returned to Magrath following the water which was "making a path for the streams that are to follow as the seasons one by one invite the husbandman to seed time and harvest."[203] Upon their return, the participants enjoyed a banquet provided by a committee of ladies headed by Mrs R.W. Bradshaw, the "mother" of the project. After the dinner she commented that "she had worked for 25 years for irrigation and the events of the day were the answer to her hopes and efforts."[204] The meal was also the occasion for a number of comments by government and CPR officials on the way the system was built and the expectation for success. P.J. Jennings, Assistant Commissioner of Irrigation observed that the Magrath district was fortunate:

> in that only a partial water right, 40 acres to the 160, has been purchased. This reduces the annual rates to around $3 an acre, well within the ability of the water users to meet on lands

Official opening of High Line Lateral, June 1927 (Magrath Museum)

> as productive as these. Under the scheme, farmers will have a block of irrigated lands supplementing their dry acreage, a most desirable combination. With an ample water supply, a well built ditch system and efficient trustees and operating staff, a new successful career is predicted for this new baby in the irrigation family of the South.[205]

FINANCING THE MID

By-law No. 3 of the MID was for the issuing of $200,000 worth of debentures. To cover the total annual payment, which included capital charges, water carrying charges and an additional sum for maintenance and operation, a rate of $1.00 per irrigable acre was charged. The contract offered to the Magrath area farmers by the CPR was different from that negotiated by the Taber Irrigation District since it required the Magrath Irrigation District to immediately begin paying both principal and interest. This amounted to a payment of $2,400.00 by 1 December 1926; $9,000.00 on 1 December 1927; $12,000.00 on 1 December 1928 and $19,142.86 on 1 December 1929. This represented a enormous financial burden for the water users in the area for which little relief would be provided until 1946. The first instalment of principal was payable on 1 December 1929 and the payment of the interest charges of 6% began on 1 December 1926.

The bond issue was designed to provide for the debt to the CPR which covered construction and the water service charge. The temporary funds required to cover the other costs of operating the system were provided by loans from the Bank of Montreal. This source of temporary funding was used as early as 1922 when the organizing committee borrowed money to cover its expenses. The MID continued this practice. By-law No. 1 authorized the board to borrow $2,000 from the Bank of Montreal at an interest rate of 8%. The MID's continued use of this source of short term credit created a significant debt for the district.

A water rate of $3.50 per irrigable acre was considered sufficient to cover the water service charge of $8,500.00, the water right payment of $9,000.00 and the other expenses of the MID in 1927. The rate remained at $3.50 for 1928 despite the fact that the payment on the debentures had increased to $12,000 but was raised to $5.00 in 1929. The increase meant that the MID had the highest rates in the entire St. Mary Project. This rate dropped in 1937 as part of a new approach to deal with its debt.

ADMINISTRATION AND OPERATIONS

In October 1926 construction of the system was underway and the MID was in a position to deal with the way the district was to be run, hiring staff and establishing procedures for the operation of the board. The board began this process with suggestions from C.L. Dodge of the the CPR on the creation of lateral associations and the delivery of water to these association. Dodge used 2½ cfs as the average rate of flow. The use of the water by the lateral association was to begin with the farmer at the end of the lateral who was allowed a certain time to use his water allocation depending on the water flow and his irrigable area. The next farmer would take the water when this acreage had had its share of the time, and so on up the ditch. The amount of time each farmer had would depend upon the amount of irrigable land he had. For instance, if there were 320 acres in the association and there were 40 acres on the farm, it would receive 40/320 or 1/8 of the total time allotted to that lateral to receive water. The lateral associations would be advised by the water master of the time allotted to each association.

The hiring of a secretary-manager began in January 1927 when an advertisement was placed in the *Lethbridge Herald* for the position commencing 1 March 1927. The notice indicted that the applicant should be familiar with irrigation and ditch maintenance and have some knowledge of book keeping. Each applicant was also requested to indicate the salary they expected. The board interviewed five of the ten people who responded to the ad. Each interviewee was shown the plan of the district and asked to give their ideas on how they would manage the system. Most of the people indicated that two men would be required for the task.

As none of the applicants had the particular combination of qualifications required, the MID divided the duties. John T Steele was hired as secretary-treasurer of the MID at a salary of $400 and Percy Perrett was hired as the water master commencing 1 March 1927 and continuing on through the irrigation season at a salary of $125.00 per month. A second ditch rider was to be employed commencing 15 May 1927 and continuing through the irrigation season. Perrett was provided with gas and oil plus $10 per month for repairs to his car. Perret remain with the district until September of 1928. It also hired people on contract to preform special tasks such as auditing the books and collecting the rates.

The MID also made use of the legal services of the Lethbridge firm of Ostlund & Clarke. In March this company requested the MID to pay it an annual retainer of $300. The firm made the request because:

> it very often occurs that clients of ours come to us with matters pertaining to your District and invariably we have to advise these clients that in times past we have acted for the district and that it would not be right for us to take matters up against you. This has a tendency to produce a loss of business to us as once you lose a client owing to your inability to act for him he very seldom comes back to you again. In times past we have served the District to the best of our ability but we have reached a point where we must either know whether you are going to retain us definitely by the year or else we shall have to accept any work that comes in against you.[206]

The board responded by expressing its satisfaction with the work carried out by Ostlund & Clarke but felt the District was not in a good enough financial position to provide a yearly retainer

The board also decided that it would maintain and control the main canals and water users associations should be formed to control and maintain all laterals and sub-laterals. The MID board was prepared, however, to undertake the replacement of all timber structures with the exception of farm bridges. The MID therefore operated in a similar way to the CPR in that it only assumed responsibility for the operation of the two laterals which constituted its main canal while the water users were responsible for the rest of the system

The water users initially indicated their willingess to create lateral associations as outlined in the Irrigation Districts Act. At the April meeting of the board, petitions were presented for the creation of lateral associations along the lines suggested by the CPR. The associations were designated by A, B, C, D, E, F, G, J, L, M, N, and O after the letter assigned to the lateral. Approval was given to most of the applications with specific individuals designated to serve as returning officers for elections purposes. The development of the lateral associations in this way was not continued. The water users later voted to reverse the process and have the board assume responsibility for lateral maintenance. The board did not however assume this responsibility thus leaving lateral maintenance to informal arrangements between the water users. This trended was regretted by Isaac Sabey who suggested that the District could be saved a lot of expense and trouble by the organization of such associations.

Like the TID, the MID went though a period of testing the system to determine its condition and see if it met the specifications of the plans. This process was begun in June 1927 when M.N Sutherland, an engineer with the Department of the Interior made an inspection. Following the inspection Water Master Charlie Perroit turned water into the Magrath Lateral and the new High Line Lateral. Many of the canals were, however, clogged with soil and in many cases the banks were too low resulting in extensive flooding of roads and adjacent property. Following adjustments by the CPR it was in full operation by the 1928.

The district also rented a rented a DNR excavator from the CPR. The terms under which the first machine was rented in 1935 required the MID to pay to the CPR only the actual cost the CPR of the maintenance and operation of the machine from the time it left the home farm near Lethbridge until it was returned to this location. The CPR supplied a two man crew with caboose in which they lived. The District furnished the necessary gasoline and oil, a light delivery car, a team to change the position of the caboose and any additional help as required. The purpose of the light delivery car was to provide transportation for the crew back to Lethbridge on the first night and into Magrath on the second night. After the DNR excavator arrived at the canal, the car was used to transport the crew between the excavator and the caboose. The car was needed because the crew was required to keep the gasoline, oil and tools with them during the transit of the machine to Magrath and during its operations on the Magrath canal. As Houston explained to Horace Ririe Secretary of the MID:

> we have found it necessary even though the machine is comparatively close to the caboose to drain the gasoline from the engine each night when finished work and remove all tools etc. from the machine and take them back in a light delivery to the caboose so that they cannot be stolen.[207]

The district was charged $16 a day, which included the wages of the operator and helper. The MID was not required to pay the $6.00 charge to cover the depreciation and interest on the cost of the machine. The equipment was rented again in 1938, 1942 and 1943.

THE BURDEN OF DEBT

The history of the MID from the late 1920s to 1946 is concerned primarily with the efforts of the board to deal with the mounting level of debt. Solving this problem involved the MID in constant negotiations with the CPR and the Bank of Montreal to develop a plan which would permit the district to continue to function. The debt was created between 1927 and 1934 and eliminated between 1935 and 1946.

The crisis had its origins in the 1926 agreement with the CPR which made it necessity for the MID to have a relatively high water rate to pay the debt created by this agreement. The MID water rate went had gone from $1 in 1926 to $5 in 1928. To provide operating funds and to pay the CPR, the MID obtained operating loans on an annual basis from the Bank of Montreal.

The first indication of the problem came on 8 December 1928 when notices were sent to all water users explaining the difficulty the district was having in raising the funds to pay the first interest payment of $12,000 due on 1 December. The district could not meet the payments because the water users had not paid their rates. The board noted in its letter to the water users that the rates had to be paid no later than 31 January 1929 or the land was would be liable to confiscation by the board.

The CPR was also aware of the financial situation of the MID and undertook an investigation in January 1929. G.N Houston, superintendent of the Lethbridge Irrigation District, and P.E. Mileson, Chief Accountant for the Department of Natural Resources in Calgary interviewed Mr. Nicol, the Manager of the Bank of Montreal at Magrath and John Steele, Secretary-Treasurer of the MID. Nicol advised them that the MID still owed the bank slightly over $4,000.00 which they expected to receive by the 29th of January since that was the date of the court for confirmation of the rate enforcement return. After this debt had been paid, the bank "contemplated advancing sufficient [funds] to pay off the debenture indebtedness to the Railway Company, and the rental of $3,375.00 payable on the 1st of last May."[208] This would bring the MID's total debt to the Bank of Montreal to $15,375.00. Nicol explained that the delay in payment of the 1927 and 1928 rates "was due to the delayed movement of the crop in 1928 as cars were not furnished in sufficient numbers by the CPR until about Christmas, and many of the farmers who had shipped their grain had not received the returns for same."[209]

John Steele indicated that the amount due to the District for 1927 rates as of 4 January 1929 was $13,376.44 of which $1,747.46 had been collected since that date. The outstanding amount owing on the 1928 rates was $17,461.00 of which $1,262.38 had been paid. Steele gave Houston and Mileson the same explanation as Nicol for non-payment and indicated that Mr. Emeno, the Inspector of the Bank of Montreal at Calgary, had approved of the arrangements made for advances by the bank to take care of the CPR indebtedness. Steele was optimistic, however, that the financial situation would be resolved. He:

> was of the opinion that the bulk of the 1927 indebtedness would be paid off before the 29th of January 1929, and these payments made would go in reduction of the indebtedness to the Bank, and if payment was not made by the 29th of January, the date for the District Court Judge confirming the Rate Enforcement Return, the lands included in the return would be subject to forfeiture pursuant to the provisions of the Irrigation Districts Act.[210]

Steele, however, did not think that rate enforcement would be necessary since "he thought the farmers in the district, as and when they received the returns for their crops for the next three months, would make payment on account of the 1928 indebtedness . . . ."[211] Houston and Mileson suggested to Steele "that they should be pressed to do this as the 1929 rate would be heavier than the 1927 and 1928 rates."[212] The optimism of Nicol and Steele was not shared by Houston and Mileson who left Magrath convinced that the CPR could not expect to receive payment on the bonds or WA 1375 from the bank or the district before 1st February at the earliest.

As Houston and Mileson expected, there was no sudden infusion of cash to the MID because of payments by the farmers. The MID also did not carry out its threat of confiscating land through rate enforcement. The Board "desired to give every one ample time to pay the rates without taking enforcement proceedings to the limit."[213]

The expectation that future economic growth would deal with the MID's debt was shattered by the fall in commodity prices in 1929. Because of the situation in the world wheat market, the MID deferred registration of the rate enforcement return for 1928. On 30 August 1930, the board also send a note to all people in arrears and advised that if payment was not received by 15 September 1930 that the 1928 return would be registered and all necessary steps taken according to the Act. This threat apparently proved successful since by October, the 1928 rates had been received with the exception of one individual who was given a further week to pay.

The good news about the 1928 rates presented at the 28 October 1930 meeting was negated by the news that many of the water uses would not be able to make payment of either the arrears for 1929 or the current rates. The MID now had no funds to meet the debenture payment of $18,714.29 due to the CPR on 1 December1930. The MID dealt with this situation by making its first contact with the CPR . John Steele was directed to write to Sam Porter, Manager of the CPR's Department of Natural Resources and explain the situation to him. The MID board also decided to sell 60 acres of land which had come into its possession because of rate enforcement. The public auction was set for 10:00 o'clock on Monday 19 January 1931.No bids were made for the land, however, on the day appointed and the sale was adjourned.

As the MID tried to make a deal with the CPR, it suffered another blow when it was advised by the Bank of Montreal in April 1931 not to write any more cheques until some agreement with respect to its debt could be made. As of 11 June 1931, the MID owed the local branch of the Bank of Montreal $17,781.00 This money had been borrowed to make the 1 December 1929 debenture payment, provide operating funds for 1930 and 1931 and to pay the interest on the debt to the bank. Without credit from the Bank of Montreal, the MID could not function.

The Board considered various ways to deal with the situation, one of which was having water users undertake canal maintenance work in exchange for credit for their work. Ditch rider Blumel was requested to interview water users to determine their interest in such an idea.

In July 1931, the Bank of Montreal submitted a proposal to reorganize the financial affairs of the MID. The MID was to assign to the Bank of Montreal all its claims to the arrears and uncollected portion of the irrigation rates for the years 1929, 1930 and 1931. The only prior claim to these funds, acknowledged by the agreement, was that of the debenture holders. The board was to take all steps that may be necessary and pursue all remedies available under the Irrigation Districts Act to collect arrears. The Bank of Montreal was no longer willing to simply loan more money without some control of the MID's affairs.

The Board of Directors did not approve the agreement but rather initiated a series of meetings between 27 and 31 August 1931 with Sam Porter of the CPR, L.C. Charlesworth of the Irrigation Council and the Bank of Montreal. At the meeting with Porter, the financial situation was reviewed. The debenture payment due 1 December 1930, amounting to $18,714.00 had not been paid with another payment of $18,284.00 due on 1 December 1931. The MID's debt to the Bank of Montreal was approximately $19,000 which was acquired to make the 1928 debentures payment of $14,410 and to provide funds for operating expenses in 1931 and 1932. The MID also owed the CPR about $11,000 for WA 1375 and WA 1374.

The MID indicated to Porter that the problem was their high water rate and that the best they could do was to collect approximately $2.60 per acre out of the assessment of $5.00 There were are approximately 4,900 acres under the new water agreements, and 1900 acres under the old water agreement making a total of about 6,800 acres liable on these assessments. Collection of a $2.60 rate was more likely over the entire area would realize nearly $18,000 if successful. Even if the collection was less than 100%, the board felt that at least $15,000 could be collected. Sam Porter did not discuss a reduction of rates but simply requested payment to the CPR of $10,000.00 and a payment of $4,500 to the Bank of Montreal. Sam Porter promised future deferments of principal and interests if the MID could make these payments. The CPR also suggested that the MID needed to improve its collection procedures.

The board immediately followed through with the commitment to improve its collection procedures by meeting with officials of the Lethbridge Northern Irrigation District on 31 August 1931. Sutton explained the methods followed by the Lethbridge Northern Irrigation District in the enforcement of payments of water rates and the forms used in such collections. After reviewing this information, the board decided to follow the LNID example and hire a rate collector. Former board member Isaac Sabey was chosen as a rate collector. The Board also decided that since the water users could only pay $2.60 an acre that should be the amount collected from all water users for the year.

Issac Sabey did not have any more success in collecting the reduced rates than the MID had in collecting the higher rates. In December 1931, the CPR suggested that the MID sell the CPR more bonds. This proposal was discussed at a rate payers meeting held on 30 December 1931. Jensen opened the meeting with an explanation of its terms and indicated his strong support for acceptance of the proposal, he felt that it was the best thing to do in the situation. Charlesworth of the Irrigation Council also reviewed the terms of the proposed agreement and recommended its acceptance. The ratepayers were not, however, willing to vote on the proposal but raised questions about the way the system had been built. They suggested that the CPR be contacted and the capital cost of the system be reduced since it was too expensive to pay for. They also demanded that Sam Porter of the CPR attend a meeting to discuss a reduction in rates. The meeting concluded with a comment by the water users that they did not want to belittle the efforts of the Board "but to get an expression from Mr Porter as to the possible reduction in rates."[214] Sam Porter's refusal to attend a meeting in Magrath forced the rate payers to approve the solution suggested by the CPR..

Under the terms of the 21 March 1932 agreement, the CPR was prevented from taking any of the steps as provided in The Irrigation Districts Act for the collection of principal and interest due under the original debentures. The MID agreed to issue and deliver to the CPR 15 new debentures with a par value of $5000 each. The debentures were payable at par at the Bank of Montreal on 1 December 1938 and on 1 December for each year thereafter up to 1952. This agreement was not a long term solution to the financial problem of the MID since it perpetuated the cycle of going further into debt to meet current expenses and made it more difficult to reduce the water rate, which remained at $5.00 per acre.

The MID began the 1933 irrigation season with its arrears problem temporarily solved but further in debt. The collection of the water rates was thus even more critical than in the previous years. In an effort to improve collections, the MID, with the approval of the CPR, introduced a crop share program to pay the rates. The CPR also agreed to "postpone their remedies" for the holders of the old water agreements, if they participated in the program. For all grain crops, hay, fodder and potatoes 12 ½ % was the required share. In the case of sugar beets, 12½ % was required of the first 100 tons grown and 10% of the sugar beets grown over 100 tons. Wheat was to be delivered to the elevator in the name of the Trustees of the MID. Other grains were to be held for delivery at any specified point, not further than the nearest elevator. Hay and fodder crops were accepted in stack in the field. Special arrangements could be made for the substitution of a larger share of one crop and a smaller share of another to allow the farmer to take advantage of market conditions. This offer was conditional upon the water user paying in cash the proportion of the 1931 and 1932 rates which was paid in those years by other rate payers.

In 1935 the MID reconsidered its approach to financing its operations and began a comprehensive program to reduce their debt to the Bank of Montreal and the CPR, to increase revenue through a more aggressive collection of rates and reduce the water rate making it easier for the water users to pay. The operating loan for 1935 from the Bank of Montreal came with various stipulations including a requirement for the express consent of the CPR. The district agreed to appoint a collector nominated by the CPR to get in as much of the outstanding arrears as possible. The bank advised selection of someone other than the water master. The third requirement was that crop share program be continued.

The CPR then offered to write off $120,000 of debt in return for full payment of $80,000 which to cover the construction of the High Line Lateral and that other creditors of the MID participate in debt write down. The MID requested $80,000 from the recently established Prairie Farm Rehabilitation Administration (PFRA). Chris Jensen wrote on 17 September 1935 to L.B. Thomson, Chairman of the Water Development Committee of the PFRA reviewing the history emphasising its financial obligations to the CPR. Jensen explained the impact of the Depression on the district:

> irrigation was sought because the settlers had learned from sad experiences that it was necessary to successful agriculture in that part of the country. We now believe that in their anxiety for water the settlers were over-optimistic as to the amount that they could afford to pay for it. To meet operating costs, CPR water

> rental, and capital charges it became necessary to levy a rate of $5.00 per irrigable acre per annum. It is possible that had the prices of farm products continued at the level which existed when the District was formed , there might have been some chance of collecting such a rate, but that was very doubtful.
> With the disastrous drop in prices, the position of the majority of the farmers in the District became hopeless, and the District authorities were quite unable to meet debenture payments Interest charges accumulated and increased the burden, and the settlers soon were in a position where they saw no hope of ever meeting the obligations which they had incurred, and consequently lost all heart.[215]

If assistance could be provided to relieve the people of their debt Jensen concluded, they would "proceed with renewed hope, and would undoubtedly develop the territory as it should and can develop under irrigation. If such an adjustment cannot be made, they will be compelled to throw up their hands and with no resources seek to re-establish themselves elsewhere."[216]

PFRA director Thomson's response indicated that the request would be investigated in preparation for the next meeting of the Water Resources Development but he did not encourage Jensen to expect any assistance. The engineering and economic studies required before a final decision could be made were carried out by W. Wotherspoon, District Engineer, Lethbridge and Mark Mann, Agricultural Adviser. Both individuals made very positive comments about the future of the MID, given the resolution of its debt problem. Wothersespoon indicted that the works were in a serviceable state of repair and that a rate of from $1.00 to $1.25 would provide for the proper maintenance of the system. If the district could be placed in a position to take advantage of the company's offer, he concluded, the settlers "will make good." Mann recommended assistance based on the fact that the scheme was feasible, irrigation was needed and that maintenance costs would be covered by a rate of $1.50. Despite the favourable recommendation, the PFRA eventually decided against providing assistance because the Minister in charge of the PFRA, Jimmy Gardiner, was not willing to contribute to a project that would benefit the CPR.

Having been denied help by the PFRA, the MID resumed its negotiations with the CPR in December 1936. After three trips to Calgary and various drafts of the agreement, an final agreement was presented to the rate payers at a general meeting on 3 May 1937. Jensen indicated that it was not as good as they would like [but] it was as good as the CPR officials would give at the present time. Despite the fact that the board was not totally satisfied, it represented a major accomplishment. This agreement put the MID on the road to financial recovery by breaking the cycle of increasing debt to cover the non-payment of rates while at the same time lowering the water rate for the first time since 1928. Under its terms, the MID agreed to impose a rate of not less than $2.50 for the years 1937, 1938 and 1939. These rates were to be collected as crop shares method. The MID was to receive a one-tenth share of the sugar beet crop and a one-sixth share of all other crops grown on the land. The revenues raised were to be applied firstly to the current operating expenditures of the district; secondly to the payment of current water rentals; and thirdly in payment of arrears of water rentals. The CPR agreed that if the MID complied with these provisions from1937 to 1939 the Company would waive all interest on the debentures and all other indebtedness of the MID due to the CPR.

OLD WATER CONTRACTS

Acting as the agent for the CPR in the collection of rates from old water contracts was appealing initially since it gave the MID an additional source of income. The terms of the agreement were a liability during the 1930s, since the MID was responsible for paying the total amount owing to the CPR even if the farmers did not pay the MID. The MID could, however deal with the problem by requesting the cancellation of the contract.

The first problem with respect to the old water rates was clause 12 of their agreement with the CPR. Steele had deducted the MID's 25% commission from the rates collected and forwarded the balance to the CPR. Under the terms of the agreement, the MID could only take its 25% portion after the CPR had received its entire share.

To ensure collections under the old water contracts, the CPR closely monitor collection activities of the MID and to take immediate action if results were not being obtained. In the spring of 1937, the CPR advised all its contract holders, including those in the MID, that failure to pay the rates could result in the cancellation of the contract. In May, an official was sent from Calgary to

Magrath to review the MID records to determine who would be permitted to receive water during the 1937 irrigation season. Following this review, Sam Porter intended to discontinue the delivery of water for the use of those who had not made suitable arrangements for payment. The CPR also tried to improve rate collection on the old water contracts by discouraging the MID from giving the holders of old water contracts the same length of time as new water contract holders to pay their arrears.

DEBT REDUCTION

By1939 the MID had dramatically improved the collection of rates and thus was able to eliminate its debt to the Bank of Montreal and all arrears on old and new water contracts. The MID's debt reduction program continued. On 1 April 1940 negotiations were resumed with the CPR .At this meeting, Porter repeated the offer, originally made in 1935, that the CPR would take $80,000 cash in settlement for the bonds. The economic situation of the MID and the economy in general had improved since 1935 to the point were this offer could now be seriously considered. The idea of making one large payment to the CPR to eliminate the debt in one stroke was brought to the 10 April 1940 annual general meeting for discussion. The meeting authorised the board to borrow $65,000 and then add to it the cash on hand, $7,500, and a further $15,000 which the board was to receive in arrears and for the payment of the 1940 rates. The board was unable to borrow the $65,000.00 by May 1941 and thus had to opt for a second agreement which dealt with the debenture debt on a longer basis. The terms of this second agreement covered the years 1941 to 1944 and required the board to continue to collect a rate of not less than $2.50 per acre. The board was to "exert every effort" to collect its irrigation rates both current and arrears and turn all revenues over and above necessary operating expenses to the CPR. These funds were to be applied to the debenture debt on the basis of a credit of $3 for every dollar paid. In the event that the MID was able to raise $80,000 at any time during this four year period, the CPR agreed to accept payment in full on the basis of $80,000.00 as at 1 December with interest at 5% after crediting all cash payments made on the debenture account subsequent to 1 January 1940. When this agreement expired on 1 December 1944 the CPR agreed to continue giving three to one credit on debenture payments but charged interest at 5% on one third of the total outstanding balance on the debenture account.

In 1945 the CPR agreed to forgo the interest on the water service charge provided that it was paid in the calender year in which it was levied. It would not, however, reduce the interest on the bonds arguing that Magrath had had more consideration than any of the other districts. In 1945 the board decided to try again to settle the debenture debt by making one large cash payment. The strategy followed in 1945 was to raise the rates to $5 and offer the CPR an amount less than $80,000. The board initially decided to offer $45,000.00 but in November, Jensen offered the CPR $37,559.23 as a cash settlement and arranged to borrow $22,000.00 from the Bank of Montreal on personal signatures pending the 1947 collections. The board supported this initiative and decided at the 2 December 1946 meeting to raise the water rate to $6.00 for 1947 to clean up the debt issue. The CPR accepted the offer, ending a 20 year struggle to cast off the burden of debt which had been the most significant barrier to the MID's progress since 1926.

THIRTEEN
RAYMOND IRRIGATION DISTRICT
1924 - 1946

The creation of the Raymond Irrigation District (RID) was the second step in the process of dividing the Southern Irrigation District. Its creation parallelled that of the MID but was delayed by complications over determining the legal description of the land to be included in the districit. Extensive subdivision of the land to facilitate its use for sugar beet production caused many complications. Many parcels were being farmed by non-owners which brought about a further delay as the owners had to be consulted in the creation of the district.

The RID, from its creation to the end of the Second World War, focussed on one primary issue, namely the construction of storage reservoirs to relieve the annual water shortage. Raymond took the lead on this issue as early as 1919 when members of the community such as John W. Evans were involved in the creation of the Irrigation Development Association. The RID continued that tradition as the most vocal advocate of the construction of storage reservoirs in the St. Mary Project.

Second only to the issue of reservoirs, as a concern for the RID board, was the financial situation of the district. This problem, which was only resolved on the eve of the CPR's departure from the St. Mary Project in 1946, involved the RID in a constant round of negotiations with the Irrigation Council concerning rate enforcement and with the CPR regarding payments on the bonds and water rental. The board was also involved in a constant effort to ensure the payment of the rates by the water users. Rate enforcement, as conducted by the RID board, was a delicate balance between applying the rules for rate enforcement as defined by statute and protecting the water users from the harsher aspects of these rules. Given the extent of this problem during the early 1930s and the implications of the rules for rate enforcement, the RID could have acquired title to a good portion of the land in the area. The board, however, felt that the water users would pay given the improvement of conditions. They were proven correct in this assumption thus avoiding the necessity of widespread confiscation of land. The RID used other means to ensure its survival during the Depression.

CREATION OF THE DISTRICT

The process of creating the RID began in July 1923 when Heber Allen and Thomas O'Brien organized and forwarded two petitions to Commissioner of Irrigation, Victor Meek. One petition requested the exclusion of certain land from the Southern Irrigation District while the other requested the formation of the RID. Allen indicated it was the intention of the Raymond water users to take over the Raymond-Stirling lateral from the CPR and assume the company's obligation to supply water. The district would collect the water rental from users and pay the CPR for this water a yearly rental of $112.50 per cfs with the balance to be retained by the district. The district was to pay the CPR $4,000 per cfs for the water right for additional land mentioned in the petition. The letter concluded by pointing out that the proposed MID and the proposed RID had paid all of the debts of the Southern Irrigation District so there could be no objection to the withdrawal of lands from this district.

Meek referred the petitions to Irrigation Council for review. As with the MID, L.C. Charlesworth raised concerns with respect to the proposed district. He questioned the need to create an irrigation district when he thought it would be easier for the landowners to simply enter into individual contracts with the CPR. He also felt that serving some users under CPR contracts and others under the terms of the Irrigation Districts Act would lead to many difficulties. His final objection was to assessment based on the number of acres owners wanted to irrigate regardless of the irrigable acreage. In his view, this was contrary to the intention of Irrigation Districts Act.

Allen immediately responded by pointing out that the CPR would not contract with so many individual farmers who wanted more water. He also indicated that the financial terms offered by the CPR were very satisfactory:

> if we get the water from the CPR now offered at $4,000 per foot and this will irrigate 170 acres the cost would be $23.53 per acre for a system already built. It will not cost more than $2.00 to $3.00 per acre to repair the ditch to take the additional water now offered so that the total cost to the farmers would not exceed about $27.50 pre acre Under present conditions this would seem very reasonable.[217]

Charlesworth's concern about the necessity of creating a separate district to provide more water to the farmers in the Raymond area was dealt with by Peter Naismith who informed him of the accuracy of Allen's information. The method of assessment was also not a problem since the federal government considered that the RID was taking over an existing project and land classification surveys were not needed.

Other problems with the petition which took longer to resolve included the fact that a number of people who signed the petition were not the owners of the land. Petition organizers had to contact the owners to obtain powers of attorney, authorising the renters of the land to act as owners' agents. The legal descriptions were also the subject of numerous letters between the provincial government and Raymond. It took a full year for these issues to be resolved.

The official notice of the intention to create the district was published in the *Alberta Gazette* on 12 November 1924. Having received no objections to its creation during the period indicated in the notice, a vote was taken and a ministerial order was issued for creation of the RID on 1 March 1925. Only minor changes were made to its boundaries after 1925.

AGREEMENTS WITH THE CPR

Under the terms of the 12 May 1926 agreement the CPR was to furnish 40 cfs of water at the headgates of the Raymond Lateral and other gates on the main canal throughout the irrigation season except when prevented from doing so by conditions beyond their control. The CPR reserved the right to shut off water for repairs upon reasonable notice to the RID and to prorate the water during drought. The district agreed to take over and operate and maintain the Raymond Lateral and all other works in the district and indemnify the company against all liability arising from the operation and maintenance of the works. The RID agreed to pay the CPR $160,000.00 for the water right and an annual carrying charge of $4,500 or 64 and 1/3 cents per acre to cover the new land brought under irrigation.. The water was to be used only for irrigation, filling stock ponds and to supply water to the sugar factory. The board agreed to deliver water to the current holders of water contracts.

The 574 existing contracts transferred to the administration of the RID were amalgamated into one water agreement between the RID and the CPR covering 8,702.75 acres. Very small amounts of water were diverted under this agreement as the 574 contracts averaged less than 16 acres. The amount of water diverted dropping as low as 4/75 of a cfs. The RID's role was to collect the water rates and deduct its fee for performing the service for the CPR. If a water user was delinquent, the RID was liable for the payment to the CPR. In the case of a water user who persisted in not making payment, the RID could notify the CPR and require that the contract be cancelled. These water rights could not be transferred to another parcel of land or water user. The amount of water being diverted by the CPR was simply to be reduced and the money owed under contract 1340 reduced. Like the MID, the RID had old and new water agreements. Unlike the contract between the CPR and the TID and the MID it did not include any construction work by the CPR for the extension of the system.

Arthur E. Fawns (Raymond Irrigation District)

ADMINISTRATION

The first Directors were Heber S. Allen, A. E. Fawns and Jesse. H. Wilde. J. W. Evans was the first Secretary-Treasurer while John Salmon was the first Water Master. Salmon was Water Master from 1926 until December 1945. Evans had come from Utah in 1903 and was employed by the Knight Sugar Company as the superintendent of field operation. He had played a leading role in the effort to save the plant from closure. He was employed by the Canadian Sugar Factories Ltd. after 1925 and carried out his duties with the RID on a part-time basis.

The operating policy of the RID was the same as the CPR and the MID in that it divided the operation of the system between the district which operated the main canal and the water users who operated the distribution system through lateral

Heber S. Allen (Raymond Irrigation District)

Jesse H. Wilde (Raymond Irrigation District)

associations. This policy was the subject of a review in 1932 when an individual who had purchased land after the creation of the district objected to paying for water rights when the district refused to build a canal to deliver water to his land. He wanted the district to either build the canal or remove his land from the assessment role. This dispute led to the passage of The Raymond Irrigation District Act in 1933. It gave a legal basis to the board's argument that it was subject to the same rules of operation as those followed by the AR&ICo. The preamble of the legislation indicated that irrigable land sold by the AR&I entitled the owner to a supply of water for irrigation from the main canal of the company upon the owner constructing and maintaining at his own expense all ditches from the said main canal. The preamble also stated that the RID had been established based on the perpetuation of this concept since it was then generally understood and accepted that the district would deliver water to the owners of irrigable lands in the district at a convenient point on the canal acquired by the district and that all ditches

John F. Salmon (Raymond Irrigation District)

J. W. Evans (Raymond Irrigation District)

necessary to convey the water should be constructed and maintained by the owner at his own expense. Following this introduction, the Act stated that:

> the district shall be under no duty to construct, maintain or operate any ditches in the district required for the conveyance of water from any canal owned by the district to any parcel of land, and the right to receive water in respect of any parcel entitled thereto shall be conditional upon the provision by the owner of the parcel of all ditches required for the conveyance thereof from the canal of the district to the parcel.[218]

This law remained in force until the 1968 Irrigation Act.

WATER SHORTAGES AND RESERVOIR DEVELOPMENT

Finding a solution to the district's water supply problem through the establishment of its own system of reservoirs was the main concern of the Board of Directors from 1928 to the early 1940s.The Raymond and Craddock Reservoir sites were developed and the sugar factory reservoir was purchased. The site with the most potential for reservoir development in the RID was the Raymond Reservoir on the north face of the Milk River Ridge in Middle Coulee about 2.5 miles south of Raymond. With the construction of a dam at this location, this site had the potential to capture runoff from the Milk River Ridge. In July 1928 Evans wrote to Commissioner of Irrigation J. S. Tempest for permission to build a dam at this site. In his response, Tempest indicated that "if you have in mind storing the natural run-off from the small creek which comes down from the Milk River Ridge, it will be necessary for you to make application for the use of this water . . . since the waters which you would be able to store in the above reservoir would probably deprive riparians on the creek lower down of their stock water supply, it is doubtful if your district would be given permission to use this water."[219] Tempest, however, did encourage the RID to provide him with full particulars as to the quantity of water the district proposed to store and the use to which this water would be put. Once this information was provided, Tempest promised that an engineer from the Reclamation Branch of the Department of the Interior would investigate the matter.

In 1935 the community resumed it efforts to develop this reservoir when the Town of Raymond requested help from the Water Development Committee of the recently established PFRA. The response from the Water Development Committee was that this project was not eligible for assistance since the PFRA's mandate was to assist the rural areas of the prairie West.

When the water shortage problem intensified in 1936, the community, through the RID, the Raymond Board of Trade, the Town of Raymond and even the local Social Credit organization increased its efforts to develop the Raymond Reservoir. Appeals for assistance were made to Premier William Aberhart, L.C. Charlesworth, Director of Water Resources for the Province of Alberta, Lethbridge lawyer, L.E. Fairbairn and the CPR. The request for assistance to Aberhart, which was made during his visit to the community on 7 October 1936, was the most impassioned. In a letter to Aberhart, following his visit to Raymond, the RID, along with other community groups, made a formal request for assistance. It was supported by background information on why this development was critical to the welfare of the community.

The RID requested $15,000 to build a dam to impound water in Middle Coulee 2.5 miles south of Raymond, the reservoir created by the dam would store 17,000 acre-feet of water. This amount of money covered the cost of bringing in 61,000 cubic yards of earth, 2200 cubic yards of rock for riprapping, pipes and headgates and other materials necessary to complete the dam. The work was to be done by men and teams of the district during fall and winter of 1936.

The runoff water conserved by the dam was to be used in the summertime for the domestic requirements of the Town of Raymond, to supplement the water available from the present irrigation system and for the stock located east of the Town of Raymond.

The irrigation farmers in the area needed the dam because irrigation of the RID's 15,000 acres required 85 to 90 cfs of water at the Raymond-Stirling Lateral from 1 May to 1 October each year:

> unfortunately almost every year during the irrigation season there is a period when the supply falls far below the full water allotment for the district. In 1936 for example during July and August the supply fell to between 50 and 60 cfs and much below that in September. The consequences of this was a loss to the farmers of approximately $150,000 this season. A similar loss has occurred during the past several years during the months of July, August and sometimes September when the crops are most urgently in need of moisture. The reason was that the CPR has no reservoirs in its system to conserve flood water and therefore can supply only the natural flow of the St. Mary River which has proven to be quite inadequate during the dry seasons to supply the requirements of the crops supposed to be irrigated under this system. This company has protected itself in its contracts to supply water by limiting its obligations to supply only the natural flow of the river at the same time requiring the water users to pay on the basis of a full supply regardless of any shortages that might occur during the irrigation season and the resultant short crops and increased difficulty in paying water rates and rental.[220]

A reservoir, the Raymond group wrote, would restore dignity to the lives of many people in the community, raise the standard of living of both rural and urban residents and permit economic expansion. Farmers would be able to pay their rates and rental and live comfortably on their farms. It would employ people in the district who needed "work in order that they might procure for themselves food, clothing and shelter."[221] The extra water would permit the expansion of "intensive agriculture . . . in the district, thus providing more employment and increasing production."[222] The Town of Raymond could use the extra water to construct an effective sewer system. The ranchers living to the east would be relieved "of the necessity of begging for water from the irrigation farmers to help them in their distress as in the past."[223]

In writing to Charlesworth, Evans recognized that the development of the Milk River and Raymond Reservoir was part of a plan for the expansion of the St. Mary Project. The RID, however, was afraid this development would take a long time and believed the smaller dam was the proper thing at the present time. If the Raymond Reservoir could be built immediately, Evans concluded it will serve the double purpose of impounding water for next year and providing a relief project that would take care of the unemployed in this district. The view of the RID that they could not expect the expansion of the St. Mary Project anytime soon was understandable given the fact that no action had been taken on the idea since the plan for the Lethbridge South Eastern Project had been prepared during World War One.

The CPR gave the most definite response to the RID's request for assistance. Sam Porter, indicated that the CPR did not consider itself obligated to construct storage reservoirs. He advised the community to look to the government for assistance in this matter.

The RID did not immediately continue its efforts to develop the dam site, leaving that task to the Town of Raymond. In 1939 the Town of Raymond built a dam at the site and obtained exclusive use of the water. Water from this source, however, was most unsatisfactory prompting the RID to make another attempt to develop the site for its own use. In December 1941, the RID undertook the construction of a new reservoir in a coulee just east of the south end of the townsite. The plan was to trade the water from the new reservoir for the water impounded in the Raymond Reservoir. The PFRA was contacted in December 1941 for assistance in the implementation of the plan. George Spence, Director of the PFRA, indicated that before any action could be taken, further investigations were needed.

The RID also acquired the reservoir northwest of the sugar factory originally constructed by the Knight Sugar Company in 1903 and which continued to be used by the Canadian Sugar Factories Ltd after 1925. The opportunity to purchase this 118.46 acre reservoir site came in March 1936 when the CPR offered it for sale to the Canadian Sugar Factories Ltd. The Sugar Company immediately referred the offer to the RID. The acquisition of this reservoir was of interests to the RID despite its small size because, as Evans pointed out, "even a reservoir the size of the factory one is considerable value when used judiciously."[224] The RID board immediately began negotiations with the CPR. Its first approach was to suggest to the CPR that the site should be transferred without any payment since "this reservoir [was] so much a part of the canal system purchased from your company that it should have been included in the transfers that were made to the board for other property."[225] The RID also indicated that despite its desperate need for storage reservoirs, the "value we can get out of it must be measured in dollars and we think, under the circumstances, you are asking too much money."[226] The RID board, therefore, made the counter offer of $1,200. Sam Porter's response to the offer disputed the RID's claim about the value of the reservoir. Porter agreed that storage reservoirs were "useful adjuncts to irrigation projects" but he could not "accept the view that [his] company [had] any responsibility for providing reservoirs for the system we turned over to you."[227] Porter viewed the sugar company reservoir at Raymond as essentially a sugar company development which was useful to the RID but not absolutely necessary and he did not considered it as part of the irrigation system as transferred to the RID. Despite these views he indicated that the CPR was prepared to accept the offer. Following its transfer, it was leased as planned to the Canadian Sugar Factories Ltd. in December 1936.

The attitude of the CPR and the lack of any definite plans by the government to begin the expansion of the St. Mary Project prompted the RID to briefly contemplate the possibility of taking legal action to force the CPR to build reservoirs. In preparation for this possibility, the RID contacted W.D. Trigo who had led the fight between 1911 and 1925 in the western section of the CPR irrigation block against what the water users there considered injustices inflicted upon them by the CPR.

The ultimate solution to the water shortage problems of the RID was the construction of the Pothole and St. Mary dams after 1946. The decision of the federal government to undertake this post-war construction program came only after all of the water users and potential water users united to lobby for it. Raymond played an important role in the South Alberta Water Conservation Council (SAWCC) which was created for that purpose.

COLLECTING THE RATES

The second main preoccupations of the RID board during this era was the financial health of the district, that is dealing with the arrears in payment of water rates and managing their financial obligations to the CPR. The obligations to the CPR included the payment of principal and interest on the bonds. The RID was also responsible for the collection of the rates on the old water contracts. The RID received 25% of the rates but was required to make up any deficiencies in the payment. This responsibility could thus be an advantage or disadvantage depending on the economic situation. The RID paid for the water delivered to the holders of new water contracts.

The administration of the rate enforcement sections of the Irrigation Districts Act was the responsibility of the Irrigation Council, under Chairman L.C. Charlesworth. He attempted to monitor the activities of the various districts and to ensure that they followed the rules. The Irrigation Districts Act gave him some discretion on the extension of deadlines for such things as the registration of a rate enforcement return with the Land Titles Office. Charlesworth did not, however, have the power to investigate the conduct of the individual districts allowing them considerable latitude in the way they implemented the rules. The RID took full advantage of this situation to deal effectively with the financial problem.

The collection of rates did not become a problem in the RID until 1930. The rate enforcement return for 1929 submitted for confirmation before Judge Jackson on 28 January 1931 included 79 water users. Given the dramatic increase in the amount of land involved, the RID board decided to delay sending the rate enforcement return to the Land Titles Office. As Evans explained to Charlesworth it was the opinion of the trustees that the return should be held for some time to give the delinquents further opportunity to pay the arrears. Charlesworth responded by extending the period by four months. The RID board took advantage of the time to advise the delinquent water users of the delay and urging them to take advantage of this opportunity which might not come again. The letter indicated that Mr. Charlesworth had extended:

> the time to June the 1st, before action should be taken by the Board of Trustees to take title to the land on which the 1929 rates were not paid. Whether he will make any further extension or not we do not know. If he does not it will be necessary for us to apply to the Land Titles Office immediately after June 1st for a title to this land.
>
> Under the circumstances the safe thing to do is to pay the rates. If you cannot pay all a very substantial reduction should be made at once and a very definite proposition as to when the remainder will be paid. Anything unpaid July 1 will have added to it a 5% penalty.[228]

The circumstances of many of the water users prevented them from following this advice and the RID request another extension to 15 November. On 12 November 1931 Evans wrote to Charlesworth requesting a further extension to 10 December in order to give the beet growers an opportunity to use the payment they would receive from the Sugar Company 1 December to pay their water rates. Most of the people on the return, Evans reported had paid and he thought that most would pay by 10 December. Charlesworth agreed to the request and extended the deadline to 15 December. By that time only eleven land owners remained on the rate enforcement list.

On 31 December 1931, Evans finally sent the rate enforcement return covering the outstanding rates for 1929 arrears to the Land Titles Office. It was returned on 4 January1932 with a request for further information. The document was retuned with the additional information and an unknown amount of land came into the possession of the RID. The RID ignored the requirement that this land be sold within one year and simply held it until such time as the original owners could redeem the land. A similar approach was taken to the land acquired in 1932. The fact that the RID had kept the land in trust for the original owner was not noticed by Charlesworth until 1938.

The preparation of a rate enforcement return and having it confirmed by a judge appears to have been discontinued altogether after 1935. L.C. Charlesworth, however, continued to remind the RID about the need to publish a notice in the *Alberta Gazette* regarding the court of confirmation.

The RID, however, did take other steps to collect the water rate arrears for the years from 1930 to 1934. These steps were less drastic than confiscating the water user's land and more effective in providing the RID with revenue. It used Section 154A of the Irrigation Districts Act which had been revised to permit a district to obtain the income received by water users from the sale of their crops. The RID advised the Canadian Sugar Factories Limited and other companies such as the Ellison Milling and Elevator Company that they should send the payments for the sugar beets or wheat sold by certain water users directly to the RID. This procedure was implemented in October 1932 and continued until at least 1937.The number of water users affected ranged 95 in 1932 to 85 in 1937.

OBTAINING RELIEF FROM THE CPR

One of the consequences of the board's declining income during the Depression was its failure to meet its financial obligations to the CPR. The 1930 obligations were payment of interest and principal on construction bonds, collection of water rates on CPR contracts and payment of the rental on the water distributed to their own water users. The RID first indicated to the CPR that it would not be able to meet its obligations in March 1931. The CPR expressed its "disappointment" at not being paid. On 24 March 1931 the RID proposed to the CPR that it relinquish 5 cfs of the new water at the same price it paid which was $4,000 per cfs. The CPR rejected the idea and advised the RID to increase their efforts to make collections. The RID promised to do this but advised the CPR that it could not promise to reduce its account any faster than arrears of rates and rental were paid. In January 1939 the RID requested concessions similar to those received by Magrath. Sam Porter offered to cut the interest rate on the bonds from 6% to 4% but would not consider any changes to the old water contracts or the outright cancellation of the interest on the bonds.

The RID's debt to the CPR was finally dealt with in 1946 when acquisition of the St. Mary Project was under consideration by the province of Alberta. The RID took advantage of these negotiations to propose that the provincial government also take over the RID at the same time. This solution was appealing because the RID water users could receive their water at the same rate as Coaldale. This rather drastic solution was proposed at a joint meeting of the RID and MID on 2 February 1946. The motion incorporating these views indicated that the terms of the agreement between the CPR and the RID did not give equality of costs and services with other water agreements in the Coaldale area or with the water rental charge to the TID. The provincial government was therefore asked to take over the operation of the RID along with the AR&ICo. irrigation system. The water users of the RID could then receive their water at $1.00 per acre. In return, the RID agreed to turn over to the provincial government all its canals, right-of-ways irrigation works and equipment and in so doing would be free from any further debentures payments to the CPR. The RID wanted a satisfactory agreement drawn up with the government of Alberta and that this agreement be incorporated into the legislation ratifying the transfer of the St. Mary Project to the government of Alberta.

This proposal was apparently taken to Edmonton in February 1946 when Jesse Wilde and Rulon Dahl along with representatives from the MID who asked the provincial government to give these districts help in their efforts to reduce their bonded debt to the CPR. The delegation was given assurance by MLA Tanner that the province would make representation to the CPR on their behalf. The delegation also asked the government to make water service charges for these districts the same as the water users in the Lethbridge-Coaldale area and in the TID when the government took over the system.

At the annual meeting of the RID on 1 march 1946 a complementary strategy was presented. The motion was passed requesting the government to deliver water to the RID at the same price the TID charges which was 50 cents per acre, and that the RID continue to distribute it throughout its own system under the direction of the irrigation district board.

Given the motion passed at the annual meeting, the new board of directors elected in 1946 had to find another solution to the CPR debt problem other than giving up the district's independence to the new crown corporation which was established to run the St. Mary Project. The new board met with the CPR on 7 May and offered to pay 50% of the debt as a final payment to cover the entire $55,000 which it owed. The CPR accepted the offer.

These developments were communicated to the water users in an open letter which indicated that the provincial government had taken over the St. Mary Project from the CPR and that the RID had been able to make new arrangements both with the government in the matter of water service charge and with the CPR on the bonded debt. The water users were advised that the CPR had agreed to accept $27,500 as final settlement of the debt as long as it was paid by 1 July 1946. Given this offer from the CPR, the Board asked for the:

> full cooperation of all water users as we feel this is an opportunity to clear the land of this district from all bonded debt on a very satisfactory basis. There is now owing $55,000 bond debt on 6,635 shares of new water or approximately $8.30 per acre which falls in ten annual installments. Our rate on new water for a number of years has been $1.50. So to clear off the bonds and pay this year's rates it ordinarily would cost $8.30 plus $1.50 equalling $9.80. By taking advantage of the reduction offered, the Board has set an assessment for 1946 of $4.00 per share. This will include the final bond settlement as well as the 1946 rate. This represents a saving of $5.80 per share on new water and the Board suggest that water users borrow the money if necessary to meet this assessment so that the district can advantage of the $27,500 reduction.[229]

The other good news in the open letter was that the annual charge for new water in the future was not be more than $1.00 per share because the RID had arranged with the provincial government for a water service charge of $7,750 instead of the $11,003 previously charged by the CPR. This reduction gave the district $3,253.25 additional revenue each year for operating costs.

FOURTEEN
HEADWORKS ADMINISTRATION AND THE LETHBRIDGE-COALDALE AREA DISTRIBUTION SYSTEM 1919 - 1946

The CPR's administration of the St. Mary Project between 1919 and 1946 was based on maintaining the existing system combined with cooperating with the farmers in their efforts to expand the project through the creation of irrigation districts. During this period, it also introduced the DNR excavator (Ruth dredger) and draglines to keep the canals free of silt and weeds as part of a trend towards mechanization. These machines replaced floating dredgers and horses by 1946 as the principle means of maintaining the system. The number of seasonal staff was gradually reduced as a cost cutting measure during the Depression along with the size of the maintenance budget. Only key structures in the system were replaced with concrete structures while wood remained the dominant construction material for checks and drops. One of the few new additions to the system was the Northeast Reservoir which complemented the role of Chin Coulee in collecting waste water and passing it on to the TID.

Operation of the St. Mary Project became a greater challenge after 1919 because of declining water supplies and the increase in demand. The annual water shortages, beginning in 1919, raised the need for reservoir construction which had been advocated as early as 1900 by George Anderson. No action, however, had been taken on his reservoir plan. The IJC also advocated a reservoir construction program as part of its ruling on the division of the water of the St. Mary and Milk Rivers in 1921. Sam Porter, the Superintendent of Operations and Maintenance, even suggested that such a program was feasible when the water shortage problem developed for the first time in 1919. The CPR, however, consistently rejected the view that it had any responsibility for their construction. The annual shortage was dealt with by rotating the water between the two divisions and the laterals depending upon the amount available. The water supply problem, like the weather went in a cycle with 1919, 1936 and 1941 being the years when the flow of water in the St. Mary River was at its lowest.

The creation of the Taber, Magrath and Raymond Irrigation Districts was encouraged by the CPR because they reduced its responsibility with respect to distribution system administration. It also made one last effort to create a comprehensive system of lateral associations in the Lethbridge-Coaldale area. The increased settlement combined with the diversification of agriculture through the growing of crops such as sugar beets made their role even more important in the efficient operation of the sub-lateral system. The failure of this attempt left lateral operation and maintenance to informal agreements between the water users along the ditch. The exceptions to this trend were the Readymade, Forbes and the Whoop-Up Lateral Associations which were formally organized under the provisions of either the Irrigation Districts Act or the Water Users' Districts Act. Even these associations had difficulties in collecting rates and ensuring the proper maintenance of their respective laterals.

The CPR's interest in land settlement continued into the 1920s through participation in the program to bring Europeans to work in the revived sugar beet industry. Its involvement in this program, which will be covered in Chapter Fifteen, brought about the further diversification of southern Alberta society with the arrival of Mennonites from Russia, Hungarians and other eastern Europeans. This was the last settlement program undertaken by the CPR which affected the development of the St. Mary Project. The ongoing recruitment of labour for the beet fields was assumed by the sugar beet growers themselves. The Japanese who arrived during World War Two were brought in at the request of the sugar beet growers.

The CPR had total control of the water supply available to the St Mary Project through to the early 1920s. This situation was the result of the federal government's plan to reserve the water of the St.Mary, Belly and Waterton Rivers for the irrigation of land southeast of Lethbridge, since it was judged more suitable for that purpose. The 1902 addition of 500 cfs of water from the Belly River to the water diversion permit of the AR&ICo. was part of their strategy to protect the water supply of the St. Mary Project given the potential loss of water to the Americans under the terms of the Boundary Waters Treaty of 1909.

This policy excluded irrigation development in the area from Cardston to Glenwood as well as on the Blood Reserve The first project in the area west of the St. Mary River, was the United Irrigation District. It went into operation in 1923, following the decision of the CPR to relinquish rights to an amount of water which would permit the district to be established. The United Irrigation District and the Mountain View, Leavitt and Aetna districts which followed, re-established irrigation in an area which played an important role in the development of irrigation in the 1890s. The United Irrigation District also continued the important role which the LDS Church played in irrigation development in Alberta beginning with the contract to construct the main canal of the St.Mary Project in 1898.

CANADIAN PACIFIC RAILWAY RESPONSIBILITIES

The administrative structure of the St. Mary Project established by the CPR in 1912, when it was integrated into the Department of Natural Resources, was retained through to 1946. The Superintendent of Operations and Maintenance from 1919 to 1924 was Sam Porter who had originally worked for the Irrigation Branch of the Department of the Interior as an inspection engineer. In 1925 he was promoted to Chief Engineer of the Department of Natural Resources. He was replaced by G. N. Houston who remained until his retirement in 1935. He was succeeded by Frederick G. Cross who held the office from 1936 to 1940 when he returned to active service in the Canadian Army. George S. Brown was promoted from canal superintendent of the Northern Division to succeed Cross in 1941.

The size of the permanent staff assisting the Superintendent of Operations and Maintenance remained small, ranging from a low of 17 in 1939 to a high of 22 in 1935. The headquarters of the St. Mary Project continued to be the old AR&ICo. building at 207-7 St. S., east of Galt Gardens. The appearance of this building was changed by the 1940s with the removal of much of its decorative trim. The residences used by the Superintendent of Operations and Maintenance and the garage foreman as well as a garage were the other buildings in Lethbridge owned by the company. The latter two structures were located in the 500 block of 4th street south.

The system continued to be divided into Southern and Northern Divisions with headquarters at Magrath and the Home Farm east of Lethbridge. With the creation of the MID and the RID, the number of staff assigned to the Southern Division was reduced. The Northern Division continued to be the major consumer of the water diverted by the headworks at Kimball. The buildings on the Home Farm site by the 1920s included a blacksmith shop, root house, bunk cars and a residence for the canal superintendent. The Home Farm also stored construction supplies, housed the company's horses, grew vegetables for the camps and feed for the horses. The size of the herd in 1919 was 46 head which increased to 56 head in 1922. In 1919 the herd was rejuvenated when sixteen of the oldest and poorest stock were replaced. By 1935 the herd had been reduced to 22 when three were destroyed because of old age and two were sold because they were past their usefulness for the work required. Six head were brought in from Brooks to bring the total to 28 by the end of the year. The herd of horses was gradually reduced until only 16 remained as of 1945.The horses were given names and numbers; Maude was 203, Pedro was 13 and Prince was 4. The vegetables grown on the farm were used in 1919 to serve 37,187 meals to the men in the camps who were charged 35 cents per meal. In 1936 the number of meals was reduced to 8,685 and in 1939 this employee benefit was eliminated altogether. In 1935 the land was used to grow timothy, alfalfa, oats and vegetables including potatoes. The surplus hay was supplied to the Galt Mines Branch and in 1935 to the Lethbridge Collieries. Expenditures on the maintenance of the farm were reduced in 1935 by reducing weed control and repairs to field structures

The number of seasonal employees declined from 120 in 1919 to 25 in 1945.This trend reflected the need to economize during the Depression, the greater use of machinery for maintenance and the impact of higher labour costs during World War Two. The wages of labourers and teamsters was 25 cents an hour up to 1936 after which it was increased to 28 and 29 cents for labourers and teamsters, respectively. It was increased again in 1939 when the minium wage was raised to 33 cents. The majority of the staff, both seasonal and permanent, were concerned with the operation and maintenance of the system. The collection of the water rates and the engineering work required for the limited amount of new construction was coordinated through the head office of the Department of Natural Resources in Calgary. The overall budget for operations and maintenance was reduced from an average of $140,000 for the years 1918 to 1929 to $80,000 for the years 1930 to 1946.

HORSE POWER AND MACHINES

A combination of horses and machines were used to maintain the system. In addition to the horses owned by the company, the CPR also rented additional horses and teams on a regular basis. In 1936 hired horses performed 23,858 hours of work. This total had been reduced to 70 hours by 1945.

Canal construction 1920s
(St. Mary River Irrigation District)

Horse power on the St. Mary Project in the 1920s
(St. Mary River Irrigation District)

The garage in Lethbridge maintained the company's fleet of motor vehicles. The motor pool in 1919 consisted of 5 cars and 1 truck which increased to 20 cars and 2 trucks by 1945. The majority of the

Lethbridge garage at 526 4th Street South
(St. Mary River Irrigation District)

vehicles were Ford products. The annual reports included detailed statistics on mileage, fuel consumption and repair costs.

The Bay City, Marion and Beatty floating dredges were used until the 1930s. In 1923 the Marion dredge was modified at cost of $5,263.73 which included rebuilding the boom, A-frame and spud control. The main barge was widened by the addition of two pontoons each 6 feet wide by 32 feet long. The boom was increased in length from 40 to 50 feet and the spuds, which were used to keep the barge in place while it was working, were changed from the bank to bottom type. The spuds on the Beatty Dredge were also changed from the bank to the bottom type to accommodate the increased width of the main canal.

Marion Dredge 1923 (St. Mary River Irrigation District)

Marion Dredge 1923 (St. Mary River Irrigation District)

The principal work of the Marion dredge in 1923 was to strengthen the lower bank of the side hill portion of the main canal near Kimball. It was difficult to do this with teams because the material below the canal was wet while the upper side was on a high steep side hill which would have required the construction of bridges across the canal if horses were used. The only feasible and economical method was to excavate the material from the upper bank of the canal with the dredge and deposit it on and below the lower bank.

Bay City Dredge working on Chin 1 Lateral, 1922 (Sir Alexander Galt Museum & Archives, Lethbridge, Alberta)

In 1919 an Austin Dragline with a half yard bucket mounted on caterpillar tracks was purchased to clean canals and raise and strengthen canal banks. Pads were used to provide a stable

Austin Dragline (St. Mary River Irrigation District)

Austin Dragline beginning the Malloy Drain 1924 (Sir Alexander Galt Museum & Archives, Lethbridge, Alberta)

base from which it could work on soft fills. Increasing use of draglines was made through to 1946.

The CPR also brought in a specialized piece of equipment for cleaning the smaller laterals which could not be dealt with by the floating barges or as effectively as draglines. The DNR excavator or Ruth dredger was invented by C. H. Ruth and first used by the Imperial Valley Water Company in California. The use of this machine came to the attention of the CPR in December of 1917. The Superintendent of Operations and Maintenance in the Western Section of the CPR's irrigation project east of Calgary, Augustus Griffin was particularly impressed with its operation. As he explained in a letter to Ray Carberry, the Superintendent of the Imperial Valley Water Company:

DNR excavator working on the Cameron Lateral 1925 ((St. Mary River Irrigation District)

> We have many miles of small ditch from 3 to 4 feet on the bottom and 2 to 3 feet deep which have to cleaned out and in some cases enlarged. The ditches get tramped in with the cattle and filled in with silt and grass grows in the ditches making a heavy sod and greatly retarding the velocity of the water.[230]

DNR excavator (St. Mary River Irrigation District)

In the past, the CPR had hired teams to deal with the cleaning of small ditches which was becoming increasingly expensive. Griffin estimated that the use of the DNR excavator would eliminate the need to hire 30 additional men during the working season.

The first DNR excavator was unloaded at Strathmore on 24 June 1918 and started working in the Western Section of the CPR block on 24 July. The machine was extensively modified for use on

DNR excavator (St. Mary River Irrigation District)

the CPR system during this first year of operation. Rather than buying additional machines from Ruth, the CPR paid him a royalty for the use of his basic idea and continued the development of the machine to best suit their purpose. By the mid-1920s, the CPR had a large fleet of these machines which were used in all of its irrigation operations.

CANAL MAINTENANCE

The headworks, main canal and the main laterals in the Lethbridge-Coaldale distribution system constructed prior to World War One remained largely unchanged through to1946.The removal of silt and weeds continued to be the major tasks in keeping the canals in shape to move the maximum amount of water. The major source of the silt was from the scouring effect the water had as it moved at high speed through the natural channels. When the water reached the canals, it slowed down and deposited its load of silt. Soil from adjacent fields blown in by the wind was a second source of dirt. The extent of soil drifting depended on the strength of the wind and the way the land adjacent to the canals was cultivated. Once the canals had been cleaned, riprap was strategically placed to reduce further erosion. Stone riprap was used in some cases while in others the protection of the bank was provided by a wall made of wooden posts, wire and brush.

Each year various sections of natural channel, main canal and the main laterals were cleaned, rock riprap placed and post and wire riprap constructed. In 1936, for example, 75 miles of ditch were worked by a combination of teams and machinery. Teams did 22 miles, the most of which was in the Northern Division while the draglines did 10 miles and the DNR excavator did 43 miles. The total length of post and wire retaining wall constructed was 8,100 linear feet and 4.2 cubic yards of rock riprap were placed.

Wire and brush riprap on main canal near Magrath (Sir Alexander Galt Museum & Archives, Lethbridge, Alberta)

Christie Brothers Circus train wreck showing employees waiting for repairs (St. Mary River Irrigation District)

First car of the Christie Brothers Circus which plunged into Nine Mile Coulee (St. Mary River Irrigation District)

The removal of noxious weeds in general and sow thistle in particular was also a regular maintenance activity, although it was not given as high a priority as in the TID. The company dealt with this problem using a diluted solution of Sodium Chloride. Weeds not only caused problems in the delivery of water but were responsible for the wreck of a circus train. At 3.10 on the morning of 26 May 1920, the Christie Brothers Great United Wild Animal Show was crossing over Nine Mile Coulee west of Raymond on route to Lethbridge from Cardston when the bridge collapsed under the weight of the first of three coaches. Circus employee Joe Boch was killed while two other circus employees, Roy Baugh and Burnard Brown were injured. The cause of the accident was Russian Thistle which diverted the flow of the water against the bank thus undermining the bridge abutment.

RENEWALS AND BETTERMENTS

Renewals and betterments were the terms used by the CPR to describe repairs and improvements to the system. Canals were realigned and increased in width while their banks were strengthened and increased in height to carry more water. The structures along the canals had to be renewed on a regular basis because the vast majority were made of wood and thus had a useful life of about ten years. The Kimball headworks was reconstructed for the third time in the fall of 1924.

In a limited number of cases, the wooden structures were replaced with ones made of concrete or metal. The wooden diversion structure at Magrath was replaced during the winter of 1919-1920. The new Magrath diversion was concrete and consisted of 10 headgate openings, each 5'2" wide, controlled by flashboards, 3 concrete sluice gates, each 11' wide, and a spillway section 120 feet wide. The Kipp Coulee flume on the Raymond Lateral was replaced with a metal structure in 1925 and the box flume across Willow Creek was replaced with a wood stave syphon in 1935 when its use by floating dredges was no longer required.

The inspection and renewal of structures was given a higher priority after1936 because various structures "went out of control"during that irrigation season. In order to avoid a repeat of the events of 1936, the company launched a program to examine all important structures on the main canal. The structure most in need of repair was located at the point where the main canal ended and the Lethbridge-Coaldale distribution system began. One combined structure was used at this location up to 1936 to direct water into the Chin 1 and North laterals as well as into the Six Mile Coulee Spillway. When this structure failed, it was replaced with three separate structures. The new Six Mile Spillway headgate was rebuilt using 11,740 FBM of new untreated lumber and 466 feet of piling together with material salvaged from the old structure. The new North Lateral Headgate was built of pressure treated material, pre-formed at the CPR plant in Calgary. The new Chin1 headgate was built with untreated lumber and salvaged material from the old structure.

In 1920 the Six Mile Coulee Spillway was constructed from the main canal at the Chin Lateral headgates through a constructed canal nearly 3 miles long which ended at Six Mile Coulee which in turn continued on to the Oldman River. This spillway was constructed with a capacity of 500 cfs or about 60 percent of the main canal capacity at that point. It was used in the regulation of the main canal, the disposal of weeds and ice in the spring and surplus water in case of accident. Spillways played a critical role in the operation of the system

New Magrath diversion structure looking from the west (St. Mary River Irrigation District)

New Magrath diversion structure looking from the west (St. Mary River Irrigation District)

New Magrath diversion structure looking from the east (St. Mary River Irrigation District)

prior to 1952 because if a break occurred in the canal or a structure "went out of control" they could be used to divert the water back to the St. Mary River thus permitting repairs to be made. The use of natural channels facilitated this process as well.

Additional drop structures were added to the natural channels and the Northeast Reservoir was built in 1935. This second reservoir was built to eliminate the erosion of the bank at the Oldman River from the water coming from the North and North East Laterals. The unused water from these laterals collected by this reservoir was taken via a new canal to Chin Reservoir for eventual use by the TID. The company also did some work on the

New Kipp Coulee Flume built in 1923 (Sir Alexander Galt Museum & Archives, Lethbridge, Alberta)

Malloy Drain in the1920s. The construction of drains was, however, considered a responsibility of the water users and thus was a low priority for the CPR.

CANAL OPERATIONS

The total length of this system by 1946 was approximately 220 miles which included 58 miles of natural channels. The longest portion of natural channel was Pinepound and Pothole Creeks. The CPR rated the capacity of their main canal at 1,200 cfs in 1945 while the main laterals varied from 350 cfs for Chin 1 to 18 cfs for the Six Mile Coulee Lateral. This system by 1946 carried water to 237 headgates which controlled the entry into the 300 miles of sub-laterals operated by the farmers. These farmer laterals varied in size from 1 to 50 cfs. The system had the capacity to irrigate 140,000 acres but seldom irrigated more than 80,000 acres in any given irrigation season. The total irrigable area of the St. Mary Project, as approved by the government, was 143,661 acres. The amount of land entitled to receive water in the entire St. Mary Project, based on water agreements, remained unchanged through this period at approximately 119,000 acres. This acreage was distributed amongst Taber, Raymond, Magrath and the Lethbridge-Coaldale areas with the latter accounting for 75,766 acres of this total in 1945.

Canal operations during any given year from 1919 to 1946 were determined by a combination of factors with the most critical one being the amount of water available for diversion from the St. Mary River at Kimball. The rate of flow would generally increase through to the end of June and then taper off through to September. Water shortages could thus begin as early as the end of June and continue through to September. A second factor was the amount of water available from snow trapped in Pinepound and Pothole Creeks and runoff into those creeks in the spring from the Milk River Ridge. Successfully capturing this water in a dry year could help compensate for any deficiency which might occur in the supply available directly from the St. Mary River. A third factor was the timing and amount of rainfall. A sufficient amount of rain in June could temporarily eliminate the need for irrigation and thus allow the company to divert the water to storage or to turn the water out of the main canal because of lack of demand. A sudden rain storm in June could, however, spell disaster if it occurred when the system was carrying its peak load. Extra water was unwelcome at this time because it could potentially damage the canal and canal structures. A fourth factor was the number and duration of interruptions caused by structure failures or breaks in the canal. A fifth factor was the type of crops being grown by the farmer. Each crop required different amounts of water and had a critical time when this water had to be received. An effort was made, therefore, to give priority to certain laterals if they irrigated sugar beets. A sixth factor was the impact on the system when a large number of water users turned water back along the main lateral system at the same time without notifying the ditch riders or water masters. This was a fairly rare occurrence but could happen in combination with a heavy rain. This circumstance had the potential to put more water into the system than it had the capacity to handle. The seventh factor was the date of freeze-up in the fall which determined how late in the year water could be run through the system. Turning water into the system in the late fall for the purpose of filling reservoirs was a risk because of the potential for damage to structures caused by ice. Ensuring that the system was drained before freeze-up also facilitated a prompt start to the following irrigation season. A final factor in the way it was operated by the CPR was the need to keep a certain quantity of water in the Chin Reservoirs in order to meet its obligations to the TID.

Each irrigation season represented a different combination of these factors thus creating dramatic shifts between success and failure in the company's ability to deliver water. The operators made the best use of the system, as it was constructed, to deliver the maximum amount of water to the farmers. When the inevitable water shortage occurred, a rotation system between the divisions, districts and laterals was instituted in an effort to mitigate the worst effects of a water shortage. Based on the CPR's own reports, the water users seemed reasonably satisfied with their efforts to operate the main canal and main laterals.

A dramatic fall in the amount of water available for diversion at Kimball first occurred in 1919. After 4 July the mean flow was only 35% of the eleven year average. American diversions of water from the St. Mary River, which began in 1919, only partially explained this drop. By the end of the season, the company had less than 40% of the amount of water it was required to deliver. The shortage amounted to an average of more than 200 cfs for nearly three months or a total of 35,000 or 40,000 acre-feet. The reduction in the water supply came at a critical time since the drought was worse than it had been for the two previous years and there was a larger area of crop under irrigation. According to Sam Porter, the staff were able to cope with the situation because nearly 30% more water was delivered during the critical month of June and the water was distributed more equitably between the various laterals. The third factor he noted in his annual report was the more cordial spirit of cooperation between the operating staff and the farmers.

This situation prompted Porter to discuss the need for reservoir development in his 1919 annual report. He noted that the problem was going to increase because of increased diversions by the Americans and increased demand for the water in Canada. All of this pointed to the necessity of providing storage facilities for the maximum utilization of the high and flood flows of the St. Mary River. Even in this year of low flow, Porter noted, there was more than enough water available for storage during May and June to make up for the shortage during the balance of the season. He recommended that the CPR retain ownership of the Ridge and Raymond Reservoir sites and develop them for the benefit of their own lands rather than relinquishing control to the irrigation districts which were proposed for this area. Porter was convinced that it was economical for the CPR to proceed with development:

> They occupy strategic positions, and it would be a misfortune for them to pass into other hands. They have a combined capacity of nearly 100,000 acre-feet. Thirty per cent of that capacity should be made available as a supplementary supply for the lands now under water agreements and the other seventy per cent disposed of for use on other lands at a price which would cover practically the entire cost of the development. The company still owns more than 8,000 acres under Chin No.1 Canal and the proposed Whoop Up lateral which can be readily disposed of at $60.00 to $100.00 an acre when water is available for them. The balance can be disposed of either to individuals or to districts organized under the Irrigation District Act.[231]

This recommendation was never acted upon thus leaving Porter and his successors to operate the St. Mary Project without reservoirs.

With the end of the drought, the difficulties involved in operating the system without reservoirs lessened and did not return until the mid-1930s. The 1935 irrigation season began three of the most difficult years in the entire history of canal operations. The decline in the amount of water available from the St. Mary River was first evident on 26 June when the headgates at Kimball had to be opened ahead of schedule and left that way for the rest of the irrigation season. From that date to the end of the season all available water in the river was being diverted. Despite their use of the entire flow of the river, the CPR was unable to meet the demand for water on the Northeast Lateral during the peak load on account of a large number of small parcels growing the same crop which wanted water at the same time. The demand for water to irrigate the sugar beets continued through to September. Because of the small amount of water available, it was found necessary to rotate not only between the ditches but between the Northern and Southern Divisions. Despite the water supply problem, the 1935 irrigation season ended on a positive note when the company was able to keep the main canal in operation through to 23 October to permit the farmers to fill their stock ponds or finish irrigating their summer-fallow.

In 1936 canal operations were affected, not only by a shortage of water from the St. Mary River, but by an usually high number of system failures when various important old structures "went out of control" at a time when the CPR could ill afford to lose the water. These two problems led to grief for the water users who took it out on one another and on the CPR. The first problem in the season's operations occurred on 20 May, four days after the water had been turned into the system at Kimball. On that day, a bad leak developed in the structure that diverted water into Chin Lateral One, the North Lateral and Six Mile Coulee Spillway. This structure was located just south of Lethbridge and was the point at which the main canal ended and the Lethbridge-Coaldale distribution system began. To make the necessary repairs, the headgates at Magrath were closed and the water was wasted from that point back into the St. Mary River via the Pothole Creek. After making the necessary repairs, the water was turned back into the canal at Magrath on the 25 May. The repairs made, however, had not corrected the problem because water continued to pass under the floor of the spillway and under the Chin and North Lateral headgates from unknown points. Water was again turned out of the system at Magrath and further repairs were completed on 7 May. This repair corrected the problem and water was successfully carried through to Chin Coulee. The second breakdown of the system occurred on 5 June at 2 p.m. when a drop approximately three miles west of Bradshaw washed out. The system was shut down from 5 to the 9 June to make repairs. The third interruption in service began at 6 p.m. on 28 June when Six Mile spillway went out of control again. Water was immediately turned out at Magrath and an earth dam was built across the channel of the spillway. This dam permitted the use of the main canal which went back into operation on the 29 June. Repairs were then completed on the spillway structure while the main canal continued to operate.

> The impact of these breaks was mitigated by rainfall which began on 6 June. As result of which farmers on irrigated land placed too much weight on the beneficence of the June rainfall and the majority requested water to be shut off immediately it commenced to rain. Not until June 18th did they realize crops were suffering, and we were faced with a sudden maximum peak demand. This was accompanied by an expression of considerable dissatisfaction at our inability to deliver water in large quantities. By the end of June some of the crops on land that had not been fall irrigated and had received no water this year were in bad shape.[232]

After 18 June the supply was insufficient to meet the demand for the rest of the irrigation season. From 31 August to the end of September, the volume of water available for diversion at Kimball dropped from 207 cfs to 92 cfs. By mid-September the Northern Division was only receiving 30 to 40 cfs which was sufficient water for only one ditch. During July, August and September a rotation system was introduced to meet the heavy demand for water.As a result of these conditions:

> complaints regarding shortages were heavy and continuous from June 18th until the end of November. Quarrelling among neighbours was rife, causing numerous court cases, and farmers were in a distressed mental condition during most of the crops growing season, through this cause during the season.[233]

The delivery of water to the farmers was terminated on 10 October while the water in the system was passed into Chin reservoirs until 5 November when weather conditions changed and it was turned out at Kimball. The company made an attempt after 5 November to meet the needs of the farmers for water:

> Due to repeated requests from farmers for water to replenish reservoirs and stock ponds an effort was made to supply this demand, but this was found to be impossible for some time as the St. Mary River was down to an all time low flow of about 40 to 50 cfs. A rise took place on the evening of the 25 November and the water was turned in again at Kimball on the 26. A few days later the weather became stormy and cold and the water froze in the canals before reaching the points of delivery and the Company was forced to close the gates at Kimball on 2 December for the season.[234]

The events of 1936 prompted great fear that 1937 would be a disaster. Superintendent Cross, in his 1936 annual report, expressed the view that:

> while the last year has been critical, one cannot banish the thought that unless Providence is on our side and we get all the favourable breaks, the peak of the crisis will come next season, because land is exceptionally dry, numerous farmers are without domestic and stock water and we anticipate a peak demand as soon as water is turned into the system. Copious rains early in the spring will, of course, change the picture, but should such not arrive and we are faced with the conditions as outlined, unquestionably the flow available for delivery to storage in Chin Reservoirs will be affected.[235]

The 1937 irrigation season began with only 2,798 acre-feet in storage in the reservoirs, the land was exceptionally dry, numerous farmers were without stock and domestic water and the canal system was plugged with ice and snow. These conditions were dealt with by making an attempt early in the year to catch every available drop of runoff water:

> On April 2 it was noticed [that] a small amount of runoff was coming down Pinepound and Pothole channels and the sluice gates at Magrath and Spring Coulee were immediately closed and we caught the runoff and carried it a short distance toward Welling where it was halted by ice and snow drifts. A crew of men was set to work to cut narrow channels through some of these drifts and better results were achieved during the following week and on the morning of April 8, 75 cfs had reached the Northern Division and was being worked through Chin #1 to the Chin reservoirs. The runoff from Pothole and Pinepound increased to 300 cfs and ceased on April 13.[236]

After delivery of water from Kimball began on 21 April "considerable propaganda work was done to encourage those who were in good standing with the company to take a maximum amount of water at as early a date as possible."[237] By mid-June the system was carrying a maximum peak load with water supplied entirely from the St. Mary River and thus any further runoff was no longer welcome. A heavy rain on 12 June:

> created emergency conditions, intensified by the fact that we could not contact certain water masters due to the faulty mutual telephone systems. It [was] estimated 4000 cfs of runoff from Pinepound and Pothole was pouring into the system within twelve hours of the commencement of the rain and we were also faced with from 400 to 500 farmers taking the initiative into their own hands and turning back water on us without notification. The staff worked day and night for nearly three days to unload surplus water and this was accomplished with only minor damage to crops, channels and structures.[238]

In 1938, Superintendent Cross reported that it was an excellent season from an operating standpoint because the water level in the St. Mary River remained high for a longer period of time. The staff were also successful in "catching every available drop of runoff water." Runoff was turned into the main canal at Spring Coulee on 8 April and into the main canal at Kimball on 12 April, the same day the ice went out on the St. Mary River.

In 1939 the operations were affected by a rare case of what the *Annual Report* for that year described as "wanton vandalism." The control gates at the north end of Stafford Reservoir were blown in two by dynamite. The matter was investigated by the CPR police in conjunction with the RCMP who concluded that the target of the dynamite was actually the fish in the canal rather than the headgates structure themselves.

RATES

The collection of the rates due under these contracts produced a constant stream of letters from the CPR to their water users reminding them of their arrears. In this correspondence the CPR constantly threatened to cut off the water supply but it seldom carried out this threat. Another method used to improve collections was the introduction of a system whereby the water users could pay rates by giving the CPR a portion of the crop. The crop share program was first introduced in 1934.

WATER USERS' RESPONSIBILITIES

The farmers were responsible for the distribution of water in and the maintenance of the 300 miles of sub-laterals. The CPR assisted the water users by providing an engineer who helped them lay out the canals. By 1918 only the Pawson Lateral Association had been created to formally carry out these tasks. The decision of one of its members to take practically all of the water in the lateral for himself, led to a court case which brought its activities to an end and raised some important legal questions about the way the system could be operated.

In 1918 interest in creating lateral associations increased because of water delivery problems in the Lethbridge-Coaldale area. These concerns culminated with the creation of Coaldale Lethbridge Water Users Association on 27 June 1918. The objectives of the organization, as stated in its constitution, were to advance the science and practice of irrigation among its members and to protect their rights under the contracts they had with the AR&ICo. To achieve this goal, the association intended to obtain reliable information on the operation of the system with respect to water supply, the capacity and state of repair of the main canal and its associated structures, the amount of land that the owners of the system had obligated themselves to irrigate and the amount of water that was actually being distributed to each farm gate. The association was also concerned with investigating any sources of contamination of the water supply which would make it unfit for human or cattle consumption. It also wanted to ensure that the canal system did not serve as a means of distributing seeds of noxious weeds.

Based on the information obtained from their investigations, the association intended to request the AR&ICo. or the Department of the Interior to deal with any deficiencies or to enforce the appropriate laws. The necessary legal action to enforce the members rights or the collection of damages for the failure of the owners of the system to comply with the obligations of the water contracts was also contemplated. The objectives of the association as listed in their constitution concluded with a general declaration that the association intended to:

> do all things of whatsoever nature which may be deemed advisable for the promotion, the welfare and the science of irrigation as practised among its members, and the protection of the rights of the membership in connection with any matter pertaining thereto, whether obtaining water, looking after drainage, guarding roads, protection from weeds and water pollution, fencing of ditches, and all things pertaining to the business of irrigation which are of common interest to the membership.[239]

Membership in the organization was open to any one who had a water agreement with the AR&ICo. Terminating ones membership was only allowed if their water contract was not in arrears. The membership fee was $8 per quarter section with full water right.

The affairs of the association were administered by a seven member board of directors from which a president and secretary-treasurer were elected. The first Secretary-Treasurer of the association, and its most active member, was W. H Pawson Jr. who had established the first lateral association in the area on his farm south of Coaldale. In order to irrigate his land and that of his neighbours, he had arranged to have the AR&ICo. expropriate some land for the purpose of constructing a ditch which was then leased back to Pawson and his neighbours who had formed the Pawson Lateral Association under the Societies Act.

The first priority was to increase membership. In a circular letter dated 5 July 1918, Pawson indicated that he wanted to include as nearly as possible every irrigation farmer and land owner in the Coaldale-Lethbridge district. Each member was encouraged to talk it over with their neighbours and get them to join. By doing so they would increase membership and the working capital of the association and "save the writer's time and expense in looking up these men."[240]

This same letter also had some instructions for those members who contemplated making claims against the AR&ICo. for damages resulting from their failure to deliver the right amount of water.

The association retained the services of Lethbridge lawyer S.S. Dunham who took immediate action to deal with these concerns. In a letter to Commissioner of Irrigation F. H. Peters dated 3 July 1918, he asked if a license had been issued to either the AR&ICo. or the CPR. He also wanted to know the amount of land which the company had the obligation to irrigate and the capacity of the canal. His final request was for information on the amount of water that was available for the Lethbridge system under the provisions of the Waterways Treaty and "such information as [he could get] as to what will be necessary to conserve and transmit that water to the point of delivery . . . ."[241]

The immediate response of the Department of the Interior was to send Ben Russell to make discharge measurements of the canals in the Coaldale-Lethbridge district. His report confirmed the problems water users had complained about. He found that the Chin 2, the Coaldale and the Cameron Laterals were in poor condition. He indicated that:

> all of these canals were running brim full at the time the measurements were made and most of them showed indications of having run over, in some cases doing considerable damage, where crops have suffered for want of water they were generally found under these canals particularly toward the lower ends.[242]

LATERAL ASSOCIATIONS

The Coaldale Lethbridge Water Users Association also took the initiative, in cooperation with the CPR and the Department of the Interior, to bring about the creation of a comprehensive system of water user districts. On 14 December 1918, Sam Porter, Irrigation Commissioner Peters along with Pawson and a number of area farmers attended a meeting with this objective in mind. Porter explain that:

> the primary purpose of such an organization was to assume control of the distribution of water in the canals which were not operated by the company and to maintain these canals in proper condition for service. [He] emphasized the idea that small districts about the size that could be efficiently managed by one good water master with the assistance, possibly in the busy season, of an extra man or two would be much easier for them to control and would involve much less overhead expense and trouble than a large district taking in the entire irrigation tract east of Lethbridge.[243]

He suggested, therefore, that separate water user districts be created around individual laterals such as the North Lateral. A smaller area would permit the trustees to become familiar with all the problems involved and would allow them "to keep in intimate touch with the work of dividing water and maintaining the ditches."[244] If the area included the entire irrigable area east of Lethbridge, the trustee responsible "for the North Lateral would necessarily have to spend much time becoming familiar with the problems around Coaldale before he could undertake to direct the work in that locality."[245] With this idea in mind, the meeting approved the creation of 6 water user districts. The first proposed district included the lands under the North Lateral; the second covered the land under the Coaldale Lateral, East Lateral and the upper part of the North East Lateral; the third included the land under the Cameron Lateral and the lower part of the North East Lateral; the fourth included land under Chin Canal No.1 below Chin No. 2 headgate; the fifth covered lands under Chin No. 2 Lateral; and the sixth was for land under Six-Mile Coulee lateral, Chin No. 1 Canal above Chin 2 gates and the small canals taking water out of the main canal in township 7 and 8. This number was later reduced to four districts which were referred to as East Lethbridge, Coaldale, Cameron and Readymade.

Porter was particularly pleased to see the farmers taking action. As he explained in a letter to Chief Engineer Dawson:

> it appears that the farmers as a rule are now awake to the necessity of taking some steps to secure proper division of water on their own ditches. Many of them have not understood previously that it was not the company's place to operate these ditches down to the individual farms. The majority of complaints for improper service in the delivery of irrigation water arise out of the unequal division and poor service on the farm ditches rather than from the company's canals. Now that the farmers are beginning to understand the situation they see the necessity of creating some organization which will have the necessary authority to compel individual water users to adhere to the proper division of water among themselves and which will have also the power to levy assessments for the maintenance of their ditches so as to provide that all concerned shall share the cost.[246]

The Coaldale, East Lethbridge and Cameron Water Users Associations elected officers and employed water masters for a short period after their creation. By 1922 their operations had been discontinued thus prompting Sam Porter to appeal for help from Commissioner of Irrigation Victor Meek, since:

> these organizations are badly needed. Some of the farmers realize their need but others object to doing anything. The payment of assessments, I think, is the basis of a considerable amount of the objection. They have been accustomed to getting their water in some way or another without paying any assessments for operation and maintaining the small ditches, and they can't bring themselves to the point of realizing that they cannot continue this haphazard system now that the country is settled up and all ditches are calling for their full share of water.
>
> The associations need some encouragement and urging to get them on their feet and to realize the problems that they have to meet. I have gone about as far as I can in the matter of urging them to take some effective measures for their own relief. More insistent urging on my part would be misinterpreted by many of them in the belief that the CPR had some ulterior motive in forcing the organization on them if you could find occasion to give them some advice and encouragement, I think it would be helpful.[247]

By 1923 Porter had come the conclusion that the effort was a failure. In his *Annual Report* for that year, he noted that:

> these districts, with the possible exception of the Readymade District, have not been as successful as might be desired. The farmers have exhibited an indifference towards them that has paralysed their usefulness. They have likewise objected to the payments of the assessments made for their operation. This is unfortunate, for the Associations are very much needed in order to get a fair and just distribution of the water in their ditches.[248]

The weaknesses he identified were the lack of hearty support and cooperation of the farmers themselves and the lack of adequate legislation giving power to the manager of the district to enforce their regulations and defining their powers more specifically for the guidance not only of the district officials but of the courts which may have to be called upon to administer them. Because the farmers were not well organized to operate their own ditches, it was necessary for the CPR in subsequent years to:

> maintain a general supervision over the distribution of water in their ditches for the sake of harmony and fairness, and for the protection of the users at the lower end of the ditches. The unofficial duties of adjusting differences between the farmers and seeing that all are treated fairly [were] often more onerous than our recognized official duties.[249]

READYMADE WATER USERS ASSOCIATION

As pointed out in the CPR's Annual Reports, the Readymade area was the only one which demonstrated an interest in the creation of a water users district in the early 1920s. The Readymade area was made up of farms built by the CPR prior to the arrival of the settlers and referred to as the Coaldale Colony. It was served by Chin 1 Lateral. The water users in the area had already begun to create an association even before the meeting on 14 December 1918 organized by the Coaldale Lethbridge Water Users Association. In September 1918, S. S. Dunham, on behalf of the area farmers, sent a petition to the Minister of Public Works requesting the formation of the Readymade Water Users Association. The petition was in order with respect to the signatures of the petitioners but the description of the land by meets and bounds did not make it clear exactly what land was in the district. The schedule of the 5,889 acres of land to be included in the district was finalized by 24 October 1915. The notice of the intention to create the district was published in the 31 October 1918 issue of the *Alberta Gazette* and the 15 November issue of the *Lethbridge Herald.* On 17 December George Parish was appointed Returning Officer for the vote required to create the district. When it was discovered that he would be away, H.P. Ober was appointed to this position on January 18, 1919 William F. Wahl, H. J. Miller, H. P. Ober were elected as its first officers. The ministerial order creating the Readymade Water Users' District. was issued in February 1919.

The first task of the Association was to request the Department of the Interior to investigate the problems in the distribution of water. In a letter to the Commissioner of Irrigation dated 15 June, George Parish explained that the water did not seem to reach the lower end strong enough. O.H. Hover was sent to the area in June 1919 to investigate. Hover found that the Association was being supplied with 17 cfs above its allotted amount. Ditch Rider R.E. Wells indicated to Hover that the supply to the system had been very liberal all season. The 23 measurements he took in the Association's ditches found that no shortage was evident. Mr T. Davis who claimed a shortage previous to Hover's visit made no complaints to him and appeared satisfied with his supply during his visit.

The next problem facing the operation of the Association was a financial crisis caused by extra expenses and by the failure of half of the water users to pay the 20 cents per acre assessed by the association. In an effort to generate some revenue and save what funds they had, the association instituted a variety of measures. No assessment was to be made on land for which the arrears had been paid. The Association also considered refusing to provide water to those water users who were in arrears. The delinquent accounts were also turned over to a lawyer for collection. The most important cost saving measure was to leave structures unrepaired until they failed.

By 1926 the Association had recovered from its financial crisis thus becoming the only association to survive out of the four created in 1920. Total receipts for 1926 were $1,091.19 with only two water users in arrears. The return of their land to the CPR solved both of these problems. Maintenance work on the system was resumed with the rebuilding of a number of structures. A ditch rider was also employed for over four months at the rate of $125.00 per month.

After 1927 the Association was able to maintain the system through the cooperation of the water users and the CPR. In 1935 the CPR rebuilt the division box and the association reconstructed the drops immediately below the new structure to prevent damage caused by erosion. In 1938 the Association also rented a DNR excavator. The Association expanded in 1940 when the former central farm for the Coaldale Colony, known as Chin farm, was sold in parcels of eighty and one hundred and sixty acres. With the sale of the land, the three service gates were replaced by one double corrugated iron pipe gate and one main high line ditch. Additional lateral ditches and division boxes to the various parcels were also necessary. A Chattin ditcher was used to construct 3.3 miles of new ditch while 7,054 FBM of lumber, 196 feet of piling and 1 salvaged pipe gate were used to build the new structures.

FORBES LATERAL ASSOCIATION

The original 1920 Coaldale Water Users District had included the farmers who obtained their water from the Coaldale, East and the upper part of the North East Laterals. Beginning in 1929, smaller lateral associations were created as substitutes for this large organization which had ceased operations by 1922. The Forbes Lateral Association, named after Forbes Murray, the water master on the Coaldale Lateral, was the first of these smaller organizations. The original Forbes Lateral, which has since been replaced with a pipeline, started north of Broxburn and ran east to Coaldale. The creation of the Forbes Lateral Association was undertaken by Jacob B. Janz with the assistance of Emmett Walker. Their efforts were strongly supported by the CPR which encouraged farmer participation in the association by promising that they would be able to use the Malloy drain to dispose of their excess water. According to G. N. Houston, Superintendent of Operations and Maintenance, this was considered a great advantage since they could regulate their water without having to change the location of their gates.

The petition organized by Janz was sent to G. N. Houston on 15 March 1929. It contained the names of 25 farmers who collectively owned or rented 7000 acres of irrigable land. The majority of these individuals, including Janz, were members of the Mennonite Community which had been established in the Coaldale area during the late 1920s. The 7 December 1929 issue of the *Lethbridge Herald* advised the public that Emmett Walker had been appointed returning officer, under the provisions of the Water Users' Districts Act, for a vote to elect 3 people to the board of managers of the Forbes Lateral Water Users' District. Anyone over 21 years of age and an owner of land in the area described in the notice had the right to vote even if they had not signed the original petition. The nomination of directors and the board election was held at Crystal Lake School on 14 December 1929. Jacob Janz, Emmett Walker and A Morton were elected by acclamation.

The by-laws of the Forbes Lateral Water Users District, which were adopted shortly thereafter, began with a forward that appealed to its members to act in the spirit of cooperation to ensure its success since it was:

> the intention of the Board to carry on the operations of the District in a business-like and economical way and to distribute the water among the members of the district in an equitable manner and as nearly as possible satisfactory to all water users. It must be remembered, however, that it is a co-operative enterprise and cannot be successful unless the majority of the members at least keep the spirit of co-operation uppermost in their minds. There must be some give and take in all such enterprise. Every member of the District should feel a personal responsibility in helping to observe the rules and keep down expense.[250]

A ditch rider was to be employed for a period not to exceed three months during the irrigation season. His job was to ensure that the water users received the amount of water to which they were entitled or distribute water based on a rotation system devised by the directors. The Managers were authorized to collect up to 25 cents from the water users for the operation of the system.

The new executive began immediately to carry out their duties by obtaining lumber at cost from the CPR for the rebuilding of structures. This assistance was provided because Houston felt that "this would be a very great help to them as they will be under pretty heavy expense for the first year, as the ditch and structures have been allowed to deteriorate for a number of years and are in very bad shape."[251] Houston went to the assistance of the Association again in June 1930 when he requested permission to make a DNR excavator available for cleaning the Malloy drain without requiring the Association to pay for the use of the machine in advance. He suggested that the machined be used on the ditch immediately because:

> it would encourage the water users to go and improve the smaller laterals running out of the main ditch. I believe it would be good policy for us to help this district as much as possible in order to have it prove a success and in this way encourage the formation of other water users districts which is to be to the advantage of the Company.[252]

The Association accumulated a debt of $164.37 as a result of the use of this machine which was not fully paid until 1942.

In 1936 Gate 2 Lateral Association was established to the west of Coaldale while Gate 4 Lateral Association was established to the east of Coaldale.

WHOOP-UP LATERAL ASSOCIATION

The Fort Whoop-Up Lateral Association was organized in 1939 by the CPR as a direct result of the sale of land by the company in the area southeast of the airport. The Whoop-Up Lateral was originally located in 1899 by Arthur M. Grace who was working with George Anderson as an engineering consultant for the Canadian North West Irrigation Company. The construction of the first ditch at this location, however, did not take place until the 1920s. In April 1930 Messrs. Henderson and Salmon, who owned farms served by this ditch, requested the CPR to undertake its reconstruction and offered to contribute $100 towards the cost of this work. The CPR's review of the situation indicted that the canal would have to be relocated to best serve the land in the area. The choice for the CPR was between merely cleaning out the present ditch for the time being or waiting until they were ready to put the land on the market before attempting to change the location. Authorisation was given to do some work to improve the condition of the ditch because the $100 contribution covered 3½ days of the six days required to complete the work. When chief engineer Dawson passed on the authorisation, he reminded Superintendent Houston to be sure to get the $100 from Salmon and Henderson. The amount of work actually done in 1930 is unknown.

In 1939 the CPR decided that the time had come to sell the land and carried on with the plan of undertaking the total reconstruction of the lateral. In March of that year Superintendent Cross advised George Brown that the Company had decided to proceed contingent upon the land owners in the area forming a water users association for its operation.

The formation of a lateral association began on 11 August 1939 when Fred Cross, in his capacity as returning officer, placed an ad in *The Lethbridge Herald* indicating under the provisions of the Water Users Districts Act that an election was to be held on held on 14 August 1939. Alexander Henderson, Burnhart Gwatkin and Michael Luco were elected as the first executive. The work undertaken included the excavation of 4.5 miles of ditch by a dragline and 4 miles of ditch by the Chattin ditcher plus the construction of 21 new wooden structures. This work was completed during the fall of 1939 and water was turned into the canal in September. According to the 1939 annual report of Superintendent Cross, the board of management of the association accepted the work with expressions of highest satisfaction and will commence its operation in the 1940 irrigation season. The Association charged the water users an annual fee of 10 cents per acre.

The formal organization of lateral associations was the exception rather than the rule in the Lethbridge-Coaldale area. The vast majority of the laterals were operated through informal agreements between the farmers. Given this situation, the CPR was required to exercise a certain degree of supervision over the operation of the distribution system. This necessity increased in the late 30s as rate enforcement became more of a necessity. As Superintendent Cross noted in his 1939 report, the company now exercises "a certain control over all farm delivery gates, due to pressure applied for the purpose of collecting rates."[253] The cleaning of the laterals and their use in the distribution of water, therefore, depended upon the degree to which the farmers along the laterals were willing to assume this responsibility.

COALDALE

The founders of Coaldale continued to be strong supporters of irrigation through the activities of the Coaldale Commercial Club. Its activities included organizing an irrigation exhibition on 16 and 17 September 1919. It also published a promotional pamphlet entitled *Coaldale "The Gem of the West"* in 1920 which chronicled the development of the community under irrigation. The booklet's tribute to irrigation's impact on the community was celebrated in a section entitled "The Song of the Lazy Farmer" which dealt, in a more poetic way, with the nature of the climate and the advantages and disadvantages of being an irrigation farmer:

It doesn't rain a lot out here,
they get a dew most every year;
on leap year, if it ain't too dry,
they have a few clouds in the sky;
and once, they say, before the war,
they had a rain that lasted for four
hours perhaps a little more,
they say it was some downpour!
the soil out here is mighty rich,
they water it from out a ditch,
the plan looks mighty good to me,
if I was home again, by gee,
and rains were coming pretty late,
I'd just start out and irrigate

Suppose your corn is getting dry
and not a cloud is in the sky;
You never need to pray for rain,
You just go out and tap the main,
or else you tell the hired hand to
turn the water on your land.
You water all the grass and grain
and then just shut it off again!

There's disadvantages to it too;
you've always got some work to do,
You never have a rainy day,
When you can sleep upon the hay,
or sit around in someone's barn
and listen to him spin a yarn,
or take your pole and bait and fish,
and catch a perch or two and wish that this
here rainy spell would last until the fishin' season's
past.
I think there'd be a lot more joy,
a farmin' back in Illinois

CROPS

Despite the various problems with respect to the operation of the system, this was an era of expansion for the St. Mary Project because of a minor rush to irrigated land in southern Alberta as a result of large scale abandonment of land in the dried out areas of southeastern Alberta and southern Saskatchewan in the 1930s. These new farmers occupied land which had not been sold by the CPR. Those farmers already on irrigated land remained despite the hardships involved in paying the rates.

The development of irrigation agriculture in the Lethbridge-Coaldale section of the St. Mary Project required the farmers to adjust to a number of unpredictable factors one of which was the water supply from the irrigation system itself. As outlined above under the heading canal operations, water shortages could develop as early as mid-June or as late as mid-August. The extent of the shortage would determine the way the water would be rationed using a rotation system. In an effort to mitigate this problem, the CPR tried to coordinate the limited water supply with the needs of the crop being grown.

Like the Taber area, the Lethbridge area was able to diversify its production of crops. One of the most successful crops grown other than wheat was sugar beets. Their production increased in terms of the number of acres grown, the productivity per acre, the sugar content and the income received by the grower. This trend was evident as early as 1936 when an all time record crop of sugar beets was produced. The production per acre averaged 11.2 tons and the average sugar content was 18.54 %. The final price paid for the crop was $6.25 which was 50 cents higher than the previous year. The quantity produced and price remained at the same levels until World War Two. Outstanding crops of wheat were also produced on irrigated land during this era with the yield being over 50 bushels to the acre compared to 20 or 30 bushels to the acre on dryland.

SHARING THE WATER

Irrigation in the area west of the St. Mary River began in the 1890s but no major projects resulted from these efforts since they were either discontinued or damaged by the floods of 1902 and never rebuilt. The resumption of efforts to develop irrigation began in 1906 when the Alberta Stake of the LDS Church purchased the Cochrane Ranch with the intention of developing an irrigation system using water from the Waterton River. The LDS Church were unable to carry through with the completion of this project at that time. Their efforts to develop the former lands of the Cochrane Ranch for irrigation purposes resumed in 1914. The intention at that time was to use the Belly River as the water supply. A survey carried out in November 1914 by the Department of the Interior showed that the project was feasible. The drought, which began in 1917 spurred interest in its construction.

Irrigation development west of the St. Mary River was contrary to the intention of the federal government to concentrate it east of the St. Mary River. The CPR were issued the right to divert 500 cfs at all stages from the Belly River in order to achieve this purpose. In 1919 the federal government also passed order-in-council 1859 which ordered that all unappropriated waters of the St. Mary, Milk, Belly and Waterton Rivers and their tributaries in Alberta be reserved until surveys then underway by the Reclamation Branch were sufficiently completed to determine in what manner and in what areas the unappropriated waters may be best and most economically used.

Despite these potential political barriers to irrigation west of the St. Mary River, new surveys were carried out in the summer of 1919. Based on these surveys the United and Lone Rock Irrigation Districts were created. The latter district was located in the Glenwood and Hill Spring area and was to be served by an extension of the system which provided water to the United Irrigation District. The Lone Rock Irrigation District was amalgamated with the United Irrigation District in 1923. The United Irrigation District was able to proceed because the CPR relinquished some of its water rights on the Belly River. The federal government also decided that the creation of the United Irrigation District represented a beneficial use of the water and thus was consistent with the objectives of the 1919 order-in-council.

The residents of Mountain View also began to consider the creation of an irrigation project as early as 1914. The first petition requesting the creation of a district was sent in 1922 . It was organized by Seymour B. Smith, a rancher in the Cardston area and an irrigation engineer who had been involved in the planning of the United Irrigation District. The petition was accompanied by a letter from Smith which indicated that they intended to use Ben Russell's 1912 survey for a canal from the Belly river to the St. Mary River as the plan for their system. Smith also indicated that they could handle the construction work themselves with very little help from the government so long as the government served as their engineering consultant. Smith concluded his letter with a note that irrigation was required to relieve the feed shortage which had developed in the area because of the drought.

FIFTEEN
REVIVAL AND EXPANSION OF THE SUGAR BEET INDUSTRY 1923 - 1946

The permanent return of the sugar beet industry to Alberta in 1925 re-established the link between sugar beet growing and irrigation originally forged by Jesse Knight in 1902 and contributed to the sustained growth of a food processing industry based on irrigation agriculture. The Raymond Board of Trade led the effort to re-establish this industry with the strong support of other boards of trade, beet growers, irrigation districts and the CPR. In addition to the new processing plant, this collaboration led to the creation of a sugar beet growers association in 1925 and the increased cultural diversity of southern Alberta as a result of the European immigration to meet the labour needs of the industry. This represented a significant change since much of the immigration into southern Alberta before 1914 had been American. The new arrivals included Mennonites from Russia, Hungarians and people from the former Czechoslovakia and Yugoslavia.

The catalyst for the return of this industry was the provincial government's effort in the early1920s to diversify crops grown on irrigated land and to develop processing industries based on those crops such as sugar beets. The newly established Irrigation Council, which took a particular interest in the development of the Lethbridge Northern Irrigation District, played an important role in promoting this objective. Its 1922 brochure entitled *A Handbook of Information about Irrigated Farm Lands Improved and Unimproved in Southern Alberta* noted the excellent results obtained from the growing of clover, grasses, alfalfa, corn, sunflowers, potatoes, sugar beets, carrots, turnips, peas, cabbage and small fruits. A second brochure entitled *Small Home Farms in the Irrigated Areas of Southern Alberta* and subtitled *Healthful Occupation, A Good Income and A Chance for Betterment* noted that irrigation would even permit a settler from the "Old country" to grow an:

> old fashioned garden with fragrant sweet alyssiums and mignonettes along the borders of the paths, with sturdy coxcombs and sunflowers and hollyhocks looking over the fence, with clumps of defiant tiger lilies and pale peonies, and sweet scented beds of phlox and petunias and clove pinks and martial ranks of blue flags and vivid rockets of scarlet sage - a delight from the time the first crocus opens in the grass until the last aster withers at the touch of frost.[254]

Crop diversification was also the theme of a conference held in Lethbridge on 20 December 1922. The conference's purpose was to establish uniform policies for the preparation of land, handling of water and farm programs for the Lethbridge Northern Irrigation District which was to go into operation in 1923. While delegates were concerned specifically with this district, the ideas discussed were considered applicable to all irrigated land in Alberta.

In December 1922, as part of its investigation of the demise of the sugar beet industry, the Irrigation Council contacted James Walker in his capacity as the Secretary of the Knight Sugar Company. The Knight Sugar Company plant had been removed from Raymond in 1917 but the company continued to exist as a corporate entity for sometime thereafter. This letter was received with great interest in Raymond where a number of former employees of the Knight Sugar Company continued to live including John Evans and Thomas O'Brien who were both involved in the development of the RID. This letter was interpreted by Walker as an indication that the provincial government was seriously considering the construction of a sugar beet processing plant somewhere in Alberta. The Raymond community immediately organized to ensure that it would be selected as the site for this new plant. The first step in organizing its bid was the revival of the Raymond Board of Trade on 8 January 1923. The 10 January issue of the *Lethbridge Daily Herald* described the meeting as "one of the most enthusiastic Board of Trade meetings held at which a large representative body of Raymond's boosters were in attendance."[255] James Walker was elected president while Louis King and John Anderson were elected as first Vice-President and Secretary-Treasurer respectively. The objective of bring the sugar processing industry back to Raymond was enthusiastically endorsed by the group. According to the *Lethbridge Daily Herald*, "from the enthusiasm displayed by those present and the 'pep' that is behind the movement, we can look forward to more tangible results later."[256] The *Lethbridge Daily Herald* also lent its support by noting that success of the movement was assured because "we have the soil and irrigation and it has been demonstrated that no better quality beets can be grown on the American continent. The sugar beets grown on Alberta soil possess more saccharine than those grown in the states of Idaho or Utah, where the sugar industry is flourishing."[257]

The new executive of the Raymond Board of Trade immediately began to prepare a proposal for the construction of a new plant and enlist the support of other communities in the project. The proposal's preparation included evaluating the site of the first plant and contacting its builder, the Dyer Company of Cleveland Ohio for an estimate of the cost of a replacement.

Organizing community support was reminiscent of the creation and operation of the Irrigation Development Association in 1919 when the Raymond Board of Trade created a central committee which worked with committees established in other centres. The first meeting outside Raymond was held on 15 January in the Magrath town hall. It was chaired by Ernest Bennion, who was active in the Magrath local of UFA and one of the founders of the MID. James Walker explained that:

> it was not known definitely that the provincial government were contemplating the establishment of a sugar factory, but from certain correspondence which had passed between himself and the Department of Agriculture, it was thought that such might be the case, and in order that the towns south of Lethbridge should not be overlooked, when the time came for selecting the site of the proposed factory, the Raymond Board of Trade had taken it upon themselves to take the initiative and place the matter before the government.[258]

The plan, as outlined by Walker, was for the provincial government to advance the money for the construction of the sugar factory and for the people to take stock in the undertaking at $10 per share. The farmers were to contribute 10% of their income from their sugar beet crop each year to pay for the stock subscribed. The meeting strongly endorsed cooperation with Raymond and Stirling to prove to the government that the logical place for sugar plant was south of Lethbridge. A local committee was created to work with the Raymond committee to promote the idea amongst the people of Magrath and district. Equally strong indications of support were received from Stirling, Cardston and Taber where each meeting was attended by an estimated 200 to 300 farmers.

In addition to meetings indicating support, petitions were also being circulated in the Stirling, Raymond and Magrath areas. According to John Anderson, these petitions were being:

> signed to a man. The farmers who in the old days were not willing to raise beets, preferring to confine their operations to wheat growing,

> are now keen for the industry. The whole spirit of the farmers has changed. They realize that the one crop system has failed, and they are looking for ways to work into intensive diversified farming. The sugar industry, therefore, looks good to them.[259]

Organizing political support for the project began in mid-January when John Anderson met with Cardston MLA, George Stringham, Taber MLA, Lawrence Peterson, and Deputy Minister of the Department of Agriculture, H. A.Craig. Craig suggested that plans for the construction of a sugar factory be commenced at once and the amount of financial support promised by the farmers submitted to the legislature.

Given this encouragement, the organizers, began a second round of public meetings to obtain public pledges of financial support for the plant, the cost of which was estimated by the Dyer Company of Cleveland Ohio at $700,000. They expected that the sugar factory, together with the money subscribed, including the pledged notes, would be held by the government as security for the bonds issued. This second round began on 1 February 1923 and proved to be another success.

The proposal was ready for presentation to Premier Greenfield and his cabinet by early February. The delegation which made the presentation consisted of James Walker, Ray Knight, and J.W. Evans of Raymond; Mayor Ririe and Chris Jensen of Magrath; Seymour Smith of the United Irrigation District; Mayor Caldwell and J.W.Card of Cardston; James Rose of Lethbridge and William Valgardson of Taber. The MLAs from Taber, Lethbridge, Cardston, Fort Macleod and Warner also attended to lend their support .

The Raymond proposal was that 500 to 1,000 irrigation farmers would subscribe $100,000 in cash and $900,000 in shares, unpaid to build a $700,000 sugar factory capable of handling 25,000 tons of sugar beets yearly. The remaining $300,000 was working capital to handle the crop for the first year. The government was asked to guarantee the bonds. Greenfield was only prepared to go to the extent of guaranteeing the bonds of the irrigation districts, leaving further investment in the industry to the private sector. "If you can get the farmers under contract to grow 25,000 tons of beets yearly" Greenfield noted "for the life of me I cannot see why you can't get private capital to take up as good a project as you claim you have."[260]

TESTING COMMUNITY SUPPORT

The organizing committee reacted to this challenge by undertaking a search for a company or individual to build the plant. It brought the same enthusiasm and level of organization to this task as they had with the development of the first proposal to the Alberta government. After Ellison declined the request to become involved, the Board of Trade used its links with the Mormon community in Utah to see if an American Company was interested in the proposal. The Raymond Board of Trade took advantage of a trip to the USA by Charles McCarty to begin this process in April 1923. He met with officials of the Utah-Idaho Sugar Company, the largest operator of sugar factories in Utah and Idaho. As a result of their initial interest, John Anderson and James Walker went to Salt Lake City to continue the negotiations. The company agreed that if the people would get behind the movement and demonstrate that beets could be grown successfully, they would go ahead with the erection of a factory in 1925. The company wanted to avoid the problems of the Knight Sugar Company by determining community support before construction of the plant.

In order to test their dedication to the plant, the Utah-Idaho Sugar Company proposed two tests, the first of which was the growing of 100 acres of beets in 1923 in 1 and 2 acre plots as a means of establishing the feasibility of sugar beet culture. The task of purchasing the seed, its distribution to the area farmers, generating interest in the project and the monitoring of the results was undertaken by the Raymond Board of Trade. The test plots campaign was a virtual repeat of the way the Raymond Board of Trade had organized the preparation of the proposal for the provincial government.

The initiation of the test plot campaign began with a public meeting in Raymond followed by a new round of meetings at the surrounding centres beginning with Glenwood on 17 April, Barnwell on 19 April, and Coaldale on the 20 April. In addition to attending meetings to generate interest, the Raymond Board of Trade also offered cash prizes for the best crop on a 1 acre plot and on a ½ acre plot. The prizes in the 1 acre contest ranged from $200 for first to $25 for fifth while the prizes for the ½ acre contest ranged from $100 for first to $12 for fifth. Former Knight Sugar Company employee Thomas O'Brien was field superintendent with the task of overseeing the seed

New sugar factory Raymond 1925 (Sir Alexander Galt Museum & Archives, Lethbridge, Alberta)

planting and cultivation of sugar beets in the different test plots throughout the entire beet growing district.

Based on these results, the Utah-Idaho Sugar Company moved to the second test of the community support for the project. On 28 August 1923, Mark Austin, the Chief Field Superintendent of the Utah-Idaho Sugar Company advised the farmers that they should prepare and put into shape 6,000 acres of land in preparation for the growing of sugar beets in 1924. If the farmers prepared 6,000 acres of land for beet production in 1924, the company promised to build a plant somewhere in Alberta to handle the 1925 crop. The 6,000 acres campaign was officially begun at a meeting held in Lethbridge on 8 February 1924. It brought together the farmers, the business community, the staff from the experimental farms and the officials of the irrigation districts. This meeting planned the 6,000 acre campaign in the same way as the test plots campaign. The Raymond Board of Trade served as the central coordinating committee with assistance from a network of local committees operating outside of the Raymond area. An advisory committee which included representatives from Magrath, Raymond-Stirling, Coaldale, Taber-Barnwell, the Lethbridge Northern and United Irrigation districts and the Lethbridge Board of Trade was added to the organizational structure for the campaign along with an educational committee chaired by Earnest Bennion of Magrath and a finance committee.

The Raymond Board of Trade monitored the progress of the 1924 campaign by sending out questionnaires in April 1924. This questionnaire asked a variety of questions including the kind of crops planted, or to be planted this year, and number of each; the number of acres under irrigation, or to be put under irrigation this season; the number of acres planted to sugar beets in 1924; and the number of acres to be planted to sugar beets in 1925. In addition to production data the Raymond Board of Trade also wanted to know the attitude of farmers toward beet growing and the number of family members available to work in the beet fields.

With the farmers willingness to support the plant amply demonstrated, the only remaining issue was its location and the freight rate structure to be charged by the CPR for the transportation of construction supplies and the beets. The Raymond Board of Trade began their campaign in January 1924 with the objective of ensuring that Raymond would be the location for the plant. The various towns which cooperated with Raymond in the campaign to bring the Utah-Idaho Sugar Company to Alberta agreed that it should be located in Raymond. There was, therefore, not the usual intense competition between the various communities which usually resulted in the issue of bonuses to attract the plant. The inspection of possible sites other than Raymond was begun by the Utah-Idaho Sugar Company in November of 1924.The final issue which had to be dealt with before the construction of the plant was assured was the negotiation of a deal on freight rates with the CPR. These negotiations were successfully concluded in early February 1925. The company's decision to locate the plant in Raymond was announced on 14 February 1925. The celebration of this event attracted an estimated 2,000 people on 12 March 1925.

ALBERTA SUGAR BEET GROWERS ASSOCIATION

With the Utah-Idaho Sugar Company's announcement on 14 February 1925 of its decision to locate a sugar beet process plant in Raymond, the Raymond Board of Trade had accomplished the goal it had set out to achieve two years earlier. A new organization was now required to continue the ongoing task of working with the company to organize the industry and meet the needs of the growers. Given the degree of organization which had already been achieved under the leadership of the Raymond Board of Trade, the beet growers were able to organize a convention within a matter of days to establish such an organization. On 11 March 1925 the founding convention of the Alberta Sugar Beet Growers' Association (ASBGA) was held in Lethbridge. At that time, Arthur Dahl of Raymond was elected president by acclamation while Earl Tanner of Magrath, William Spackman of Stirling and I. B. Roberts of Raymond were elected to the other positions on the executive. Representatives were also appointed to the executive from the United and Lethbridge Northern Irrigation Districts as well as from Cardston, Welling, Lethbridge, Coaldale, Barnwell and Taber. The important positions held by people from Magrath, Raymond and Stirling was a reflection of the earlier role of these towns in procuring the sugar beet processing plant. A constitution and bylaws similar to the Utah Growers Association was adopted at the first meeting of the executive committee held on 13 March 1925.

The ASBGA's primary role was to negotiate the annual contract with the Canadian Sugar Refineries Limited and to deal with any other matters affecting the relationship between the company and the growers. The participation of the officials of the company in the affairs of the ASBGA ensured maximum cooperation as did the hiring of many people who had been involved in the effort to establish the plant such as Earnest Bennion of Magrath. This degree of cooperation between the growers and the sugar beet company was a innovation in the industry since no such organization existed before the First World War.

The labour issue, which was first raised by the Raymond Board of Trade in the questionnaire sent out in April of 1924, was one of the first issues dealt with by the ASBGA. The Association moved quickly because the failure to procure an adequate labour supply to thin and harvest the crop had been a major reason for the industries failure to become established before the First World War. At the first meeting held on 11 March 1925 various motions were passed to determine the labour needs of the growers and ways of supplying these requirements. The field men of the company were requested to canvass the farmers. Possible solutions to these needs included shortening the school year to permit children up to the eighth grade to work in the beet fields during the first two weeks of June and bringing in Japanese workers from Vancouver. At the meeting of the executive on 13 March, Indians, Mennonites, Hindus and the 150 miners who had seasonal employment in the mines in the Lethbridge and Taber districts were identified as a potential sources of labour. At this same meeting, a committee was also created to investigate the creation of a labour bureau in cooperation with the provincial government.

IMMIGRATION AND SUGAR BEETS

The most ambitious program to meet the labour needs of the industry in the late 1920s was undertaken by the CPR through the efforts of James Colley, Assistant Superintendent of the CPR's Department of Colonization and Development, Secretary of the Western Canada Irrigation Association and active in the ASBGA. He was a firm believer in combining sugar beet farming, land settlement, irrigation and European immigration. In his 1926 Annual Report, Colley noted that the beet sugar industry was a valuable aid to colonization and development providing hand work at wages for the new immigrant until they were able to secure land, as renters or owners, upon which to grow beets on their own account. It was possible, in Colley's view, for all the members of the family to engage in this work and by their combined earnings have a substantial income which would enable them to acquire land. His immediate objective was to:

> get as many people as possible located - particularly families - and to have farmers demonstrate for themselves the value to them of the beet crop and thus to bring about such an increase of acreage as to tax the capacity of the present factory and make additional factories necessary. By the time a second factory is established in southern Alberta - somewhere between Lethbridge and Taber probably - the possibilities of the sugar beet industry as an aid to development of irrigable lands and the advantage to farmers of growing beets will be so convincingly and overwhelmingly demonstrated that the way for a third factory . . . will be paved.[261]

Starting in 1925, he organized Colonization Boards in a number of southern Alberta communities in association with boards of trade, irrigation districts and sugar beet growers associations. These organizations determined the labour needs of their area, located land which could be eventually acquired by the settlers and assisted these people when they arrived. The recruitment of the settlers was the responsibility of the Canadian Colonization Association and other organizations such as the Mennonite Land Settlement Board. These included the Taber, Magrath and Irrigated Farms Colonization Board which was located in the Coaldale area. The Raymond Sugar Beet Growers Association and the Raymond Hungarian Colonization Board worked together to bring immigrants to that community. .

The development of the Mennonite Community in Coaldale beginning in 1926 was one example of this program in action. The first Mennonite families occupied land provided by the chairman of the Irrigated Farms Colonization Board O. T. Lathrop who turned over his land, together with horses, stock and equipment to five Mennonite families. In the fall of 1926 the CPR instituted a program to provide the Mennonite immigrants with land in return for the growing of beets. The settlers were provided with the land in return for turning over the gross proceeds of 10 acres of sugar beets to the company each year until the land was paid for. The CPR advanced $400 worth of building supplies to each settler. Thirty families participated in this program. Mennonite settlers also took over farms in the Van Horne Colony which had been established during the First World War as part of a soldier resettlement plan.

The Taber Colonization Board was originally established by the Taber Board of Trade in 1926. Ted Sundal and William Valgardson from the TID served on the Board of Directors. In 1928 when this organization failed to provide continued support, Ted Sundal and William Valgardson took over its day to day operations. Sundal handled the paper work involved with the program while Valgardson dealt directly with the settlement of the people. As the area representative of the Canadian Factories Ltd,. he was in the best position to know who required help. The TID was a strong supporter of the program since "our local sugar beet industry is growing rapidly and we anticipate an increase of 1000 acres [this] coming spring which will require a substantial increase of new imported labour."[262]

EXPANSION OF PRODUCTION

Sugar beet production in the St. Mary Project had increased to 19,000 acres by 1944. One of the factors in this success was the decline in the price of wheat and the increase in the price paid for sugar beets. One of the main areas where production increased was the TID. Production went from 752 acres in 1925 to 7,239 in 1945. The sugar content increased as did the gross value of the crop which went from $37,072.50 in 1925 to $1,283,422. in 1945. By 1929, sugar beets had become the foundation crop of the TID. The use of commercial fertilizers after 1929 and the inclusion of sweet clover in the crop rotation was Sundal's explanation for this success. A beet growers association was formed in 1925 shortly after the Alberta Association was established. Its activities included ensuring adequate loading facilities were provided and access by Taber beet growers to their share of the beet pulp and molasses.

ROGERS SUGAR

In 1930 Canadian Sugar Factories Limited was purchased from the Utah-Idaho Sugar Company by the Rogers Sugar Company. The Rogers Sugar Company changed from trying to stop the production of beet sugar to building new plants and promoting its use. The first new plant was built in Picture Butte in 1936. The effort to obtain a plant in Taber was a joint undertaking of the beet sugar growers association and the TID. The TID and the beet growers association worked closely together because Ted Sundal was Secretary-Treasurer of both organizations and William Valgardson, chairman of the TID, was the field man for the Canadian Sugar Factories Limited in the Taber area. In 1928 Ted Sundal wrote to the Dominion Sugar Company of Chatham, Ontario extolling the virtues of the Taber area as a possible location for the company to establish a new plant. The company indicated that it was not interested in expanding to the West. Following the purchase of the Raymond plant by Rogers Sugar, the Taber farmers made an agreement in February 1939 with this company for the construction of a third plant.

Rogers not only expanded the number of plants but also brought the benefits of a research department and the publication *Silver Sunshine* to the growers. *Silver Sunshine* which began publication in 1940 was intended to promote the industry and to educate the water users on various issues such as amount of water to maximize sugar content.

SIXTEEN
REVIVAL OF THE LETHBRIDGE SOUTH EASTERN PROJECT 1935 -1946

The first attempt to expand the St. Mary Project was made between 1902 and 1904 by the CNWICo. in partnership with the territorial and federal governments. The company built 16 miles of a canal system designed to take water from the Milk River to reservoirs on the eastern and northern slopes of the Milk River Ridge. Water from Ridge, Verdigris and Brunton Reservoirs was to be used to irrigate 500,000 acres northeast of the Milk River Ridge. The CNWICo. had elaborate plans to supplement the water supply provided by the Milk River with water from the St.Mary River. A dam was to be built on the St. Mary River close to the American border along with a canal system to take water to the Milk River via Whiskey Gap and north to a network of storage reservoirs south of Spring Coulee.

The CNWICo., and its successor the AR&ICo., abandoned all of these plans including the reservoir system. After 1904, the AR&ICo. concentrated its efforts on expanding the headworks of the existing system, completing the distribution system in the Lethbridge-Coaldale area and operating with the water available from the St. Mary River at Kimball. The AR&ICo. did not pursue its ambitious expansion plans because of objections from the Americans and potential seepage problems on a portion of the Milk River canal. This approach was continued through to 1946 by the CPR after it took control of the St. Mary Project in 1908. The only exception to this approach being the construction of the Chin Coulee and Northeast Reservoirs which were only of benefit to the TID.

The second attempt to expand the St. Mary Project began in 1909 when the federal government initiated the preparation of a plan called the LSEP to make use of Canada's portion of the St. Mary and Milk Rivers as provided for in the International Waterways Treaty ratified that year. This plan, as it evolved through to 1918, extended irrigation east to the shores of Pakowki Lake and Forty Mile Coulee. In 1919 the Irrigation Development Association was created by the farmers to encourage the government to build this system. Both levels of government rejected this idea and encouraged the creation of farmer-run irrigation districts as an alternative approach to irrigation expansion. The farmers in the Taber, Magrath and Raymond areas, with the cooperation of the CPR, were the only ones who successfully created irrigation districts in the St. Mary Project during the inter-war period. The TID, MID and RID expanded irrigation, east of Chin Coulee, added the High Line Lateral in the Magrath area and expanded the irrigated land being served by the existing Raymond Lateral in the area between Welling and Stirling.

The 1936 water shortage combined with American plans to build dams on the lower Milk River was the catalyst for the third and successful attempt to achieve the goals for the project as established at the turn of the century. This third attempt was very similar to the second in terms of the people involved and the organizations created to make the farmer's opinions known to government. The events of 1939, in particular, were a case of deja vu for people like Ben Russell and William Buchanan who had gone from an MP to a senator since 1919. The fundamental difference between these two attempts was the willingness of government after the Second World War to invest, not only in planning, but construction of irrigation projects. The PFRA, the 1936 Ewing Commission, the Engineering Institute of Canada, the South Alberta Water Conservation Council, and the 1942 St.Mary and Milk River Water Development Committee were all part of the process which brought about this policy revolution.

THE MILK RIVER DIVERSION PROPOSAL

The revival of the LSEP began in 1934 when the residents between Milk River and Raymond including settlers in the vicinity of Tyrrell Lake, Verdigris Lake and Etzikom Coulee requested assistance from the provincial government to complete the canal from the Milk River originally started by the AR&ICo. in 1902 and abandoned in 1904. These people wanted to extend this canal to both Verdigris and Tyrrell Lakes where the water was to be stored for stock water purposes in Etzikom and Verdigris Coulees. The Town of Raymond was willing to contribute $10,000 to its construction if the canal could be used to provide water for the Raymond Reservoir. Raymond was then in the planning stages of a new $30,000 water supply system. The use of the water for irrigation purposes was not considered at that time.

This proposal was brought to the attention of the PFRA soon after its creation in 1935. This organization was created in April of that year by an Act of the federal government as a temporary measure to rehabilitate the drought strickened areas of Manitoba, Saskatchewan and Alberta. The PFRA included a large number of former Department of Interior, Irrigation Branch staff who had been actively involved in water management planning before 1930. These people were familiar with the LSEP which remained as a blue print ready for use when the necessary funds were available for its construction.

The PFRA Act established an advisory committee to inform the Minister of Agriculture on the best methods to deal with the drought and soil drifting problems in the West. This committee was also to advise on the systems of farm practices, tree culture and water conservation that would promote the greater economic security of its residents. The first chairman of the Advisory Committee was Dr. E. S. Archibald, Director of the Dominion Experimental Farms. The Act also provided for the appointment of temporary technical and other support staff and provided $750,000 on an annual basis for five years to carry out its work.

Water management issues were handled by a sub-committee, established at Swift Current, Saskatchewan. The committee chairman was L. B Thomson, Superintendent of the Dominion Experimental Station, Swift Current and the Secretary was William Jacobson, an Irrigation Specialist at the Dominion Experimental Station in Lethbridge. This committee's responsibility was to provide engineering and financial assistance to farmers and ranchers in developing small scale stock watering and irrigation projects.

The Water Development Committee was assisted by Water Development Committees in each of the prairie provinces. The Alberta Water Development Sub-Committee was made up of L. B. Thomson, L.C. Charlesworth, Dr. William H. Fairfield, Augustus Griffin, Ben Russell and William Jacobson. Ben Russell was the senior technical advisor to the committee.

These committees initially focussed the PFRA's limited resources on small scale projects such as farm dugouts and small dams which stored up to 50 acre-feet of water from the spring runoff. It provided funds up to a maximum of $50 for dugouts and $350 per dam. The PFRA wanted to delay the construction of large irrigation projects until a detailed investigation of all proposed projects had been made to determine their economic as well as engineering feasibility in relation to their overall value to the community. Despite this policy, it did undertake the Rolling Hills project in the Eastern Irrigation District and made an $80,000 contribution to the Canada Land and Irrigation Company for repairs to the headworks of the Bow River Project.

The Alberta Water Development Committee considered the Milk River Diversion proposal on 4 January 1936 after it had been reviewed by its own staff. Their review indicated that Verdigris was not satisfactory as a storage basin owing to its shallow depth and the excessive losses that would be incurred through evaporation. The cost was also high because a large capacity canal was required to divert Canada's share of the Milk River which was 14,000 acre- feet during the short time the river was at the flood stage.

The committee recommended that less expensive alternatives be investigated and that any development be approved by the IJC. It also referred the proposal to the general Water Development Committee for discussion at their February 10 meeting. The Water Development Committee expanded the discussion of the Milk River proposal to include a review of storage reservoirs on the St. Mary River and the implications of the developments given the terms of the International Waterways Treaty of 1909. The committee was anxious to initiate a reservoir construction program without giving the Americans an excuse to renegotiate the International Waterways Treaty since Canada had failed since 1921 to comply with the IJC's recommendation that each country build storage dams to make beneficial use of their share of the water. As Fairfield pointed out, the underlying principle of the treaty was beneficial use. It was therefore:

> important that storage reservoirs to utilize flood water from the St. Mary River be developed a soon as possible before the Americans establish priority through beneficial use with the storage development proposed on the American side.[263]

The collection of information requested by theWater Development Committee included a meeting at the Marquis Hotel in Lethbridge on 27 October 1936 organized by Dr. E. S. Archibald and chaired by Dr. William Fairfield. The other people presented were John Vallance, Ben Russell, L.C. Charlesworth, A. Griffen, G.N. Houston, L. Fairbairn, Roi Risinger, and F.G. Cross. This group included farmers, project administrators and government officials.

The objectives identified for development of the St. Mary and Milk Rivers were to provide a dependable water supply for stock and for the irrigation of small acreages in order to supplement feed supplies. A second objective was to provide full summer irrigation for the crops intensively grown in the present irrigated area such as sugar beets, potatoes, alfalfa and canning crops. The total demand for water was increasing, particularly from late July to September. The normal flow of the river gave reasonably adequate service for grain crops (the peak demand for which is in June) but was wholly inadequate for other crops. The plan also had to extend the benefits of irrigation to additional territory by the development of storage sites on the St. Mary River in Canada. A dam on the St. Mary River between Spring Coulee and the International boundary was recommended because it would not only provide adequate storage but the water could be diverted into the existing AR&ICo. system.

On 10 November 1936 Russell submitted a report on the St. Mary Project which reviewed the water supply available from the Milk and St. Mary Rivers, the extent to which Canada was making use of its share, the feasibility of the construction of a dam near Spring Coulee and the federal government's responsibility for the construction of this dam. Russell reported that an additional 82,000 acre-feet was available, even in the low flow years, for storage purposes. Russell felt that the federal government had a responsibility to intervene because the rights granted in 1899 to the AICo on the St. Mary River, were partly made valueless by Article VI of the International Waterways Treaty. He further argued that this responsibility had been recognized in 1912 when surveys were made for a canal from the Belly River to the St. Mary River. He recommended further investigation of the Spring Coulee dam site, the Ridge and Raymond Reservoir sites and the feeder canal required to link them together. He provided preliminary plans for a dam near Spring Coulee and for canals to take water from the St. Mary River to the AR&ICo. canal and from a point on the AR&ICo. canal near Bradshaw to Ridge Reservoir.

The investigations of the Milk and St. Mary River by the PFRA and the staff of the Water Resources Branch and the Alberta Department of Agriculture by 1937 provided various options for the development of both rivers. The options for the Milk River included a dam at the forks of the North and South branches, a 150 cfs canal to the Ridge Reservoir or a 350 cfs canal to Verdigris Reservoir. With respect to the St. Mary River, two potential sites for a dam had been identified as had the possibility of simply diverting water from the existing AR&ICo. to the Ridge and Raymond Reservoirs.

EWING COMMISSION

The review of the water storage problem by the PFRA in 1936 coincided with the investigation of the Ewing Commission into the financial problems of farmers in southern Alberta's irrigation projects. Its members included Justice Ewing of the Supreme Court of Alberta, Dr. W. A. Wyatt, Head of the Soils Department at the University of Alberta and Lethbridge area farmer, Roi Risinger. Risinger was closely associated with the St. Mary Project having been active in the Irrigation Development Association in 1919 which lobbied for government funds to build the LSEP and having been involved in the attempt to create an irrigation district in the New Dayton area during the early 1920s.

The objectives of the Ewing Commission, when it was created, were to determine the value of irrigated land based on the assumption that it would be used to grow a variety of crops and raise livestock. A second objective was to determine if a farmer could pay for the land, based on the assumption that he had the ability and knowledge to be a efficient irrigation farmer. The farmer, in the commission's opinion, had the opportunity to affect his success by shifting away from straight grain farming to specialized crops, carrying out the proper rotation of crops and maintaining high soil fertility by using recognized cultivation methods. The commission also recognized that the ability to make irrigation agriculture a success was dependent on factors beyond the farmer's control such as weather and market trends. The third objective was to examine the collection procedures for the land and water purchase contracts and the problem of debt which appeared to be the inevitable result of these contracts. The water storage problems of the St. Mary Project were added to its list of concerns because of the influence of Risinger and the views expressed at the public hearings held in Raymond, Magrath, Taber, Coaldale and Lethbridge.

The Ewing Commission held its first public meeting at Raymond in August 1936 where it focussed on the affairs of the RID. Secretary-Treasurer Evans, along with RID water users expressed satisfaction with the financial situation of the district, the water rate and the collection of water rates under the present provincial laws. Their real concern was the lack of water for irrigation requirements. This point was also made at the hearings held by the commission at other locations in the Lethbridge area. The submission of the CPR dealing with their Lethbridge-Coaldale area was given by Major F.G. Cross, Superintendent of the project who provided data from a selection of land and water contracts rather than giving a comprehensive summary of the affairs of the project. When pressed by the commissioners to tell it how many of the farmers were meeting their obligations under the crop share program, he indicated that in his opinion "the contracts will probably be in fair average standing. There will be a certain amount of lame ducks and a certain amount of good farmers."[264] He only admitted that there was a water shortage problem which was affecting crop production when pressed by the commission to discuss this issue. He felt that an increased water supply:

> would be a material benefit, but as to what percentage, [he could not] say. While there [had] been a general percentage of loss in crop due to shortage of water [this year, he could not] very well estimate it. Certainly if more water had been available there would have been much more production.[265]

The water shortage problem in the Lethbridge-Coaldale area was, however, emphasized by other individuals such as A.G Virtue who represented the Irrigated Farm Co. Ltd. He estimated the loss of income from the lack of water at $1,000,000. Other speakers complained that the water shortage problem extended to the ability of the CPR to deliver the water which was available.

Based on these submissions, the Ewing Commission concluded that the water storage problem had developed after re-establishment of the sugar beet industry:

> That was a crop demanding an assured amount of water, failure to have it applied meant greatly diminished yields and great monetary loss to the farmer who has an outlay in money of about $40.00 per acre that [had] to be repaid before he can begin to realize any profit. Then too, the effective and proper use of water resulted in such an increase in yield and net money returns, demonstrated by the better farmers, that with such examples water users generally in this district have come to a realization of true value in its use and are consequently making fullest use possible of their allotment. That in the aggregate requires delivery of much more water than was used in the first fifteen or twenty years . . .of development.[266]

The commission also passed on the view of the water users that, while the CPR and the government had no legal responsibility to solve the problem by building reservoirs to capture the spring run off, they did have a moral responsibility to solve the problem because the settlers had been guaranteed an adequate water supply when they settled the land.

ONGOING INVESTIGATIONS BY THE PFRA

The PFRA continued detailed planning for the expansion of the St. Mary Project in June 1938 when its first Alberta regional office was established in Lethbridge under the supervision of Wally Foss. Under his direction, the evolution of the LSEP from1909 to1936 was reviewed and new investigations made in order to prepare a revised version. Detailed specifications for a dam at the Spring Coulee site were part of this work which was completed by April 1941.The Foss expansion plan included: a 25,000 acre-feet reservoir on the Waterton River; an 1,100 cfs canal from the Waterton to the Belly River; a 1,700 cfs canal from the Belly to the St. Mary River Reservoir; a 270,000 acre-feet reservoir on the St. Mary River created by the dam at the Spring Coulee site; and a 3,000 cfs canal from the St. Mary Reservoir to 345,000 acres of irrigable land to the east. Chin Reservoir was to be expanded to 101,000 acre-feet and an 80,000 acre-feet reservoir constructed at the Ridge site, along with a 16,900 acre-feet reservoir at the Raymond Reservoir site and the conversion of Verdigris Lake into a 140,000 acre-feet reservoir. Both Russell and Foss emphasized the dam on the St.Mary River as the key structure in the revised LSEP. Foss provided a 10 year construction schedule which anticipated that 20,000 to 30,000 acres would be added to the system each year. Development work carried out by Foss and his staff included some soil testing but this was curtailed because of World War Two.

ENGINEERING INSTITUTE OF CANADA

The first non-government agency to formally study the water problems of the West in general and the St. Mary Project reservoir situation in particular was the Engineering Institute of Canada. In 1935 the Council of the Institute appointed a committee, under the chairmanship of G. A. Gaherty, to study the problem of water distribution in the drought areas of the western provinces. In 1939 the activities of this organization were focussed specifically on the St. Mary Project as a result of the initiative of the Lethbridge Branch. At a joint meeting with the Association of Professional Engineers of Alberta on 28 October 1939 a resolution was passed drawing attention to the urgent need for water conservation in the southern portion of the Province of Alberta and recommended to the governments of Alberta and Canada that immediate consideration be given to the creation of suitable storage reservoirs so that the economic and cultural stability of this area may be maintained for future generations. The Lethbridge Branch offered its assistance in implementing a resolution which was sent to the headquarters of the Engineering institute of Canada for suitable action and to the Premier of Alberta and the Prime Minister of Canada.

The Lethbridge Branch was encouraged in its efforts by F.G. Cross superintendent of the CPR Lethbridge Irrigation Project who indicated in a December 1939 letter that:

> while the International Treaty gives Canada priority right of the use of the water of the St. Mary River, over the United States, Canada is actually in a weak position. In 1921 the International Joint Commission . . . recommended the Governments of the United States and Canada proceed to construct reservoirs to obtain the maximum benefits from the flood waters of these streams. The United States have forged ahead and at the present have an advantage in Shelbourne Lake and the Fresno Reservoir, the latter brought to completion and dedicated about two months ago. Canada has taken no action and if the United states call for an opening up of the Treaty, they will be able to present a case showing beneficial use of the waters and unless quick action is taken by Canada a valuable birthright will be lost forever.[267]

The suggestions of the Lethbridge Branch were given immediate attention by the Western Water Committee. A sub-committee of western members of the Engineering Institute of Canada was established to assemble material on this issue. It was made up of the senior officials from government and industry concerned with the administration of water resources in Alberta. It also had access to other resource people and virtually all of the documents in the form of reports and surveys which had been prepared or carried out over the last forty years. The objective of the committee was to review the proposed LSEP plan and the general topic of Canada's utilization of its share of the waters of the St. Mary and Milk Rivers as well as the Belly and Waterton rivers. The committee focussed on the international aspects of the situation, economic development of irrigable land and water supply.

The report of this sub-committee was presented to Council on 18 January 1941. The plan for the LSEP as envisaged by the Committee included a reservoir of 270,000 acre-feet capacity on the St. Mary River at a point near Spring Coulee. A 1,100 cfs capacity canal to connect the Waterton River with the Belly River and a 1,700 cfs capacity canal to connect the Belly River with the St. Mary River was suggested to increase the water supply. A third major canal with a capacity of 3,000 cfs would take the accumulated water from the St. Mary River to the Ridge and Raymond Reservoirs and into the existing system for transport to Chin Reservoir.

SOUTH ALBERTA WATER CONSERVATION COUNCIL

The second public group to take action on the issue of storage reservoirs for the St. Mary Project was the water users in the St. Mary Project and potential water users located east to Medicine Hat. Their concerns were made known after the public hearings of the Ewing Commission by the MID, RID and the Alberta Sugar Beet Growers Association. In July 1937, following a joint meeting of the MID and the RID, Chris Jensen chairman of the MID indicated to the *Lethbridge Herald* that:

> Every dry year, about July 1 the water in our ditches begins to go down and down. When one sees all the water going to waste in the river in May and June it seems foolish that we cannot hold that in store somewhere so that our crops may get the water they need at critical time. We feel that this is a problem which requires national aid. Every foot of water available should be stored and carried out over the dry land area in southern Alberta. It would add to the national income and do away with a lot of relief called for every drought period. And we are going to keep pounding away at this problem until it is finally solved satisfactorily.[268]

The Alberta Sugar Beet Growers Association began to lobby for assistance to build reservoirs in the St. Mary Project in the fall of 1937 in a letter to Jimmy Gardiner, the federal Minister of Agriculture. Gardiner's response was that established irrigation projects were not eligible for funding from the PFRA and that a separate budget appropriation would be required if any work was to be done. Gardiner's rejection of assistance to existing irrigation projects was not, however, supported by the Alberta Water Development Advisory Committee which was dismayed at the response sent to the Sugar Beet Growers Association. It took the opportunity to request Gardiner to obtain the funds to thoroughly investigate the matter of water storage and suitable reservoir sites and to construct the reservoirs. When this request was made by L. C. Charlesworth he also noted that the request had the support of the Province of Alberta.

A further indication of the split between the politicians and the staff of the PFRA over this issue came in July 1939 when Jesse Wilde and Christian Jensen reported, after a trip to Regina to meet PFRA Director George Spence, that the PFRA seemed anxious to promote the reservoir idea but were not inclined to go far unless the people were pressing the case.

The creation of an organisation to press the case came shortly thereafter as a result of a tour in September of the Taber-Barnwell area organized by the Alberta Sugar Beet Growers Association which included John W. Evans, Secretary-Treasurer of the RID. Upon his return, Evans wrote a letter to the editor of the *Lethbridge Herald* in which he described how the TID had become a Garden of Eden because of stored water from Chin Coulee Reservoir. After noting the obvious benefits of operating a system with reservoirs to store the spring run-off, Evans expressed the hope that some organization would provide the leadership to deal with the issue.

The call for action was answered this time by the Lethbridge Board of Trade which immediately convened a meeting on 4 October 1939. This meeting led to the creation of the South Alberta Water Conservation Council (SAWCC). This organization was virtually identical to the Irrigation Development Association established in 1919. Established organisations such as the Burdett Agricultural Improvement Association were incorporated into its activities. In other areas such as Coaldale, new organizations were created as chapters of this organisation. The council executive was made up of president, Phil Baker from Raymond, and Vice - Presidents J. I. McDermott from Coalhurst, Seymour Smith from Mountain View and Samuel Ashton of Burdett. The Honorary President was Senator

William Buchanan and Dr. William Fairfield as Honorary Vice president.

The council immediately arranged meetings with Jimmy Gardiner and George Spence in the fall of 1939 and prepared a submission to Prime Minister W. L. MacKenzie King which was submitted on January 26, 1940. This organization continued its efforts through to 1946 at which time it was superceded by the Alberta Irrigation Projects Association.

ST. MARY AND MILK RIVER WATER DEVELOPMENT COMMITTEE

The efforts of the SAWCC, the Engineering Institute of Canada, support within the PFRA and the dam construction program of the Americans led to the creation on 17 February 1941 of the St. Mary and Milk Rivers Water Development Committee by the federal government. The members of the committee were Victor Meek Controller, Dominion Water and Power Bureau. George Spence Director of the PFRA and W.E Hunter Accounts Branch, Department of Finance. The members of the Alberta Water Development Committee were designated by Alberta to work with the committee. The members of this Committee were D.B. MacMillan, the Minister of Agriculture, N.E. Tanner Minister of Lands and Mines and P.M. Sauder, the director of Water Resources.

The Committee prepared a report which consolidated the work done since 1909 by the Department of the Interior, the PFRA, the Alberta Water Resources Branch and the Engineering Institute of Canada. This document determined Canadian use of its share of international streams in southern Alberta, the development of a plan to put it to use, selected lands to be irrigated, an estimate of the cost of storage reservoirs and the costs of the other works required for the completion of the development and provided a yearly breakdown of the expenditures.

The Meek report went beyond the engineering aspects of the project to suggest how to deal with the current owner of the system, the benefits which this water development would confer on Canada, the Province of Alberta and the residents of the district affected, the allocation of costs between Alberta and Canada, methods of financing, administrative control after completion and the colonization of the irrigable lands. The committee did not hold public hearings but rather invited and received representations in person or in writing from a variety of experts.

The Meek Report made an important departure from previous ideas on government assistance to the building of irrigation infrastructure. Up to the 1940s the financing of irrigation infrastructure construction was based on the idea that the total costs of an irrigation system should be charged to the lands immediately benefited. The Meek report changed this view and suggested that to insure completion and successful operation of the project, there must be governmental financial assistance.

The activities of the committee drew the attention of the CPR whose interest in irrigation development was rapidly declining. The CPR indicated that it would turn over the St. Mary Project to any responsible organization. The CPR had already divested itself of the Eastern and Western sections of its irrigation block east of Calgary in 1935 and 1942 when the Eastern and Western Irrigation Districts were created. The transfer of the Western Section was negotiated by a water users committee who were advised by P. M. Sauder, the Director of Water Resources for the Province of Alberta and by Edmonton lawyer and former Premier of Alberta , John Brownlee. In both cases the CPR paid the new owners an amount of money to cover rehabilitation.

The ongoing losses from the operation of the Lethbridge section, which increased during the 1930s, ensured that this policy would be extended to this project as well. Problems in the operation of all three of its irrigation projects contributed to this decision. The western section of its irrigation project east of Calgary was a particular source of problems since the farmers in the view of the CPR were unwilling to make the best use of irrigation to develop a diversified agricultural economy. The fact that the CPR did not try and change the terms of the water delivery contracts was the main reason that its water users in the Lethbridge-Coaldale area were less rebellious than in its other projects. These contracts protected the water users from any rate increases by giving them a perpetual supply of water at the rate of one dollar per acre. Given the fixed rate of return on their investment, the CPR were not prepared to spend money on major improvements such as a reservoir system. The terms of the contract thus ensured that the system would operate but at a low level of efficiency in terms of its ability to make maximum use of the water supply from the St. Mary River.

POTHOLE DAM

With any federal commitment to the expansion of the St. Mary Project postponed until the end of World War Two, the initiative shifted back to the Alberta government, the water users in the St. Mary Project and the CPR. The two issues dealt with through to the end of the war were the construction of the Pothole Dam and the transfer of the St. Mary Project from the CPR to the provincial government. The provincial government moved first with respect to these issues. In 1943 it revised the Water Resources Act permitting it to construct, acquire, purchase or take over and operate irrigation works. The Pothole dam was the first project to benefit from the new willingness of the provincial government to take a pro-active role in the St. Mary Project. The Raymond and Magrath Irrigation Districts requested its construction in 1943. In November1944 Premier Manning indicated that Alberta would provide $350,000 to build the Pothole dam and canals leading thereto provided that credit was given to Alberta on the final allocation of the cost of the project when the entire plan to expand the St. Mary Project had been agreed upon. Manning also indicated that, in return for this contribution, the water users would be expected to pay a water rate of $3.00 per acre as their contribution to its construction. The offer was subsequently endorsed by the board of directors of the RID and by the water users at a special meeting on 25 September 1944.

On 9 December 1944 a joint meeting of the Lethbridge-Coaldale and Magrath-Raymond committees was held to discuss the issue of support for the Pothole Dam project. This meeting served to illustrate the division between the two groups since Raymond and Magrath were prepared to deal with the issue of water supply even if it required a change in rates while Lethbridge-Coaldale was not prepared to change the rate from $1.00 even to solve its water supply problem. In the spring of 1945 the provincial and federal government completed their negotiations for the construction of the Pothole dam.

GOVERNMENT ACQUISITION OF THE ST. MARY PROJECT

The CPR first indicated its willingness to relinquish control of the St. Mary project in 1942. On 27 June 1944 it repeated the offer to Christian Jensen, chairman of the MID, Jesse Wilde, chairman of the RID board and Phil Baker also representing the RID during a meeting to discuss their respective debts. Upon the delegation's return, a joint meeting was held between the Magrath, Raymond and Coaldale water users on 10 July 1944 in the Coaldale Community Hall. Phil Baker, John Evans and Christian Jensen attended the meeting to outline the offer made by the CPR. The meeting revealed a split between the two groups since the Raymond and Magrath people were anxious to make an immediate decision on the offer while the Coaldale farmers felt that more study was required.

The Lethbridge-Coaldale study committee considered the current and future cost of operating the St. Mary Project. In their efforts to gather information, the committee found out from the CPR that the capital cost of the St. Mary Project was about $2,500,000.00 with the total acreage with full water rights being 101,710 acres. The operating costs averaged over the ten years from 1933 to 1943 was approximately $90,000,000.00. A second issue considered by the committee was the extent to which it would receive government support for the construction of the necessary reservoirs and other irrigation works for adequate water storage to supply the district at all times. The results of this research were presented toa meeting in Coaldale on 29 November 1944 with the result that the decision was made not to accept the CPR offer.

The decision not to accept the offer by the Coaldale farmers was followed by the decision of the province to take control of the St. Mary Project and operate it as a Crown Corporation. This decision was the result of the influence of P.M. Sauder, the Director of Water Resources. He was able to convince the Minister of Agriculture that provincial control of the St. Mary Project would ensure its post-war expansion as laid out by the St. Mary and Milk River Water Development Committee report.

The negotiations leading to the transfer began on 9 October 1945 when Sauder, on behalf of Alberta offered to take over the AR&ICo. provided the CPR paid the Province $200,000. This amount was based on the use of $100,000 to cover a year's operation and maintenance costs of the system, plus $100,000 to cover overhead cost not included in the Company's charges to the Project, probable increased costs under Provincial management and other contingencies. The CPR indicated that $200,000 was an unreasonable amount. On November 10, 1945 Sauder made a revised offer which reduced the cash payment from the CPR to $100,000.

The drafting of a formal agreement was completed in December 1945. With the successful completion of these negotiations the provincial government passed the necessary legislation creating the St. Mary & Milk Rivers Development Corporation.

During these negotiations, Sauder made a point of keeping the water users informed On 29 November 1945 Sauder, on behalf of D. B. MacMillan, the Minister of Agriculture called a meeting of various representatives from the irrigation districts and the Lethbridge-Coaldale district. Sauder outlined the terms of the agreement which had been arrived at by the CPR and the province. Sauder suggested that the water users should have a solicitor examine the agreement. The water users immediately appointed Edmonton lawyer, John E. Brownlee to act as solicitor for the farmers. Brownlee had been premier of Alberta from 1928 to 1935 during which time the agreement transferring ownership of the natural resources of Alberta from the Dominion government to the Government of Alberta had been negotiated. This agreement, along with a subsequent agreement in 1936, shifted the responsibility for the administration of water resources to Alberta. He also had a long association with the water users of the western section of the CPR irrigation block having served as the legal advisor to the water users committee which negotiated the transfer of this irrigation project to the water users in 1944 The principal concern of the water users was the effect this transfer would have on the status of the water rental agreements.

The decision to proceed with the construction of the St. Mary Dam was made in 1946 after the federal government was persuaded not to delay construction until an unemployment problem developed in the Lethbridge area.

SEVENTEEN
GOVERNMENT ADMINISTRATION OF THE ST. MARY PROJECT 1945 - 1968

In 1945 the St. Mary Project began an unprecedented period of change during which time the expectations of the previous 55 years were realized. This process began with the construction of Pothole Dam (renamed Jensen) on the eastern arm of Pothole Creek which was connected to the existing AR&ICo. system by an 800 cfs canal from Spring Coulee. The Jensen Dam was built prior to a federal government commitment to build the St. Mary Dam which was only made the following year. The St. Mary Dam became the key structure in the expanded St. Mary Project when it was completed in 1951. The headworks were extended from the St. Mary Dam west to the Waterton River and east, via Jensen Reservoir, to Ridge and Raymond Reservoirs. A new main canal was built north from the Raymond Reservoir through to Medicine Hat. New distribution systems were built from Grassy Lake to Medicine Hat.

The full impact of the cooperation between the federal and provincial governments to expand the St. Mary Project, however, was not felt until 31 May 1950. At that time, an agreement was signed between Alberta and Canada outlining their responsibilities with respect to the project's expansion beyond these two structures. This agreement finally permitted the full implementation of the plan originally prepared by the Irrigation Branch of the former Department of the Interior during World War One and revised by the PFRA, the Engineering Institute of Canada and the St. Mary and Milk Rivers Development Committee between 1935 and 1942. The expansion program was divided into headworks and distribution system components funded by Canada and Alberta respectively while overall project management was provided by the PFRA.

After the St. Mary Dam went into operation in 1951, the canal system south of Spring Coulee, along with that portion of canal between the Magrath diversion structure and Nine Mile Coulee, were abandoned. The RID was the district most altered by the construction program, since the new main canal cut through the district from the south. The Lethbridge-Coaldale area was the least altered because the old main canal was left intact to serve the water users in this area. The new canal was added to the network of canals and laterals already built in the area.

Speed was of the essence in the expansion of the St. Mary Project since Canada had already delayed the construction of dams to make use of its full share of water from the St. Mary and Milk Rivers. The American dam development program on the lower Milk River had the potential to make use of Canada's unused portion of water from these

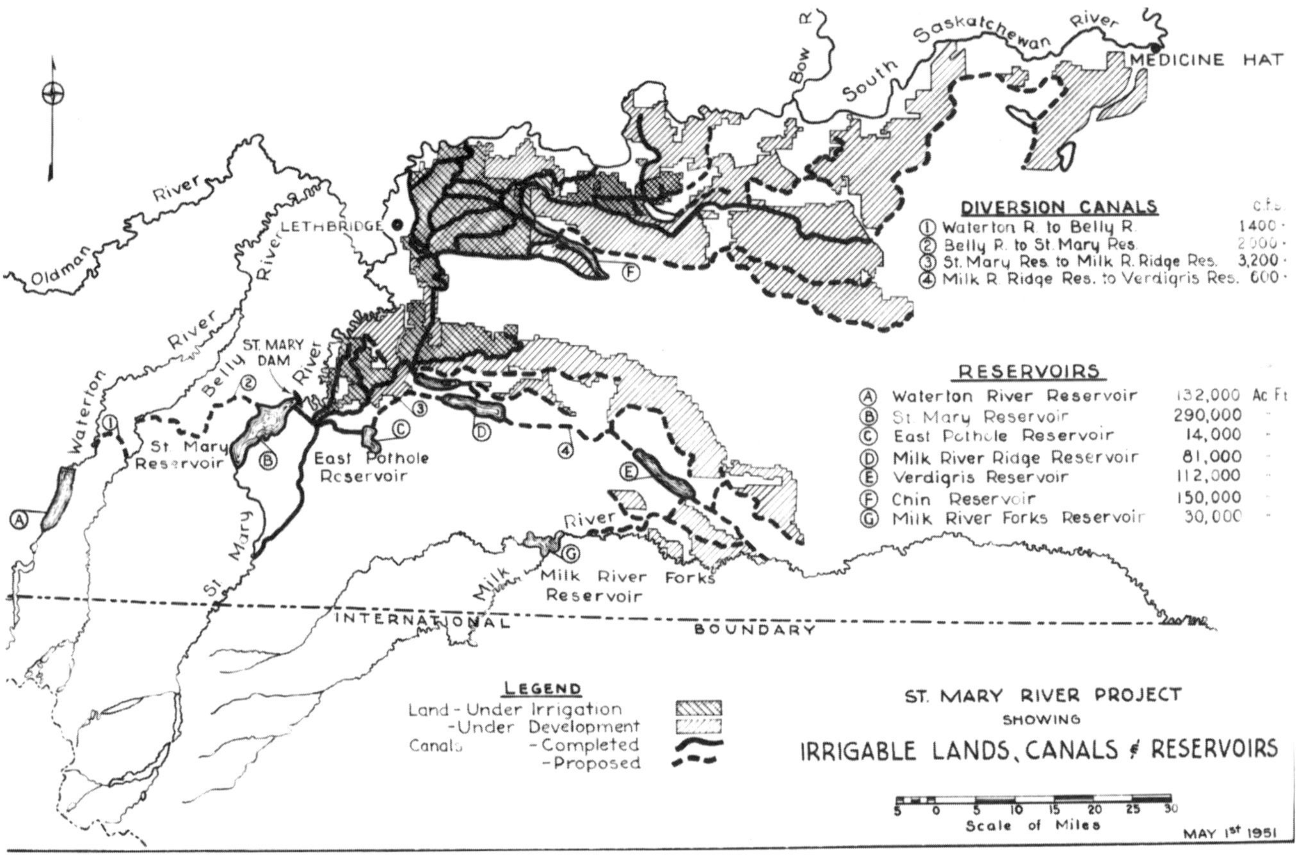

Map of proposed St. Mary Project showing irrigable lands, canals & reservoirs as of May 1, 1951 (PFRA-7216)

two rivers. The pace of construction was also driven by the need to take advantage of the favourable political climate for public investment in irrigation development. The drive of the people supervising its construction such as Wally Foss, Ben Russel and P.M. Sauder was a third reason for the haste with which it was built. All of these individuals had been involved for a number of years with planning the extension and thus were anxious to proceed without any delays.

P.M. Sauder, who was appointed Manager of the St. Mary and Milk Rivers Development in 1946, was the dominant figure in its operation up to his retirement in 1960. As Manager, he introduced a number of progressive ideas such as the delivery of water to the land and the termination of the old water contracts. He saw government ownership as a stage in the transfer of the system to the administration of the farmers. His creation of advisory committees was part of his attempt to maximize farmer involvement and thus prepare them for their ownership role. Sauder, however, lacked the financial resources to achieve many of his objectives which would have to wait until after 1968 and the increased funding provided by the provincial government. Budget restraints prevented the rehabilitation of the Lethbridge-Coaldale area and the delivery of water to the land throughout the entire system.

The physical transformation prior to 1968 was complimented by changes in its administration. With the departure of the CPR, the St. Mary Project was jointly administered by the PFRA, a provincial crown corporation, the three organized irrigation districts and the water users. The PFRA operated the headworks as far east as Ridge Reservoir. The crown corporation administered the main canal through to Medicine Hat, the main laterals in the Lethbridge-Coaldale area and the distribution systems added to the project between Taber and Medicine Hat. The three organized irrigation districts continued to operate their own systems but were given an opportunity for the first time to participate in the administration of the entire project through their membership on the Joint Advisory Committee. Lethbridge-Coaldale area water users, like those in the RID, retained responsibility for the operation of the sub-lateral systems.

Along with the new dams and reservoirs, which provided a stable water supply for the first time to all the water users in the project, other changes were made which disrupted well established procedures and agreements dating back to 1902. Water, under the original contracts, was provided at a constant price of $1 per acre and at a rate requested by the water users rather than being determined by the amount of land classified as irrigable through an independent survey. Sauder wanted to introduce a flexible approach to the determination of the water rates based on the amount of assessed irrigable land and the costs of operations rather than on a fixed charge for the amount of water delivered. This change affected both the individual water users in the Lethbridge-Coaldale area and the three organized irrigation districts. Financing the St. Mary Project in this way was second only to the construction of the St. Mary Dam in terms of changing the way it operated. The water users in the Lethbridge-Coaldale area were affected most by this change and mounted an effort to prevent it from being made. The change in the rates generated a storm of protest beginning in 1955 which overshadowed the engineering accomplishments of this period. The change in water rates was not anticipated by the water users who felt that the agreement which transferred the St. Mary Project from the CPR to the Alberta government guaranteed no changes in the way it was operated. The result was an extended period of negotiation to resolve the issue of the rates to be charged for water supplied to the districts and the individual water users in the Lethbridge-Coaldale area.

The underlying issue in these negotiations was the legal right and necessity of the SMRD to introduce a more flexible approach to determining the water rate which would reflect the actual cost of operation. P.M. Sauder knew that the old water contracts had to be eliminated if the St. Mary Project was to be both self sufficient and able to undertake badly needed improvements. Sauder, better than anyone else, also knew that the provincial government had no intention of subsidizing the operations of the St. Mary Project as had been the case with the CPR. This new approach to determining rates was successfully introduced in 1960 with the passage of the St. Mary and Milk Rivers Water Agreements (Termination) Act.

The concerns of the farmers in the new Eastern Division were different from those in the west since they were not subject to the terms of the old water delivery contracts. Their concerns were the way land was classified and the sign-up procedures carried out for their inclusion into the St. Mary Project. The SMRD was accused of forcing people,

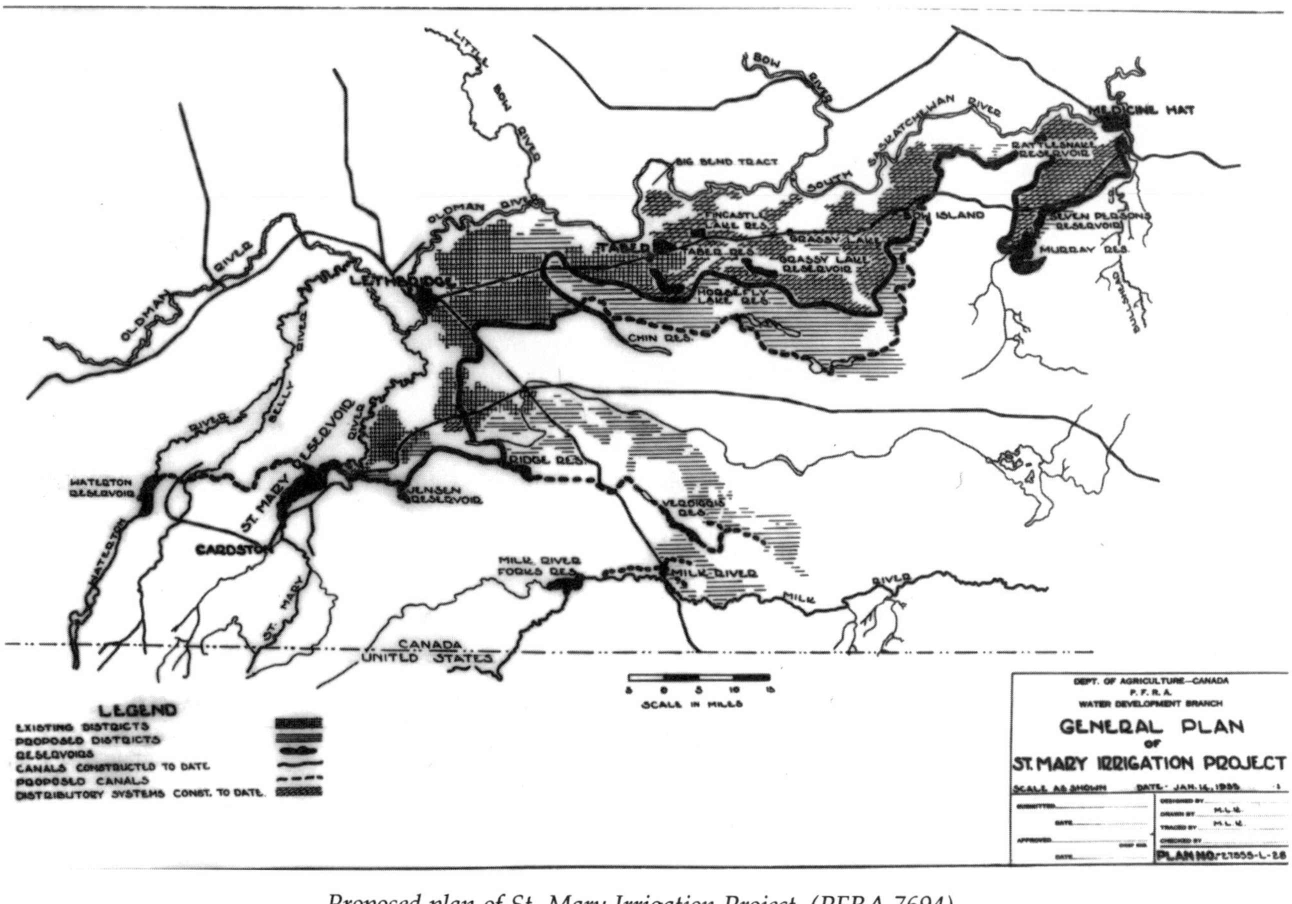

Proposed plan of St. Mary Irrigation Project. (PFRA-7694)

through various methods, to sign water supply contracts against their will. The concerns of the farmers in the new areas being served by the St. Mary Project were echoed by the farmers in the Enchant area west of Vauxhall, which was also being built by the province and the PFRA. The provincial government's response to the criticism was the 1958 Irrigation Study Committee chaired by W.R. Hanson. The creation of this committee initiated a period of review of the irrigation industry in Alberta which was completed at the initiative of Harry Strom in the 1960s under the terms of the Agriculture Rehabilitation and Development Act (ARDA).

Waterton Dam under construction. (PFRA-22383)

These reviews laid the basis for the post-1968 era since many of the recommendations for change were incorporated into the 1968 Irrigation Act and the multi-million dollar rehabilitation program funded by Alberta, Canada and the water users themselves.

During the post-1968 era the emphasis shifted to rehabilitation and rationalization of the existing system and expansion based on the more efficient use of the existing water supply. The rehabilitation of the main canal from Ridge to Medicine Hat was the most comprehensive and well planned period of construction in the history of the project. Unlike the 1950s, it was not driven by the need to beat the Americans or by outdated methods and plans. This program gave the system the capacity, for the first time, to reach the objectives for the project first stated in the 1890s.

P. M. SAUDER AND THE LETHBRIDGE-COALDALE AREA

The St. Mary and Milk Rivers Development Crown Corporation (SMRD) took formal possession of the St. Mary Project on 1 April 1946. Its first manager, P.M. Sauder, began his long association with western Canada's water resources in 1904. In that year, he joined the Department of the Interior, following his graduation from the University of Toronto with a degree in engineering. Employed first in making topographical surveys, he moved on in 1906 to become an inspection engineer with the Department. In 1908, he began the first systematic watershed surveys on the eastern slopes of the Canadian Rockies. This survey program was continued by the Hydrographic Survey Branch of the Department of the Interior which Sauder established the following year. He was the Director of this branch from 1909 to 1920. Prior to his departure from the Department of the Interior in 1920, he served for a brief period as Commissioner of Irrigation. In 1920, he joined the Lethbridge Northern Irrigation District eventually becoming Manager of that project in 1923. From 1940 to 1944, he was Director of Water Resources for the Province of Alberta during which time he played a leading role in the creation of the Western Irrigation District and the decision of the provincial government to take over the St. Mary Project from the CPR. He became the first Manager of the Western Irrigation district in 1944. In 1946, while still serving in this position, he was appointed Manager of the SMRD. He held both positions until 1950, when the increased pace of development on the St. Mary Project, required his full time attention to this task. George Brown, who had been Superintendent of the St. Mary Project prior to 1946, was Assistant Director. Few other changes occurred in the organization and operation of the St. Mary Project at the time of the transfer. The Northern and Southern divisions were retained while the SMRD took possession of the former office building used by the CPR. The water users expected that there would be no changes in the water agreements.

Sauder's administration of the SMRD was based on his view that it was a stage in the eventual transfer of the St. Mary Project to the water users as had been the case with the other CPR projects. He was, therefore, anxious to ensure that the SMRD was run as much as possible like an irrigation district and not as a branch of the provincial bureaucracy. In his contacts with the water users, he asked

Drilling to test depth and composition of soil at Belly River Diversion (PFRA-12959)

for their cooperation to ensure that there would not be any need for control from Edmonton.

MEETING THE WATER USERS

The first meeting between the water users and the new administration was held on 26 May 1947 at the community hall in Coaldale. It was attended by over 100 water users along with SMRD senior staff. This was the first time in the history of the St. Mary Project that the water users were consulted about the operation of the irrigation system. Drainage in general and Eight Mile Lake in particular, the status of the old water contracts and soil drifting were the principal concern of the water users. The SMRD staff took the opportunity to point out that it had limited resources and that many things would stay the same despite the change in ownership. As George Brown explained, the problem in dealing with drainage was the method of financing the work since the SMRD could not legally spend money for the construction of drainage works on its own. The SMRD could only assist the water users if they wanted to provide the funds for the SMRD to undertake the work on their behalf. The water users were advised that the Eight Mile Lake problem could only be solved if the farmers themselves were prepared to create a drainage district which could assess the members for the cost of the work required which was estimated at $15,000. The SMRD, like the CPR, was limited by the present water rate of $1.00 per acre which provided an income sufficient only to deliver water but not rehabilitate the system.

While Sauder and Brown made it clear that the SMRD did not have the resources or the mandate to assume responsibility for the rehabilitation of the Lethbridge-Coaldale area, they suggested a number of initiatives to improve the operations of the project. The most important of which was the creation of an advisory committee, as provided for in the Act, in order to improve the service and public relations between water users and the officials of the district. The Lethbridge Northern Irrigation advisory committee was the model Sauder wanted to introduce to the St. Mary Project. Sauder also indicated that he wanted to introduce a system of water delivery to the land and not the lateral. The was a major change in policy since the CPR had organized their entire administration of the St. Mary Project based on the use of lateral associations.

LIMITED SUCCESS

After the meeting, the farmers and the SMRD cooperated in the construction of drains in the manner suggested at the meeting. An Eight Mile Lake Drainage Association was created and construction carried out by the SMRD using money provided by the members. The drain in the south Coaldale area was also built in cooperation with the water users. The SMRD provided equipment to the lateral associations to clean out existing drains.

Sauder moved quickly to establish an advisory committee. The SMRD prepared the necessary forms to conduct the election and Sauder served as the returning officer Francis W. Richards, Harry W. Rowley, William Fairfield and Charles R. Daniel were the first members of the Western Advisory Committee. Fairfield, while retired as the Director of the Dominion Experimental Farm, still owned the farm in the Coaldale area he had purchased in 1906. The SMRD covered the expenses of the committee at the rate of $5 per day for meals and 10 cents per mile for travel mileage from their home to Lethbridge. The Western Advisory Committee met two or three times a year rather than monthly as originally intended.

Given Sauder's interest in wanting to improve the operation of the St. Mary Project, he was far more diligent than the CPR in monitoring the water supply contracts between the SMRD and the irrigation districts and the SMRD and water users in the Lethbridge-Coaldale area. Starting in 1949, the TID was charged an additional 50 cents for every acre-foot of water used above their contract amount of 34,000 acre-feet. This decision was prompted by the discovery that the TID had diverted 19,000 acre-feet more than allowed under

Dragline working at Belly River Dam. (PFRA-11163)

its contract. Sauder also monitored the on farm use of water which the CPR had also neglected. Sauder wanted to make water users pay for water used to irrigate land in excess of the irrigable area as shown on SMRD records and as agreed upon when the purchase of the land was made. This situation was the result of partial water rights which gave the land owner the opportunity to decide how much land to irrigate. The CPR had tolerated partial water rights and had not taken much interest in how the farmers used the water they received. In 1949 the SMRD collected an additional $1.00 per acre from all persons irrigating land in excess of their water right. It was suggested by the Advisory Committee in 1952 that land classification surveys be made with the view of being able to have a fair and equitable annual assessment for water delivery to better regulate on farm use of water.

The next change came in 1951 when the cost of new water rights was increased from $1 to $10. This was the first increase in the charge for a water right in the St. Mary Project since it had been established. This increase was made when 5,000 acres was added to the area served by the existing system through the conversion of partial water rights to full water rights. An additional 8,000 acres of new land was also made available in the Cameron tract through the extension of the Cameron Lateral. The Lethbridge-Coaldale water users were thus being charged the same water right as the new water users in the Eastern Division.

William Fairfield tried to stop this change because he felt that the people in the Lethbridge area, who had a partial water rights, should not be charged this amount to acquire a full water right. Fairfield, however, failed to convince Sauder and the Advisory Committee. In March 1953 the majority of the Western Advisory Committee approved the policy that all water rights granted in the Lethbridge-Coaldale area for additional water should be at the rate of $10.00 per irrigable acre irrespective of the type or wording or phrasing of the original water agreement.

Sauder's plan to deliver water to the land and not the lateral was less successful. The Advisory Board decided that the farmers wishing to acquire a new water right for additional irrigation acreage or for parcels or areas not then under contract for water should make his own arrangements for service through the farm ditch from which they were to be served.

Sauder also tried to implement this approach to water delivery by working with the Forbes Lateral Association. In November 1947 the Advisory Committee met with Messrs Wilmot, Roberts and Regier and proposed an agreement to the Forbes Lateral Association to give service to individuals on the Forbes lateral for a three year term at an additional rate of 40 cents per acre. The proposal was accepted and Sauder requested the SMRD's legal advisor to draft an agreement incorporating this approach. Sauder was subsequently advised by Virtue and Russell that they considered it extremely doubtful if the Manager of the Forbes Lateral Association had the right to enter into a contract of the nature suggested. In December 1947, an alternative form of agreement was decided upon which preserved the existence of the Forbes Lateral Association while relieving it of much of its responsibility. The new agreement provided that the officers of the Forbes Lateral Association collect 40 cents per acre from each member of the Association and turn it over to the SMRD. In return the SMRD would clean the first part of the Forbes Lateral ditch from the headgate to the drop in the SW - 12 - 9 - 21-4 and in addition would repair five structures at points where divisions of water were made. A ditch rider was to be engaged for a three month period starting on May 15. He would supervise the division and delivery of water on the main Forbes Lateral canal. The Forbes Lateral Association approved the agreement which was subsequently renewed each year up to 1960. In March 1949 the SMRD entered into a similar agreement with the Whoop-Up Lateral Association. Given these legal problems, Sauder made no further effort to implement his policy of delivering water to the land.

IMMEDIATE POST-WAR CONSTRUCTION

The construction program undertaken by the provincial and federal governments from 1945 to 1949 was based on a specific agreement to build the Jensen Dam as well as their mutual acceptance of the recommendations of the 1942 Report of the St. Mary and Milk River Development Committee. The cooperation of the federal and provincial governments was further guided by the passage in 1947 of an order-in-council which outlined in general what Canada was prepared to contribute to a joint irrigation venture with one of the prairie provinces. Under the terms of this document, Canada agreed to proceed with construction, if the province concerned agreed to transfer the water rights required for the operation of such a project. The province also had to make available to Canada the necessary provincial crown lands required for the construction of dams, reservoirs and canals. With these conditions met, Canada, at its own cost, would proceed with the construction, maintenance and operation of the main reservoirs and any connecting canals in the project. Having completed construction of the headworks, Canada agreed to deliver to the province all the water it could utilize in a distribution system which the province had been developing at the same time. The fee for the delivery of water using the new headworks was to be determined by negotiations. The federal government was free to sell any water not used by the province to other parties. The point which required clarification was the meaning of the term "main reservoirs and any connecting canals." Alberta passed the Irrigation Projects Act which permitted it to enter into an agreement with Canada providing for the developement of irrigation projects and the construction and maintenance of irrigation works within Alberta and provided for the payment by Alberta of a portion of the costs incurred under any such agreement.

Aerial view of Belly River Diversion and canal to the St Mary Reservoir (PFRA-18057)

Despite the lack of an agreement between the senior levels of government specifically dealing with the entire St. Mary Project as recommended by the St. Mary and Milk River Development Committee Report, considerable progress was made immediately after World War Two. This progress was possible because of the commitment to expansion by the PFRA, especially its senior official in Alberta, Wally Foss; Ben Russel, who was director of water resources in Alberta; and P.M. Sauder, the manager of the SMRD. These officials, like Commissioner of Irrigation, Frederick Peters in 1919, were well ahead of the politicians in Ottawa and Edmonton in terms of their willingness to pursue development. The fact that preparations had been made and that these people were anxious to proceed is very clear in a 16 September 1946 memorandum by Ben Russell in which he pointed out that:

> lands in the Project with the exception of certain areas in Township 9,10,11 and 12, Ranges 6 to 11 inclusive, or those lands to the east of Forty Mile Coulee depression were all plane tabled surveyed and classified years ago, either by the Dominion Reclamation Service or the Canadian Pacific Railway Company. Plans were prepared at that time showing the irrigable parcels in each quarter section. Those lands not originally surveyed as part of the project have since either been or are now being plane tabled and classified. The maps referred to are all in this office and are available for any purpose desired. Furthermore these plans are the basis of all estimates since prepared by the PFRA and are also the basis of the report prepared by the St. Mary and Milk River Development Committee. The PFRA in recent years has actually staked out a number of large canals originally surveyed by the Reclamation Service and I understand are continuing this work.[269]

Given the availability of all of this information, Russell concluded that what was required "now more than anything else [was] some understanding between Governments and Departments of Government with respect to financing further construction. This office, in cooperation with the PFRA will be able to look after the rest."[270]

Construction of the Pothole Dam began in 1945 when a contract was let to the Western Construction and Lumber Company of Edmonton. This company built a rolled earth filled dam with a crest length of 850 feet and a maximum height of 130 feet. A reservoir and balancing pool was created with a useable storage of 14,000 acre-feet and connected to the AR&ICo. canal at Spring Coulee by a 800 cfs canal. Construction was financed by the province of Alberta and supervised by the PFRA. The canal and the dam was completed during the 1948-49 construction season. The water which filled the reservoir was used during the 1949 irrigation season. It was appropriately renamed the Jensen dam in honour of Christian Jensen, the chairman of the MID board of directors from 1929 to 1946 and a strong advocate and campaigner for storage reservoirs for the St. Mary Project.

The St. Mary Dam was constructed at a site 40 miles southwest of Lethbridge and 4 miles northwest of Spring Coulee. A construction camp was established at a cost of $38,000. It required the building of three miles of all weather road from highway No.5, an 18 mile power line from Magrath, a pipeline, cistern and pump to provide water from the St. Mary River and a 16 mile

Members of Blood Tribe passing over bridge on way to site of official opening ceremonies St. Mary Dam, July 16, 1951.

R.C.M.P. Band leading procession to site of official opening ceremonies, St. Mary Dam, July 15, 1951

Ceremonial dance, official opening of St. Mary Dam July 16, 1951

Crowd at opening of St. Mary Dam on July 16, 1951

telephone line to Magrath. Two military buildings at the Lethbridge Airport were purchased from the War Assets Corporation and moved to the camp. One building was renovated for use as an office, soils laboratory and sleeping quarters and the other building was used for a kitchen, dining room and additional accommodation.

The St. Mary Dm is a zoned rolled earth-fill with a maximum height of 186 feet and crest length of 2,600 feet. The length at the base across the stream is 300 feet and required 4,000,000 cubic yards of fill. It was the largest earth fill dam built in Canada to that date. The construction was divided into three separate contracts. The first contract covered the construction of the diversion and irrigation tunnels and was awarded to Bennett and White Limited of Calgary in 1946. The second contract was awarded to W.C. Wells Construction Company, Limited for constructing the earth-fill dam in the fall of 1946. A third contract was awarded in 1947 for the construction of the spillway and control works. The total estimated cost of the dam in 1949 was over $6,000,000.00.

The diversion tunnel is 20 feet in diameter, 2,115 feet long and lined with reinforced concrete. Its function was to take care of the flow of the St. Mary River during construction of the dam and was completed in June, 1948. Assiniboia Engineering Company of Calgary constructed the irrigation tunnel. This tunnel is 2,500 feet in length and located 90 feet above the stream bed of the St. Mary River. It is used to supply water from the reservoir to the main supply canal. Eight control gates were installed in this tunnel to give flexibility of control over the quantity of water to be diverted from the reservoir into the distribution system. The original spillway was constructed at the south end of the dam. It was cut into the sandstone and lined with concrete. The flow of water over the 60,000 cfs capacity spillway was controlled by 12 stop-log openings, 20 feet wide by 18 feet high. The reservoir created by the dam was 17 miles long, 6 miles wide and 176 feet deep and had a storage capacity of 320,000 acre-feet.

Five separate contracts were let by the PFRA for the construction of the main canal between the St. Mary Dam and Ridge Reservoir. This work also included the building of a wood stave syphon through Pinepound Coulee.

The PFRA carried out topographical surveys on the proposed location for the dam on the Waterton river. The location of the canal from Pothole Reservoir to Horsefly Lake Reservoir was determined before 1950 along with classification of the land east of Forty Mile Coulee which had begun in the early 1920s when the LSEP was first proposed.

The expansion of the distribution system was also undertaken prior to 1950 largely through the initiative of Wally Foss,. This initiative was taken in 1948 because water was available to irrigate an additional 3,000 to 4,000 acres courtesy of the new Pothole dam which was to go into operation in 1949. The Big Bend area was a prime destination for this water because of its potential for immediate use to grow sugar beets for the new processing plant which was to be built by Canadian Sugar Factories Ltd. in Taber. In July 1948 Wally Foss met with Ted Sundal Secretary-Treasurer of the TID, N.M. Sutherland of the office of the Provincial Director of Water Resources, W.B. Thomson, irrigation engineer with the PFRA to review the plans.

The plan for the Big Bend area called for the construction during 1949 and 1950 of a new main canal with a capacity of 1,700 cfs from Chin Reservoir to Horsefly Lake with a possible extension to Bow Island, the conversion of Horsefly Lake into a reservoir with a capacity of 10,000 acre-feet, the enlargement of Taber Lake from 3,500 acre-feet to 7,000 acre feet and the construction of a main lateral and distribution system in the Big Bend area for about 12,000 acres of irrigable land. Foss included Taber Lake in his plan because water wasted from the TID which went into this reservoir could be reused in the Big Bend area. The construction required to irrigate 3,000 acres of the Big Bend area with a view to its expansion included about 6 miles of main lateral with a capacity of 150 cfs and 5 or 6 miles of smaller ditches all for the cost of $40,000.The PFRA prepared the plans and supervised the work while the actual construction was handled by Alberta. The contract was awarded to Shannon Brothers of Didsbury in September 1949.

COMPREHENSIVE DEVELOPMENT AGREEMENT

The construction of the distribution system in the Big Bend area raised the issue of the need for the clarification of the responsibilities of the federal and provincial governments. This need was pointed out as early as September 1946 by Ben Russell, but it did not become an issue until water was available from the new dams. In September

1948, Mark Mann, a PFRA Administrative Officer, pointed out that the report of the St. Mary and Milk River Development Committee was:

> now six years old and in this day and age many changes can take place in that length of time, it would be as well probably to not take too much for granted. The thoughts expressed in 1942 may have undergone some change in the intervening years. It would seem well, therefore, to more or less make a fresh start toward the whole setup.

Alberta entered these negotiations with the expectation that the federal government would construct the system at their own expense as far east as Horsefly reservoir. The province was prepared to accept a compromise plan in which the federal government would build the system, at its own cost, as far east as Chin Coulee. The federal government, on the other hand, entered these negotiations with the view that it was only committed to the construction of the Waterton and St. Mary dams and the necessary connecting canals between these dams. These differences began to affect the project in September when Jimmy Gardiner, the federal Minister of Agriculture, advised his deputy on 28 September 1949 to suspend all work on the Project pending the outcome. The Alberta government also considered the possibility of cancelling the project when it became aware of these differences. Following a meeting on 3 January 1950, Premier Manning advised Jimmy Gardener that:

> in the light of this development I had no alternative but to advise that if this was to be the full extent of the Dominion Governments's participation it could only mean that the further development of this entire project would be

Aerial view of St. Mary damsite. (PFRA-7001)

> stopped in that the provincial government could not undertake to complete a project of this magnitude without a substantially greater measure of assistance from the Dominion Government.[271]

Over the next five months a compromise was agreed upon which established Ridge Reservoir as the point between the federal and provincial portions of the St. Mary Project.

A second issue that had to be resolved during these negotiations was the extent of canal and reservoir development east of Ridge Reservoir using the Milk River. On this point, both the federal and provincial governments agreed very quickly that there would be no development making use of the Milk River and that the canal going east of Ridge Reservoir to Verdigris Coulee should be eliminated from the St. Mary Project.

The decision, at this time, not to include the Milk River and the region east of the Milk River Ridge was the most significant change in the plan for the St. Mary Project which had been developed over the last fifty years beginning with George Anderson and carried on by the 1942 Report of the St. Mary and Milk River Development Committee. As late as December 1949, Wally Foss had been planning for expansion based on the use of Verdigris Reservoir.

The final agreement was signed on 31 May 1950 and incorporated the provisions of the 1947 Order-in-Council while being very specific about what was meant by the phrase "main reservoirs and connecting canals." It also added provisions for a land settlement program. Alberta made commitments to the federal government to permit it to build a new headworks system and then divert sufficient water to permit the irrigation of all of the potentially irrigable land in the St. Mary Project which was estimated at 500,000 acres. The federal government was thus given Crown land and a license to divert water from the St. Mary, Belly and Waterton rivers plus all of its tributaries. Additional land required for construction purposes was to be acquired by Canada at its own expense. Alberta assumed responsibility for the development of the rest of the expansion program which included all of the dams, reservoirs, canals, distributaries, and other works required for the conveyance and distribution of water to all irrigable lands located below Ridge Reservoir. This commitment included four reservoirs one of which was Chin which was to be increased to a capacity of 150,000 acre feet. Horsefly, Grassy Lake and Whitla (renamed Sauder) all of which were to have a capacity of 6,000 acre feet were the new reservoirs included in the system. The new canal system was to include the "highline" canal east of Chin Coulee Reservoir which was to be used to supply a distribution system in the Foremost area. Water for this system was to be provided directly from Ridge Reservoir based on the plan first suggested by the St. Mary and Milk Rivers Development Committee. Alberta was also responsible for the land settlement program which was to be carried out upon the completion of this construction program. In this program first preference was to be given to veterans of World War Two and to the re-establishment of farmers from sub-marginal lands.

The commitments of Canada under the terms of the agreement were to build the headworks system at its own expense and serve as the overall project manager by completing all surveys, engineering design and cost estimates required for the construction of the entire project and to furnish all engineering, legal services and supervision required in the construction of the project and to file all such plans with the appropriate land titles office. Once the headworks system had been built, Canada agreed to maintain and operate this system and deliver water to Alberta at cost. The charge for this water was not to exceed 25 cents per acre-foot per year and was to be reviewed every five years.

The PFRA did not require any time to begin the implementation of the agreement, since Wally Foss was well ahead of the politicians in his planning of the project. The province was also ready to go, since it had enacted a new St. Mary and Milk Rivers Development Act on 5 April1950. This new legislation provided the machinery for the management and operation of the expanded St. Mary and Milk Rivers Development. The new Act changed the SMRD into an agency equipped to aggressively pursue the expansion of the St. Mary Project. The most significant change was the establishment of a Colonization Branch within the Department of Agriculture and the appointment of a Colonization Manager and Assistant Colonization Manager. P.M. Sauder, who already held the position of Manager, was appointed to the former position on 1 April 1950.

The new Act also introduced a method of financing the operations of the St.Mary Project which was based on the procedures then in use in those irrigation projects subject to the Irrigation Districts Act. The Manager of the St. Mary Project was now authorized to collect water right payments and water rates. The Manager was required to make an estimate of the amount required to pay the water service charge which was to include all the expenses connected with the administration of the district and the diversion, distribution, and drainage of water, and the maintenance of the irrigation works for the year. The Crown Corporation was thus given the right to levy a service charge to cover the actual cost of maintenance and operation plus a contribution to the maintenance fund. The maintenance fund was to provide funds for the rebuilding of works destroyed or worn out. No provision was made for the Legislative Assembly to vote funds for its operation since it was expected to be self sufficient. Provision was also made for the recovery of rates through the seizure of property owned by the water user. This new Act was thus another important step in the transformation of the St. Mary Project from a government run project into a water user run irrigation district operated on the basis of the Irrigation Districts Act.

Despite the fact that the development of the St. Mary Project did not include the Milk River, the Crown Corporation name remained the same. The Milk River for the first time in fifty years had been separated from the St. Mary Project during the negotiations with the federal government.

ON TO MEDICINE HAT

The signing of the agreement in May of 1950 was followed by an unprecedented period of construction in the history of the St. Mary Project which continued through to 1955. Main canal construction east of the St. Mary Dam included the enlargement of the existing canal between the St. Mary and the Pothole Dams and the construction of a entirely new canal 200 miles east to Medicine Hat. The task of enlarging the canal from 800 cfs to 3,300 cfs was carried out in 1952 while the construction of the new canal was begun in 1951 when 50 miles were constructed between Ridge and Chin Coulee. This canal did not replace but rather paralleled the existing canal and Chin One lateral for much of this distance. A portion of the old main canal and Chin One lateral built in 1904 were left to serve the Lethbridge-Coaldale area. In 1952 three further contracts for work on the main canal were completed on the section between the Pothole Coulee Dam and Ridge Reservoir while sixty-five miles of main canal were constructed between Chin Reservoir and Forty Mile Coulee. The water was taken across Forty Mile Coulee by a 2,640 foot wood stave syphon completed in 1952.The section of the main canal east of Forty Mile which had to go around a large hill and approach Seven Persons Creek from the north was completed through to Medicine Hat in 1954.

The reservoir development program began with the enlargement of Chin Reservoir and the commencement of the construction of Ridge Reservoir in 1952. The construction of the four new reservoirs to the east of Chin was started in 1951. The Sauder (originally called Whitlaw) and Murray Reservoirs were completed and in operation by 1953. . The Grassy Lake Reservoir was finished in 1953 along with its main connecting laterals serving tracts around Burdett and Grassy Lake. This lateral was syphoned across Yellow Lake Coulee south of Grassy Lake and distribution ditches were also constructed between Burdett and Grassy Lake. A distribution system for 23,000 acres located between Grassy Lake Reservoir and Forty Mile Syphon was built in 1953 along with a system to serve a further 16,000 acres in the Bow Island tract. A total of 56,000 new acres were added to the St. Mary Project by the 1953 construction program thus making a total of 96,000 acres of new land ready to receive water in the spring of 1954. In 1954 and 1955 twelve contracts were awarded for the completion of the Bow Island and Medicine Hat distribution systems. These contracts brought the total of new acres added to the project to 187,000. The provincial involvement in this construction program was under the control of the Manager of SMRD until 1954. With the creation of the Water Resources Branch of the Department of Agriculture, these duties were taken over by Ron Francis of that branch on 1 July 1955.

The St. Mary Dam was officially opened on 16 July 1951. After the official opening the concrete spillway was completed and the site cleaned up. The new system began partial operation in 1952 when 186,000 acre feet of water was delivered to approximately 130,000 acres of land from the newly constructed St. Mary Reservoir. The greater part of the water was applied to 120,000 acres in the older irrigation districts of Magrath, Raymond, Lethbridge-Coaldale and Taber. The remaining 10,000 acres irrigated was located north of Taber in

Aerial photo of St. Marys Dam (PFRA-7215)

the Big Bend area. During 1954 water was carried entirely by the new canal from the St. Mary Reservoir to the Medicine Hat for the first time. This canal constructed during the last five years was 220 miles in length and had a capacity which varied from 3,300 cfs at the upper end to 6 cfs at the lower end.

CONSTRUCTION DELAYS

One of the first problems to affect the pace and the ultimate extent of development was the difficulty in acquiring a right-of-way through the Blood Reserve for a canal to connect the proposed Waterton Dam with the St. Mary Reservoir. This situation temporarily caused the PFRA to look for an alternative to the dam on the Waterton river and affected the construction program in the east. In their search for an alternative to the dam on the Waterton river, the PFRA re-examined the idea of building a dam on the Oldman River immediately up stream from Lethbridge. A dam at this location had been suggested by Ted Sundal in 1941 but the idea was rejected at that time because of the excessive cost of having to build a dam with a height of 300 feet, a top length of 4,200 feet a bottom length of 1,500 feet and requiring 32,000,000 cubic yards of fill. The $6,500,000 in additional costs to build this dam instead of a dam on the Waterton river could not be justified in 1941.

The search for an alternative site began in April 1955 when the PFRA's Chief engineer G.L. MacKenzie asked Wally Foss to provide him with information on the proposal to construct a high dam between Lethbridge and the mouth of the St.Mary River for the purpose of diverting water by gravity to the old AR&ICo. canal. The combined flows of the Oldman and Waterton River also raised the possibility of its use for power generation. When this investigation confirmed once again that this site was unsatisfactory, negotiations were resumed with the Blood Tribe. When an agreement was reached, construction resumed on the headworks in 1956. They were completed in 1968. The eastern extension of the headworks includes the Waterton Dam, the diversion dam on the Belly River and the canals connecting both structures to the St. May Reservoir.

The delay in the completion of the headworks, in turn, had an affect on the pace of the construction program in the east because it temporarily reduced the water supply. Because there was now less urgency to complete distribution systems, the decision was made in 1957 to defer the construction of the high line canal and its associated distribution system.

IRRIGATION AND SOUTH EASTERN ALBERTA

The main canal was completed to Medicine Hat 42 years after the first attempt to bring large scale irrigation to the area had begun. In 1906 the Southern Alberta Land Company acquired two blocks of land located north of the South Saskatchewan River in the vicinity of Bow Island which contained approximately 144,323 acres and south of the South Saskatchewan River in the area south west of Medicine Hat which contained 236,250 acres. The Southern Alberta Land Company proposed to irrigate the land south of the South Saskatchewan River by means of a pumping station on the Oldman River located north east of Barnwell. A six mile pipeline was to carry the water to Horsefly Lake from which a canal was to carry the water further east to Forty Mile Coulee. The Southern Alberta Land Company also intended to use the natural run off collected by Forty Mile Coulee and water from Seven Persons Creek to irrigate this tract. The Company's application encountered immediate objections from the Department of the Interior which pointed out that the low water flow of

Aerial view of St. Mary Dam & Reservoir (PFRA-10729)

Seven Persons Creek had already been granted to prior applicants and that Forty Mile Coulee, was only a chain of alkali lakes with no flow except in spring time when there has been snow enough to provide run-off. In 1908 the Southern Alberta Land Company abandoned this first plans in favour of an alternative plan which abandoned any further attempts to irrigate land south of the South Saskatchewan River in favour of irrigating land in the Suffield area and west of the Bow River. In 1921 the successor to the Southern Alberta Land Company abandoned the attempt to irrigate the land north of the South Saskatchewan and east of the Bow River.

The second proposal for a large scale irrigation project in the Medicine Hat area before 1954 was prepared in 1920. The plan for what was referred to as the Medicine Hat Southern Irrigation Project was virtually identical to the one developed by the SMRD in the valley of Seven Persons Creek in the mid-1950s.

One of the projects which was actually built before the arrival of St. Mary main canal was the Moore Project. It was established in the 1940s on the northern slope of the Cyprus Hills.

LAND DEVELOPMENT AND COLONIZATION IN THE EASTERN DIVISION

While construction of the main canal and distribution systems in the Eastern Division were underway, other activities with respect to its development were taking place. These included incorporating the farmers into the project who were already resident in the area, the organization of a land settlement program, land classification surveys, land levelling, farm ditch surveys and educating the farmers on how to operate an irrigation farm. These activities were undertaken jointly by the SMRD as well as the Extension Services and Colonization Branches of the provincial Department of Agriculture.

The incorporation of the farmers east of Taber into the expanded St. Mary Project began with a canvass of the area by individuals hired by the Colonization Manager to determine their interest in irrigation. The farmers indicated their interest or lack thereof by signing or not signing water purchase contracts. The majority of the owners in an area had to sign such contracts before construction of the distribution system began. This procedure was a dramatic change from the pre-Second World War approach to the expansion of irrigation which placed the responsibility on the local residents to initiate irrigation development. The Irrigation Districts Act required that the farmers themselves prepare petitions which indicated community support. These petitions had to indicate that the owners of over fifty per cent of the land in the area supported the creation of an irrigation district. This was the procedure followed in the creation of the TID, RID and MID.

The contract contained various provisions one of which was a water right charge of $10 which was the farmers contribution to the cost of construction. This amount was based on the recognition that the farmer could not bear all the costs of an irrigation project and that increased production on irrigation projects not only increased business in the locality of the project but throughout Canada. To ease the transition from dry-land to irrigation farming, the government provided free water to the farmers for two years after they had signed the agreement.

Based on reports by Sauder, close to 100% of the farmers agreed to receive water. The only residents, according to Sauder, who offered resistence were a few of the elderly people over 65. When the land changed hands, however, he expected that the new owners would sign. The SMRD could also force land owners to pay the rates if the majority of the owners in an area had signed contracts. By 1955, 710 owners had signed contracts to irrigate 1,400 parcels of land. The size of these parcels ranged from 20 acres to 1,600 acres with average size being 1.96 quarter sections or 313.6 acres.

The organization of a land settlement program began with P.M. Sauder's appointment as Colonization Manager. Assembling land for distribution began in August 1950 when tax recovery land was transferred to the Colonization manager from the Municipal District of Eureka No.14 (now Taber No.14) and Bow Island No.13 (now Forty Mile No.8) Additional land for settlement purposes was obtained from the Minister of Lands and Forests.

The preparation of the regulations, terms and conditions for its sale were prepared in 1951. These regulations gave preference to veterans and secondly to farmers from sub-marginal lands such as the dried out areas of Alberta and Saskatchewan. Veterans could not, however, participate in the program if they already rented or owned other land. The successful applicants were to be granted a maximum area of one-quarter section at a nominal

Tractor and BR 2 Roller compacting embankment at St. Mary Dam (PFRA-14946)

purchase price. Settlers were required to present a program of their proposed farming operations for the next year and a statement of production for the current year to the Advisory Committee for approval. They were also required to establish a proper home on the land and other buildings required to house farm implements and livestock. Purchasers were required to pay all taxes, water rights and other charges assessed against the land. The $10.00 per acre for water rights was payable either by cash or by assigning a share of the crop. Interest on the unpaid balance was to be three and one half per cent per annum. The Advisory Committee could refuse an application if it considered the applicant not qualified for irrigation farming or because he did not have sufficient cash, capital, livestock or equipment. After ten years, the settler could apply for title to the land and if all the requirements had been satisfied, would receive title to the land.

The distribution of land based on these regulations began in 1951 when 8,000 acres was available for settlement in the Big Bend area. Sauder and the Advisory Committee examined and rated 15 veterans and five non-veterans at Lethbridge and Taber between January and April for their suitability as settlers. As Sauder explained in his *Annual Report* "each man, and if married his wife also, was carefully examined and rated for age, health, agricultural experience, education, capital resources, personality and character, service record, wife's personality, wife's aptitude and children."[272] By April 14, 14 quarter sections were allotted to veterans and one to a non- veteran. Several others were found qualified but land with water was not available for them at that time. Eight sections of land was also allotted in 1951 to farmers as compensation for land lost because of the construction of the Horsefly Reservoir. The review and allotment procedure was followed through to 1955 by which time 117 new settlers had been established. The constant problem was meeting the demand for land which was greater than the supply.

ADMINISTRATION OF THE EXPANDED ST. MARY PROJECT

The SMRD took over the operation of the main canal through to Medicine Hat in 1956. Given the physical separation of the land added to the project from the Lethbridge-Coaldale area and the administration of the headworks by the PFRA, it was necessary to alter the administration of the St. Mary Project for the first time since 1926. Between 1954 and 1957 the Southern and Northern Divisions were replaced with a Western, Eastern and Main Canal Divisions. The Western Division consisted of the main canal and laterals in the Lethbridge-Coaldale area along with Chin Coulee Reservoir. The sub-laterals continued to be owned and operated by the farmers thus a smaller staff was required for its operation. In 1961 the Whoop-Up Lateral Association voted unanimously to join the Lethbridge-Coaldale area. The future of the Forbes Lateral was resolved in 1965 when it was incorporated into the SMRD. The Readymade Water Users Association , was far more reluctant to give up its independence and made an attempt to establish its self as a separate irrigation district. This request was made to Minister of Agriculture, L.C. Halmrast on 10 February 1960. In the letter from Secretary-Treasurer Thomas Ober, it was pointed out that as an irrigation district such as Taber and Magrath the Readymade area could be run more efficiently. Halmrast referred the letter to Sauder who took the view that dividing the Lethbridge-Coaldale into small districts was a step in the wrong direction.

The creation of a separate administration for the Eastern Division with its headquarters at Bow Island, began with the appointment of an Advisory Committee by 1954 and a full time staff by 1957. It included Fincastle, Grassy Lake, Bow Island, Seven Persons and Medicine Hat. The reservoirs operated in this division were Horsefly, Fincastle, Grassy Lake, Sauder, Murray and Seven Persons. It provided service directly to the farm gate thus a larger staff of ditch riders and water masters was

Moving concrete form, diversion tunnel at St. Mary Dam. (PFRA-808)

required. The headquarters for the St. Mary Project remained in the former CPR building until 1965 when it was moved to a new building on Mayor Magrath Drive. P.M. Sauder retired in December 1960 to be replaced by Jim Graham who was the superintendent of the Eastern division at the time of his promotion.

NATIONAL AND INTERNATIONAL IMPLICATIONS OF EXPANSION

The expansion of the St. Mary Project had international implications which were taken care of by the IJC. The international implications were the result of Canada's use of the Waterton and Belly Rivers which was not covered by the International Waterways Treaty of 1909. In 1955 the IJC rejected the American request that it review Canadian use of these rivers based on the argument that they were international.

With the post-Second World War boom in irrigation development in Alberta, the new demands came from Saskatchewan and Manitoba for their share of the water from the eastern slope of the Rockies. The creation of an agency to carry out this task was first established in 1930 when Alberta and Saskatchewan received control of their natural resources and the Department of the Interior no longer provided for the management of the water resources of the West. In 1930 Alberta, Saskatchewan and Manitoba formed the Western Water Board. Its failure to operate effectively led to the creation in 1945 of the Prairie Provinces Advisory Water Board. Before this organization could play any role on the management of inter-provincial water it was succeeded by another organization which reflected the renewed interest on the part of the federal government in the development of prairie water resources. In 1948 the governments of Canada, Alberta, Saskatchewan and Manitoba signed the Prairie Provinces Water Board Agreement. The board created by this agreement was to make recommendations for the best use and allocation of interprovincial water resources.

CONTROVERSY

The completion of the main canal through to Medicine Hat began a period of controversy and re-evaluation during which time both the PFRA and the SMRD were severally criticized for the way the St. Mary Project was built and operated. This controversy was sparked by the charge levied by the PFRA for the delivery of water by the new headworks. Under the terms of the 1950 development agreement, Alberta had agreed to pay the federal government for the water supplied by the new headworks at a rate not to exceed 25 cents an acre-foot. This charge was not levied, however, until 1953 when the SMRD received a bill for $48,000. At that time Sauder was able to persuade the PFRA that the SMRD did not have to make payment. In 1954, however, the PFRA sent a bill the SMRD with the expectation of being paid.

The implications of the imposition of the new charge was first discussed by the Western Advisory committee at their 3 December 1954 meeting. At this meeting, Sauder reviewed the financial

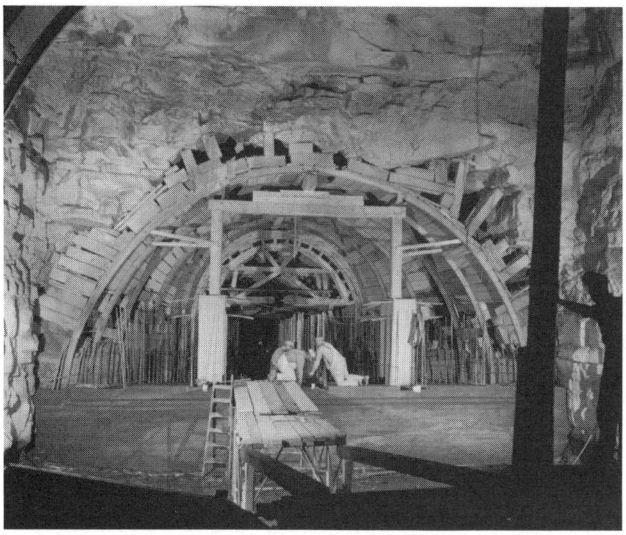

Concrete slab below control conduits in irrigation tunnel at St. Mary Dam (PFRA-1183)

history of the Project since 1946. He indicated that by the end of 1946 revenue exceeded expenditures by $20,509.00 and until recently the annual surplus had averaged about $13,342.00 per year. This provided an accumulated surplus of $106,734.51 by 1953. If the SMRD had absorbed the cost of this water in 1953 instead of a surplus of $2,021.00 the SMRD, Sauder noted, would have had a deficit of $36,800.00. If the increase was passed on to the water users, it would have increased the rates in the Lethbridge-Coaldale district to $1.37 while the rate for the farmers in the Taber, Magrath and Raymond area would be increased to 87.5 cents to meet the cost of giving service without drawing on the maintenance reserve to make up the deficits. Sauder suggested two ways of dealing with this issue, one of which was to call a meeting of the water users and ask them to agree to a voluntary raise of rates. The second way suggested by Sauder was to ask the Legislature to outlaw the old agreements and order the water users to pay the actual cost of giving service.

Given the lack of support for a voluntary payment, Sauder proceeded with the second option. The decision was strongly opposed by the Western Advisory committee which immediately began a campaign to rally water users support to prevent this from taking place. Part of this effort was a brief to the Minister of Agriculture, in January 1956. The purpose of the St. Mary Project according to this document:

> was to attract settlers and sell land with water agreements at a water rate that would make it possible to operate profitably and retire the capital and get on to the business of railway operations. The works were poorly designed and engineered without storage or drainage with an eye to economy and not to permanency.[273]

Construction camp, St. Mary Dam (PFRA-838)

Despite these circumstances, the brief continued the "farmers in this area through hardship, sacrifice, trial and error and without relief or government assistance or reduced payments, honoured their obligations in full."[274]

The brief then dealt with the transfer of the St. Mary Project from the CPR to the government of Alberta. According to the Advisory Committee:

> the irrigation officials of the CPR guaranteed the water users of this area that the $1.00 an acre water agreement then in existence would be maintained in perpetuity and were insistent that the statute respecting the transfer would conclusively protect the farmers in this regard. We wish to point out that during the meetings with Government officials . . . there was never at any time any suggestion that the water users of this area would be expected to pay for water in storage, it came as a complete bombshell that our rates might be changed. The term "increased water service" was always presumed to mean delivery of water to the farm.[275]

The brief then cited section 3 of the 1950 St. Mary and Milk Rivers Developement Act which supported their case. Beyond the legal requirement that the existing rates be retained, the brief indicated that "a change in the water rentals would constitute a breach of faith as serious in principle as the abrogation of the Crow's Nest Agreement."[276] The brief conceded that only "at such time as it becomes imperative to provide irrigation works and drainage necessary to conserve productivity of the soil and adequately secure the future of this irrigation area some consideration of water rate revision may be deemed mutually desirable."[277] This brief was consistent with the position the water users in the Lethbridge-Coaldale area had taken since 1944 which was that they were not prepared to pay for a stable water supply or the rehabilitation of their system, if it meant a change in water rates.

The next step of the Western Advisory Committee was to organize a meeting of the water users which was held at the Coaldale Community hall at 2:00 p.m. on 27 March 1956. It began with an opening statement by chairman C.E. Parry who indicated that the purpose of the meeting was to present a report of the work of the Advisory Committee in their effort to forestall an additional charge for water. The position of the Advisory Committee was presented by Fairfield and his fellow members of the Committee. The arguments used by Fairfield and his associates against the change were that there were no improvements in service, no consultation and the additional water made available by the construction of the St. Mary dam and the new headworks was not intended for use in the Lethbridge-Coaldale area.

Brown countered by indicating that the delivery of stored water represented an added service as did the improvements to various canals which had been made to deliver the additional water made available by the St. Mary Dam. Sauder began his response by expressing surprise at the view that providing stored water was not an improvement in service. He then pointed out that the provincial government had the legal right to make a change in rates. Sauder also tried to acquaint the water users with the reality of the provincial government's attitude to irrigation development. Sauder concluded his presentation with the comment that the water users should concentrate their energies on the improvement of their system.

The resolutions passed following these presentations showed that Sauder's views had made no

Pinepound syphon from the east side. (PFRA-1216)

impact. The first resolution was that the water users of the Lethbridge-Coaldale area in the St. Mary and Milk River development held "sacred our water agreements as being binding upon the Government of Alberta the same as they bound the Canadian Pacific Railway Company and our selves and that the Alberta government is morally and legally bound to meet the conditions regardless of loss or profit."[278] The second solution requested the provincial government to arrange for the appointment of an independent committee.

The brief and the resolutions had immediate short term effect since the provincial government temporarily withheld the revision of the Act pending a review. This review was begun by members of the Irrigation Committee of Cabinet which visited the area in November 1956.

The next step taken by the water users was the creation of the Lethbridge-Coaldale Irrigation and Water Conservation Association. The first meeting of the Association was held on 12 December 1957 and was attended by 168 farmers, the senior staff from the SMRD plus R.S. Lee, MLA for Taber. The meeting was brought up to date by C.E. Parry who indicated that there was nothing binding in the agreement following the transfer of the irrigation system to the Province of Alberta that the rates would remain the same. He also acknowledged that the district was in bad shape and something had to be done about drainage. He asked that the meeting make some decision as to the next step to be taken. He indicated that Sauder was their best representative to the government. The absence of Fairfield may explain this sudden change.

The offer from the SMRD was then formally considered which was that the SMRD was prepared to improve service in return for a rate of $2.00 per irrigable acre. The response of the meeting was to request the government to put the capital works into shape comparable to the new area and for drainage. Given this expenditure, the water users were prepared to pay the operation and maintenance costs.

Following the meeting, Sauder sent a letter to the Executive of the Lethbridge-Coaldale Irrigation and Water Conservation Association on 20 January 1958 which dealt in more specific terms with the willingness of Sauder to seek increased funding from the provincial government. In this letter Sauder repeated the point about the condition of the system which was inferior to every other system in Alberta. He attached an estimate of the cost

Jensen (Pothole) Reservoir & Taylor chute. (PFRA-10751)

of rebuilding and rehabilitating the distribution system and putting it in the same shape as the distribution system in the Eastern Division. The estimate included paying the water users for a few new culverts put in recently by them, one delivery for each quarter of land and canals and ditches of sufficient capacity to provide satisfactory service to all water users. Government willingness to undertake this work, noted Sauder was dependent on outlawing the old water contracts and payment by the water users of the levy set by the Manager and approved by the Irrigation Council. This meant that the Manager will submit his estimate of expenditures each year to the Advisory Committee elected by the water users and if the advisory Committee considered the levy higher than necessary it can pass its views on to the Irrigation Council which had the final say regarding the water rates levied each year.

Sauder concluded his letter with the suggestion that if a majority of the Committee was in favour of asking the government to provide the necessary funds and have its construction branch purchase the necessary materials and equipment and build the works, that a small sub-committee made up of water users and SMRD officials could take the proposal to the Minister. If the government gave the request favourable consideration, the association could take it to their members. If sufficient support was not received the plan could be dropped.

PROJECT AND INDUSTRY REVIEW

The request of the farmers in the Lethbridge-Coaldale area for an independent committee to investigate their concerns, combined with the concerns raised by the farmers in the Enchant area west of Vauxhall, prompted the first re-examination of irrigation development in Alberta since the Ewing Commission of 1936. This review was conducted by a special Irrigation Study Committee created by the province of Alberta on 6 May 1958. This committee was chaired by W.R. Hanson, Chief Forester for the Eastern Rockies Conservation Board. Having grown up on an irrigation farm in

First scraper full of earth of the main canal of the Big Bend, Sept. 12, 1949. (Taber Irrigation Impact Museum)

P.F.R.A. building headgates back into Taber Lake for Big Bend area. (Taber Irrigation Impact Museum)

the Cardston area was his only direct association with irrigation. The other committee members were A.T. Johnson, an irrigation farmer and rancher in the Brooks area; William Mackenzie, an agricultural economist from the University of Alberta; Doug Paxman, an irrigation farmer in the Raymond area and a member of the RID Board of Directors; and C.L Sibbald, the Western Manager of Catelli Food Products.

The Irrigation Study Committee investigated the organization and administrative set-up of present districts to determine if a standard for irrigation districts might be desirable and what, if any, amalgamations or divisions of existing units were required to obtain better results for the water user. The Committee's report, which was presented on 18 September 1958, made a number of general observations about the development of irrigation through to the post-Second World War era. The most important of these was the role played by the need to protect Canada's share of the water from the St. Mary River which was the catalyst for the return of the federal government to active participation in irrigation development in Alberta. Two problems resulted from the sense of urgency created by this situation, the first of which was the province's willingness to use water as soon as the PFRA was in a position to deliver it. By making this commitment, the province was "launched on a rate of development which might possibly have been reconsidered if the agreement had not been signed."[279] The second was that construction of the system was carried out by engineers who were not responsible for its operation.

The impact on the St. Mary Project, given this haste and division of responsibility, was the lack of a comprehensive plan which incorporated recent land classification survey information and an evaluation of the potential markets for the products of the expanded St. Mary Project. These two factors also led to a failure to keep the water users informed and involved in the development process. The failure to take advantage of the expertise of established farmers and irrigation district administrators was also noted. The controversy over the charge for stored water was cited as a consequence of this lack of communication. The third general criticism was that a crown corporation was not the appropriate agency to deal with the needs of the individual parts of the St. Mary Project. The failure to create an agency which could deal effectively with such problems as the rehabilitation of the Lethbridge-Coaldale area was illustrated by the decision not to specifically outlaw the old water contracts.

Two recommendations of the Irrigation Study Committee had important implications for the St. Mary Project. The first was that the provincial government should take the initiative, and not the federal government, in the future development of irrigation in Alberta. Having taken control, the province should reconsider all plans for future expansion. The second recommendation was that the Lethbridge- Coaldale area, along with the other parts of the St. Mary Project, should be established as separate farmer administered irrigation districts. The future role of the SMRD was to be reduced to that of an administrator of the main canal.

The recommendations of the Hanson Committee pleased the Lethbridge-Coaldale Water Users Association but dismayed Sauder since he believed in the further consolidation of the Project. In an effort to restore the strained relations caused by the rate issues and to decide a future course of action a meeting was held on 8 January 1959. It included water users, SMRD officials and the Western Advisory Committee. This meeting reviewed the role played by the Advisory Committee. Sauder's conclusion was that the committee could be a great help to him and the St. Mary and Milk Rivers Development if it would forget the past and think about the future. With respect to the future, the Lethbridge-Coaldale Water Users Association was concerned that the recomendations of the Hanson Committee did not address the issue of the rehabilitation of the Lethbridge-Coaldale area. The water users were not anxious to take over its administration without a prior commitment by Alberta for rehabilitation funding. A detailed review of the idea of creating a separate district was begun by the Alberta government at the request of the water users. The issue of the water rate was resolved in 1960 with the passage of legislation terminating these agreements. The establishment of a water rate based on this new system was first made in 1961.

St. Mary and Milk Rivers Development office. (Sir Alexander Galt Museum & Archives, Lethbridge, Alberta)

Forty-mile syphon south of Bow Island (PFRA-13877)

DISTRICT NEGOTIATIONS

The negotiations between the districts and the SMRD continued on a separate basis through to 2 January 1960 before a final agreement was negotiated. In 1957, 1958 and 1959 the three districts withheld the extra amount pending the finalization of an agreement Under the terms of the final agreement, the TID, MID and RID agreed to pay 10 cents

Staff of the SMRD, LNID and the Colonization Branch of the Department of Agriculture in the SMRD Building, circa 1950. The SMRD staff included Alfred Kress (back row, far left), George Brown (back row, third from left), P.M. Sauder (back row, far right, Mary Slevin (front row, third from left) (Sir Alexander Galt Museum & Archives, Lethbridge, Alberta)

per irrigable acre for 1957 and 1958 and 19 cents per irrigable acre for 1959 as a canal cost and maintenance reserve from Ridge to Chin Reservoir and pay interest on the unpaid balance of the revised water supply accounts for 1957, 1958 and 1959.

Another important result of these negotiations was the establishment of a permanent organization to manage the main canal in 1962. It was made up of seven members drawn from the various areas based on acreage. Magrath-Raymond and Taber received 1 member each while the Lethbridge Coaldale area had 2 and the Eastern Division of the SMRD received three. The function of the committee was to advise the Manager on the operation of that portion of the main canal from Ridge to Chin Reservoirs which had to be operated for the benefit of all the units in the St. Mary Project. The specific concerns dealt with by the Overall Advisory Committee were the disposition of the maintenance fund created through contributions by its member organizations and the annual charge for water supplied and transported for irrigation and domestic purposes by the SMRD. Despite the delays in arriving at this approach to managing the main canal, it had resulted in an agreement which recognized the need to provided for the

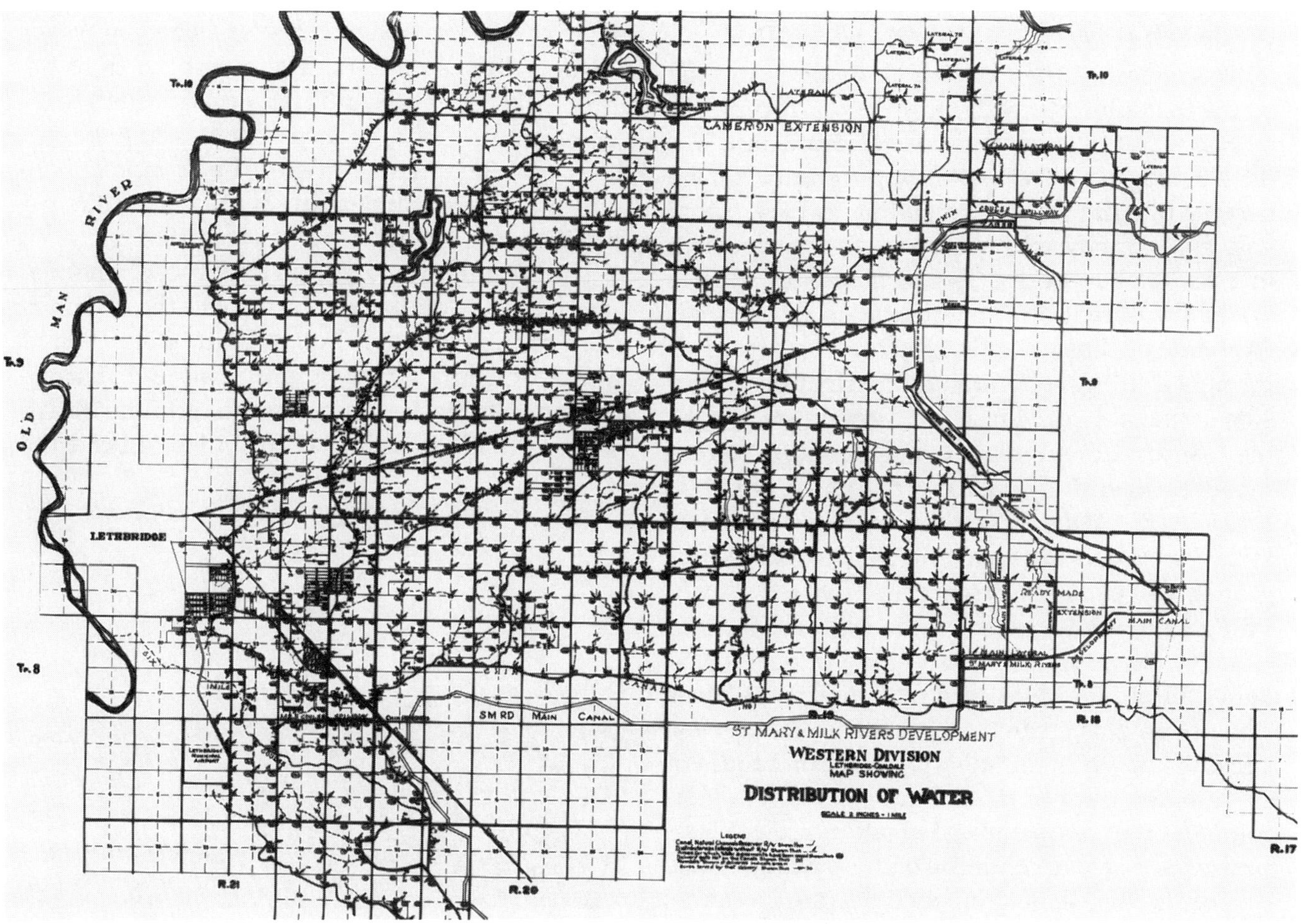

Plan of the Western Division Lethbridge-Coaldale Tract (PFRA)

Flooded area in Sec. 14-9-20, North of Coaldale, June 1947 (SMRID)

changing cost of the operation of the system rather than basing it on a fixed cost which had been the approach of the CPR. In 1962 the rate set for the delivery of water to the headgates of the districts was set at twenty-five cents per acre-foot and an additional charge of ten cents per acre was to be assessed to create a maintenance fund. This rate remained unchanged through to 1967.

ALBERTA IRRIGATION STUDIES

The review of the industry undertaken in 1958 by the Irrigation Study Committee was completed by Harry Strom through the Agricultural and Rural Development Act (ARDA) program after he became the Minister of Agriculture in 1962. Strom was first elected as the MLA from the Burdett area in 1955. He had been active in the South Alberta Water Conservation Council and in the Burdett Agricultural Improvement Association. He thus brought a strong commitment to irrigation expansion to the office of Minister of Agriculture. Unlike the Hanson Commission, which based its recommendations entirely on briefs presented at public hearings, the ARDA irrigation studies were undertaken by experts in various fields. The ARDA program involved the first comprehensive effort to investigate the point made by the Wilson Commission that other interests besides the farmer derived benefits from the industry in the context of Alberta. The documents created as a result of this review became known as the Alberta Irrigation Studies.

The objectives of these studies were to: determine the cost of operating irrigation projects; appraise the physical condition of the existing works; determine the economic benefits from irrigation; and to assist in the development of a government policy on irrigation which was equitable to all projects and which would serve to enhance the effectiveness of irrigation production in the province. One study was designed to quantify the input and output costs and benefits of irrigation. This was achieved through a case study of one irrigation district from which the necessary conclusions could be drawn for application to the entire province. The Eastern Irrigation District was chosen for this purpose because it was a well-established district. It was originally the eastern part of the Canadian Pacific Railway Irrigation block, construction of which had begun in 1910. It was also the largest of the irrigation projects then in existence in Alberta. It made up 30 percent of the irrigated acreage in the province. It was also a clearly defined economic unit.

The task of collecting data was shared between various agencies. The PFRA determined the costs of irrigation. The additional income and costs of agricultural production attributable to irrigation was determined by the Economics Branch of the Canada Department of Agriculture. The magnitude and distribution of the benefits and costs of irrigation were determined by the University of Alberta. The data obtained was analyzed by Walter B. Rogers and Travis W. Manning both from the University of Alberta and Herbert W. Grubb, Texas Technological College. Their conclusions were outlined in a document entitled "The Economic Benefits and Costs of Irrigation in the Eastern Irrigation District of Alberta" released in 1966.

The formulation of overall policy recommendations based on the data collected was the responsibility of the Irrigation Policy Committee chaired by Cy McAndrews. His report made recommendations on financial arrangements, administrative structure and construction. In order to fund the

Excavating drift soil on North-east Lateral Extension in Spring 1946 (SMRID)

capital works of irrigation projects, his report recommended a cost-sharing program in which Canada would pay 45%, Alberta 41%, water users 14%. It further recommended that the cost sharing agreements be revised in 1970 with the distribution of costs being Canada 35%, Alberta 32%, municipalities 22% and water users 11%. This was the first time that a cost-sharing formula for irrigation development in Alberta had been suggested.

The changes introduced by this review process served to shift the initiative for irrigation expansion from the PFRA to the provincial government and resolved the issue of the future of the Western and Eastern Divisions of the SMRD through their establishment as a single farmer operated irrigation district. This era of review also produced the first comprehensive, on going financial plan for the funding of irrigation development in Alberta. The time had truly come for the idea, first suggested in the 1890s, and repeated many times there after, that public funds be used for the irrigation infrastructure construction.

Water Users Combined Drop and Division Box in the Lethbridge-Coaldale area. (PFRA-28052)

Structure in the Lethbridge-Coaldale area in need of rehabilitation. (PFRA-28051)

EIGHTEEN
ST. MARY RIVER IRRIGATION DISTRICT
1968 - 2000

The 1968 Irrigation Act created the St. Mary River Irrigation District (SMRID) and provided the legislative authority for an irrigation rehabilitation program as recommended by the Irrigation Policy Committee. These two developments brought about the final stage in the evolution of the St. Mary Project since planning for its construction had begun in the 1890s. The transfer from the Crown Corporation to the water users also brought to fruition the original objectives of P.M. Sauder who had administered the SMRD with the view to its eventual transfer to the water users. Between 1969 and 1973 the provincial government replaced the federal government in the management of the headworks and took the lead in providing funding for its on going development. The changing role of the federal and Alberta governments was the result of the 1973 Alberta Irrigation Agreement which ended over 75 years of association between the federal government and the St. Mary Project This transition to water user control was made with relative ease since preparations had begun in 1946 with the creation of the first advisory committee and continued with through to 1960 the termination of the old water contracts.

In 1974 the St. Mary Project, as expanded after the Second World War, reached the limits of its ability to provide service thus initiating a long-term planning program to deal with its deficiencies. The construction program which emerged out of this planning process was implemented in the 1980s through the ongoing assistance of the Lougheed government who provided funds from the Heritage Trust Fund for irrigation expansion. The rehabilitation of the main canal from Ridge Reservoir to Medicine Hat and the addition of Forty Mile Coulee Reservoir to the system were the largest components of this work. The completion of these tasks brought about a second construction boom and necessitated the relocation to a new office which was required to accommodate the additional engineering staff. Other improvements to the system included the introduction of the overshot gate to replace the radial gate, the use of pipelines rather than concrete lined canals to transport water beyond the main laterals and the use of computers which further automated the system. Peter Langemann, irrigation farmer, a past chairman of the Board of Directors and inventor contributed the LOPAC Gate which automated the flow of water on smaller laterals and the Langemann Gate which applied the overshot principal to increase the efficiency of larger canals.

The rehabilitation of the system necessitated a larger staff to direct construction and to provide water to the land both in the Western and Eastern

Divisions. The end of the era when the Western Division depended upon lateral associations was symbolized in 1969 when the SMRID took over management of the laterals formerly maintained by the Readymade Water Users Association. The North Lateral Water Users Association #2 is one of the last surviving examples of this type of organization

With the issue of rehabilitation and expansion successfully dealt with by 1992, the SMRID, in partnership with the RID and the TID, moved on to other tasks which included the development of the hydro power potential of the system. Other recent developments include the establishment of the Blood Tribe Agricultural Project and improvements to the headworks through the construction of a new spillway at the St. Mary Dam and the replacement of the wood stave syphon across Pinepound Coulee with a buried syphon made of concrete.

CROWN CORPORATION TO IRRIGATION DISTRICT

The 1968 Irrigation Act was the most comprehensive irrigation legislation passed since the North-west Irrigation Act of 1894. It re-organized the Irrigation Council and created an Irrigation Secretariat to carry out its duties, standardized the administration of irrigation districts and consolidated provincial irrigation legislation dating back to the passage of the original Irrigation Districts Act in 1915. Some of the legislation repealed included the Raymond Irrigation District Act of 1933 and the Taber Irrigation District Act of 1934. The former statute stated that the responsibility of the RID Board of Directors extended only the operation of the main canal while the latter statute authorized the TID to invest a portion of the district's funds in the construction of a cannery.

The 1968 Irrigation Act transferred the Lethbridge-Coaldale area as well as the entire Eastern Division from the SMRD to the SMRID. The authority of the Manager of the SMRD was transferred to the Board of Directors. The SMRID was thus the successor to the Crown Corporation in terms of its responsibilities for the overall coordination and operation of the main canal from Ridge Reservoir to Medicine Hat. The Act did not go as far toward rationalizing the St. Mary Project as was recommended by the Irrigation Policy Committee. The MID and the RID were not amalgamated with the Lethbridge-Coaldale area to form one large district west of Chin Coulee. The TID and the Eastern Division of the SMRD were not amalgamated to form one district east of Chin Coulee. The alternative to amalgamation has been the cooperation of the districts through a main canal agreement which provides for the independence of the districts as well as the efficient operation of the main canal.

As in 1908 and 1946, the transition from one administration to another was relatively smooth since it was not accompanied by any changes in organization or major turnovers in staff. Continuity was also ensured by the election of a board of directors made up of people who had taken an active role on the advisory committees. The first Board of Directors included Jack Brewin, Stan Tiffin, Peter Shologan, Gifford Woodcock, Ted Eiserman and Paul Edlund. Jack Brewin had been a member of the Eastern Advisory Committee since its creation and had been active in the review of the irrigation industry during the 1960s. The first election for board members was held on 6 April 1970.

The organizational structure of the SMRD, as revised in the mid-1950s to accommodate the construction of the St. Mary Dam and the expansion of the main canal to Medicine Hat, was retained. The Western Division was operated in 1968 by a superintendent, six water supervisors, one equipment operator and a utilityman. Additional staff was not required because the delivery of water beyond the main laterals continued to be to the responsibility of the water users. Bow Island continued as the operating headquarters for the Eastern Division. It had a staff of 22 in 1968 which consisted of a Superintendent, 14 Water Supervisors and 2 Field Superintendents. The larger staff was required for the Eastern Division because the SMRID delivered water to the land. The operation of the main canal was the responsibility of Bob Fuller the Superintendent of the Western Division and the two water supervisors.

One of the first tasks of the board was to hire a new Manager since Jim Graham left in August 1968 to become Manager of the newly established Irrigation Secretariat. He was succeeded by Jake W. Thiessen, a graduate in Agricultural Engineering from the University of Saskatchewan. Following graduation in 1961 he spent three years working on an irrigation project in India. Upon his return to Canada, he was employed by the consulting firm of Underwood, Mclellan and Associates (UMA) of Saskatoon. In November 1965, he joined the Alberta Department of Agriculture, Water Resources Division in Lethbridge where he served as an irrigation and design engineer. He became manager of the SMRID on 1 August 1968.

MAIN CANAL AGREEMENT

Another task which required immediate attention was the resolution of the revenue surplus dispute which it had inherited from the SMRD and the negotiation of a new agreement between itself and the TID and RID for the joint administration of the main canal. The revenue surplus dispute went back to 1955 when the SMRD had raised the price it charged the districts for their water supply. The view of the TID, RID and MID was that they were entitled to a portion of the surplus created through the collection of this payment.

The distribution of this money which the SMRID was forced to pay, and the other aspects of the operation of the main canal were incorporated into an agreement dated 1 January 1969. Magrath, Raymond and Taber received one time payments of $17,890.77, $44,381.08 and $44,179.49, minus the amount they had refused to pay prior to 1969. The SMRID was to retain all future payments made by the districts. The remaining portions of the agreement outlined the way the future costs were to be divided. The agreement also had clauses which dated back to the agreements of 1902 which allowed the SMRID to shut down the system for repairs and protect it from damages caused by a drop in the water supply beyond the control of the SMRID. The provisions in the agreement concerning the ongoing operation of the main canal did not include the MID since its water supply was provided by Alberta Environment directly from the headworks.

ALBERTA IRRIGATION AGREEMENT

The SMRID begun their administration of the project in anticipation of a federal- provincial agreement, similar to the one negotiated in 1950, incorporating the cost-sharing recommendations of the Irrigation Policy Committee. It had recommended that Canada, Alberta and the irrigation districts pay 45%, 41% and 14% respectively towards the cost of rehabilitating the irrigation system. It also suggested that this distribution be revised in 1970 with the municipalities assuming 22% of the cost thus reducing Canada's share to 35%, Alberta's share to 32% and the water users' share to11%.

The municipalities were the first to indicate their unwillingness to be part of any cost-sharing agreement arguing that they could not afford the cost. The newly elected government of Pierre Elliot Trudeau was also reluctant to participate. The extent to which the SMRID had based its planning on this program and the effect the federal government's attitude was having on the willingness of the provincial government to participate was communicated to the Minister of Agriculture Bud Olsen in a letter dated 21 October, 1969. Jack Brewin reminded the Minister of the research that had been undertaken to arrive at the cost-sharing formula and that the districts had been warned that unless the government of Canada entered into a cost sharing agreement there would be no provincial funds available for 1970. This decision jeopardised the SMRID's plan to:

> complete [the] renovation of the irrigation system in the Lethbridge-Coaldale area which has been in operation for over sixty years. We are convinced that any construction which we undertake now will serve this area for at least the same period of time projected into the future. With this in mind, we cannot afford to ignore new technology and materials which are available to us.[280]

The letter went on to point out that plans and economic feasibility studies were underway for an 8,000 acre pilot project to test a buried pipe irrigation system in the Lethbridge area. Olsen was invited to attend a meeting of the Board when the results of this pilot were available for discussion. Olson referred the letter to the newly created department of Regional Economic Expansion which now had responsibility for the PFRA. This department indicated that a new round of policy reviews had begun and that no immediate commitment could be made until they had been completed.

The SMRID continued its effort to bring about an agreement by working with its fellow members in the Alberta Irrigation Projects Association (AIPA). It assisted in the preparation of a briefs and it was part of a delegation which went to Ottawa in 1970 to meet with Olsen and Jean Marchand, Minister of Regional Economic Expansion. The delegation included; Farrell Nelson, chairman of the RID and the AIPA; R.T. White, Manager of the EID; Les Toth, Manager of the LNID; and Jake Thiessen, Manager of the SMRID and secretary of the AIPA. Following this meeting, Olsen announced at the AIPA annual meeting that several million dollars would be made available for irrigation rehabilitation but no further details about the program were released.

The negotiations remained stalled until Eugene Whelan succeed Olsen as Minister of Agriculture in 1973. He met with the directors of the AIPA in Lethbridge. The main topic of discussion was the state of disrepair of Alberta's irrigation system and the need for cooperation between the two senior levels of government to finance a program of rehabilitation. Shortly after this meeting, the Alberta Irrigation Agreement of 1973 was signed. This agreement transferred administration and control to Alberta of the headworks of the Bow River and St. Mary Irrigation Projects. It also provided a one time contribution towards the rehabilitation of both projects as well as for other irrigation projects in Alberta. Canada agreed to pay Alberta $4,200,000 as compensation for accepting the transfer.

The transfer took place between 1 April 1973 and 1 April 1974. Canada and Alberta agreed to make every effort to employ present staff. An Implementation Board made up of three provincial and three federal representatives was created to oversee this process. This Board had the responsibility of planning the transfer and of ensuring that all the appropriate records were included. The1973 agreement thus ended the dominant role which the federal government had played in the St. Mary Project since 1895 when it prepared the first plan to divert water from the St. Mary River for the purpose of irrigation. The PFRA's only ongoing involvement after 1973 was with the development of the Blood Tribe Agricultural Project. This agreement brought to an end the partnership with the provincial government which developed following the passage of the Irrigation Districts Act in 1915. The Department of the Interior had provided technical assistance to the province in the administration of this legislation. This partnership continued into the 1950s through the joint federal-provincial agreement to expand the St. Mary Project.

PROVINCIAL INITIATIVE

The provincial government, unlike the federal and municipal governments, moved immediately to implement an irrigation rehabilitation program in partnership with the irrigation districts. This response was consistent with its commitment to irrigation expansion which went back to the construction of the Jensen dam in the late 1940s. It had also initiated the irrigation industry review undertaken in the mid-1960s. A number of individuals who had participated in these studies already held or rose to positions of influence within the provincial government in the late 1960s and 1970s. This group included Harry Strom who continued to be an advocate of irrigation development in his role as minister of agriculture from 1962 to1968 and as premier from 1968 to1971. Cy McAndrews went from being the colonization manager for the St. Mary and Bow River Projects to Assistant Deputy Minister of Agriculture He was instrumental in developing the irrigation extension service of Alberta Agriculture which worked with irrigation farmers. This service provided assistance and advice with respect to land levelling, sprinkler system design, irrigation scheduling, advice on crop rotations and other services. Other knowledgeable irrigation specialists who at one time worked in the Colonization Branch and then moved to more senior positions in the Department of Agriculture in the late 1960s included Norman Thomson, Doug Harrington and Lloyd Sadler. Jim Graham, who had served as Manager of the SMRD from 1961 to 1968, moved to Edmonton to head up the new Irrigation Secretariat.

Preparations for the implementation of the irrigation rehabilitation program began with the drafting of the Irrigation Act 1968 which made provision for irrigation districts to enter into agreements with the provincial and federal governments for the purpose of cost-sharing on irrigation rehabilitation programs. Later revisions to the Irrigation Act also encouraged participation by providing an exemption from seepage claims provided that the district filed a ditch rehabilitation plan with the Irrigation Council.

The Alberta government first made funds available in 1969. This money was intended to be a one year program only since the provincial government in 1969 did not intend to enter into longer term agreements until the federal position on participation was known. The provincial government assumed responsibility for the total government contribution in the proposed funding formula. For the 1969/70 fiscal year $583,630 was distributed to the districts on the basis of the 86 - 14 formula. This contribution from general revenue was increased to $965,630 in 1970/71. In the spring of 1971 the provincial government stopped waiting for any contributions from the federal government and began planning a long term program based on no federal assistance. This decision was made to permit the districts to facilitate their plan of the rehabilitation programs. Funding for 1971/72 and 1972/73 was one million per year.

THE NEED FOR REHABILITATION

The SMRID inherited a project that was in need of rehabilitation for a number of reasons. The Western Division had not received the benefit of any of the funds spent on construction during the 1950s. A portion of the original main canal and the entire lateral system constructed before World War One was still in use as late as the 1970s. The CPR had spent a decreasing amount of money on maintenance through to 1946 and the water users associations lacked the funds and support to properly maintain the sub-lateral system. A system of drains to take care of return flows was virtually non-existent. The Crown Corporation between 1946 and 1968 lacked the resources and the mandate to undertake a rehabilitation program. Rehabilitation had been further delayed by the water rate controversy in the 1950s during which time P.M. Sauder had great difficulty in convincing the water users to adopt a more flexible approach to establishing the water rate. The PFRA had begun to prepare a rehabilitation plan but it had not been implemented. The Western Division thus had a number of wooden structures which dated back to the 1930s, if not earlier. Sauder described the Lethbridge-Coaldale area as being in the worst condition of any irrigation district in North America. It was estimated in 1969 that the rehabilitation cost for the Western Division was $15,000,000.

The Eastern Division ,which was fifty years younger that the Western Division, was in need of repair because of the way it was built. It was designed in such a way as to minimize its initial cost and speed construction. The idea being that once built, it could be repaired and the acres irrigated from the system expanded. It made use of the side hill technique which involved using the side of a hill for one side of the canal. The result was the accumulation of water in any coulees or depression along the side of the hill. The canals were constructed on the contour of the land with no particular emphasis on selection of construction materials. Seepage, thus became a problem shortly after water was turned into the new canals thus reducing the productivity of the adjacent land because of logging, salinity and alkaline conditions. The construction in the 1950s had been based on lower construction standards and the wide spread use of untreated timber for drop structures and bridges. The life expectancy of these structures was only 10 years and sometimes less if they caught fire when the maintenance crews were burning weeds along the canal banks. The new system had been superimposed on the existing system rather than replacing it. It was estimated in 1969 that the rehabilitation cost for the Eastern Division was $4,000,000.

The most dramatic demonstration of the deficiencies of the existing system occurred in 1974 when a major break in the main canal near Barnwell occurred on 1 August. Water began to flow around a turnout structure which delivered water to a TID lateral. From the time of initial failure until the break was closed approximately 12 hours later a section of canal bank over 100 feet in length had been eroded .Property damage claims amounting to approximately $250,000 were received. The compensation paid to land owners was only partially covered by the insurance. This incident prompted the SMRID to conduct investigations into the cause of the failure and possible preventive measures that may be taken so to avoid a similar occurrence in the future. The results of this investigation indicated that the turnout was initially undersized when it was installed in 1967.

INITIATION OF REHABILITATION PROGRAM

When the SMRID took over the system, Manager Jim Graham had already initiated an investigation of the use of pressure pipelines for irrigation water conveyance. This evaluation was continued by the SMRID Board since it wanted to investigate new methods and technology such as buried pipe or concrete lined canals which were then in use in the USA to irrigate millions of acres. The Readymade area was chosen for a pilot project to evaluate three different types, all of which covered the same area and did not create seepage problems. The first option was a complete pressure pipeline system, with a central power unit and valve for each 160 acres located in the middle of the quarter section. Other valves were to be located at different locations along the line at the land owners expense. The second system was a gravity buried pipe scheme, located along the roadside with one turnout along the fence line on the high side of the quarter section. The third system made use of concrete lined laterals. Piped irrigation systems recently installed in the western USA and the

Above and below: *Trenching equipment of the type used by Getkate and Dyck Bros. to excavate the earth section of canal prior to placing slip form concrete lining. (early 1970s)*

Completed concrete lateral (St. Mary River Irrigation District)

New concrete drop structure in Lethbridge Coaldale area. (N.E. lateral early 1970s)

Electrically driven pumps to take water from Yellow Lake and place it into the Grassy Lake syphon for irrigation downstream (circa 1970)

Okanagan valley of B.C. were reviewed as part of this test. The cost of pipelines was generally considered excessive for southern Alberta before the time of major cost sharing programs by senior levels of government. The use of concrete lined laterals was thus given preference.

The first improvement made to the system had begun in 1967 and involved the deepening of the channel in the lower Stafford Lake area. The shallow depth of the canal at this location was a bottleneck in the main canal system between Ridge and Horsefly Reservoirs. The cost of this work was paid for by the Main Canal Maintenance and Contingencies Reserve Fund. Between 1969 and 1973 the SMRID spent $2,044,597.22 on rehabilitation work with 86% of this amount coming from the provincial government while the remaining 14% was provided by the water users. These funds were used primarily to line laterals with concrete. Other laterals were relocated, deepened and polyethylene lining added to reduce seepage. Open and tile drains were constructed as part of this program and new precast concrete checks, drops and bridges replaced structures which had previously been constructed of wood. These funds were also used to install a pumping station at Yellow Lake the function of which was to stabilize the level of the Yellow Lake and provide additional water for the North Grassy and North Burdett distribution systems. An agreement was arranged with the County of Forty Mile to provide for joint funding of its operation. The County paid half the operating costs of the three pumps up to a maximum of $3,000.00. After 1971 the expenditures on ditch rehabilitation were based on a 5 year ditch revitalization plan submitted to the Irrigation Council in accordance with Section178 of the Irrigation Act. The submission of this plan exempted the district from seepage damage claim settlement procedures. This plan committed the SMRID to capital works expenditures totalling approximately $350,000 per year up to and including 1975. The savings realized in operation and maintenance expenses as a result of this plan was about $.15 per acre annually.

LONG TERM PLANNING

The first indication that the capacity of the system was being strained came in 1969 when the *Annual Report* pointed out that the main canal along certain areas during the peak irrigation season flowed to capacity. Consideration, it noted, should be given to decide how many more permanent acres to be irrigated can be added to the assessment roll. This point came in 1974 when for the first time since the construction of the headworks and main canal in the early 1950s there have been major restrictions placed on the expansion of irrigable area to include new lands within the district. Since 1 May 1974 the Board adopted a policy of opposing new applications for permanent water rights and terminable agreements in all areas except those where the irrigation works had adequate capacity to serve additional areas. This means that new water right applications were not being processed for most of the area lying below and served out of the Chin Reservoir headgates. Existing water rights can be expanded within a given parcel of land either by increasing the permanent "acres to be irrigated" or by processing a terminable agreement.

Planning to deal with this problem began in 1973 when UMA were hired to prepare feasibility studies on various projects, one of which was to investigate the elimination of the old AR&ICo. main canal and most of Chin 1 lateral and make use of the new main canal to serve this area. The second Project was to determine the approximate cost of construction and the benefits of constructing a storage reservoir at Forty Mile Coulee.

The long-term planning process continued in 1975 with the preparation of a 10 year rehabilitation program for the district. The program proposed the rehabilitation of the Lethbridge-Coaldale area to provide a district owned and operated delivery to each parcel of land. The installation of canal lining and the construction of drainage to relieve a number of the worst seepage problems was also recommended. The third feature of the plan was the enlargement and upgrading of the main canal system to supply more water to the area lying east of Chin reservoir . This was evaluated in detail in 1976 by UMA who conducted a main canal system study. The objectives of the study were to determine the available water resources, potentially irrigable lands in the project area, and relative costs of increasing water delivery capabilities. The 1977 UMA report indicated that water and land resources were sufficient for a project 500,000 to 550,000 acres in size within the present district. The key item in efficient supply and control of water in the project both now and in the future was storage at Forty Mile Coulee. Substantial canal up grading, erosion protection, seepage control and replacement of inadequate or deteriorating structures were also required to serve the long term needs of the project. This report was the first comprehensive review of the St. Mary Project since the St. Mary and Milk Rivers Committee report was prepared in 1942. Like the 1942 report it prepared a plan which provided a way to achieve the expectations of the St. Mary Project as defined by the LSEP prepared during the First World War. The review undertaken by UMA coincided with the completion of phase one of a review of the water resources and their use in the Oldman River Basin by the Alberta Government. The second phase of the study was initiated with the creation of the Oldman River Study Management Committee. It was appointed on 6 May 1977 and included Jack Brewin, chairman of the SMRID board. The Committee recommended the construction of a dam on the Oldman River.

NEW FUNDING

One of the results was increased funding for irrigation expansion. The 2 million dollars in funding for irrigation rehabilitation available from general revenue had been supplemented after 1973 by funds from the federal government. These funds were further supplemented in 1975 with funds from the Alberta Heritage Trust Fund. Between 1976 and 1980 31.6 million heritage fund dollars were expended on irrigation works rehabilitation based on the 86 - 14 cost sharing formula. In 1980 the Alberta government announced a decision to expand it's water resources and irrigation development program for southern Alberta. This announcement was contained in a ministerial statement jointly released by Jack Cookson, Minister of Environment and Dallas Schmidt, Minister of Agriculture. The program allocated $234 million dollars to Environment for the development of major conveyance and storage systems and $100 million dollars to assist irrigation districts in the rehabilitation and upgrading of their water distribution systems.

This policy statement was a long term strategy for the further development of renewable resources in southern Alberta and closely followed the recommendations of the Oldman River Management Study Committee. This announcement was considered the most important event in the history of irrigation in southern Alberta since the decision was made to irrigate land in southern Alberta. The benefits of this program for the SMRID included a commitment to reconstruct the main canal from Ridge Reservoir to Murray Reservoir and the construction of Forty Mile Coulee Reservoir.

NEW CANAL AGREEMENT

The SMRID's access to these funds was contingent upon the negotiation of a new main canal agreement to replace the 1969 agreement which had expired on 31 December 1978. A new agreement was signed on 17 December 1981. The new agreement provided for the sharing of the main canal costs from Ridge Reservoir to Horsefly Check. The new agreement perpetrated the Advisory Committee to advise the SMRID Board of Directors on the operation of the main Canal. It consisted of the Chairman of each district and the Manager of each district. In 1982 the Committee dealt with the main canal budget for Ridge Reservoir to Horsefly Check, the main canal reconstruction priorities and schedule as they affected the TID and RID, main canal turnout financing, cost of additional right-of-way for the main canal and the transfer of certain lands to the RID as provided for in the agreement

THE SECOND POST -WAR CONSTRUCTION BOOM

Agreements for undertaking the main canal enlargement and construction of Forty Mile Coulee Reservoir were negotiated and finalized in December of 1981 and early 1982. The rehabilitation of the main canal was divided into a number of separate contracts and came into force as the work was undertaken. The Forty Mile Coulee Agreement on the other hand was comprehensive and covered the entire project from start to finish. The construction schedule was revised in 1983 with the result that the construction of Forty Mile Coulee Reservoir was delayed by three years.

The enlargement and rehabilitation of main canals presented a whole new set of challenges. One of these challenges was keeping the system in operation while rehabilitation was carried out. This meant that the bulk of construction had to be carried out in cold weather and under difficult wet soil conditions. These factors had to be taken into consideration during the design of the new systems. This type of work had never been done on existing main canals in a climate that had below freezing temperature for most of the time available for construction. This called for the development of new design criteria as well as new construction techniques. Before rehabilitation began on the St. Mary main canal an agreement was negotiated between the government of Alberta and the Board of the SMRID A negotiation committee was established headed by Peter Melnychuk for the Province and Jack Brewin for the SMRID. This negotiating team quickly determined the responsibilities for each party. Alberta Environment would provide project management, engineering and funding for all capital works construction. The board would be responsible for obtaining right of way and would remain responsible for operation and maintenance of the system.

Alberta Environment appointed a project manager for each of the major main canal and storage projects. The project manager reported directly to the chairman of the Irrigation Implementation Committee and was responsible for the management of all aspects of design and construction as well as providing liaison with the district board

Lining of main canal in 1985 using automated gravel placement on PVC liner

and management. George Forsyth and Pat Duffy served as project managers of the St. Mary main canal Project. John Campbell was project manager of the Forty Mile Coulee project. The SMRID canal was designed by private sector consultants. Because of the size of the SMRID Project it was broken down into reaches of about 40 km. each with a separate consultant or consortium of consultants for each reach.

New inlet pipes leading to pump station at Forty Mile Coulee reservoir (circa 1985)

In order to ensure consistent design standards throughout the entire project main canal design criteria had to be developed. Under Thiessen's direction a series of meetings were held between consultants and government design engineers. Input from the irrigation districts was obtained from staff and board members. UMA Engineering was retained to write the standards and prepare a suitable document for approval by the Board and Alberta Environment. Ken Craig of UMA coordinated this work. This manual covered all components of main canal systems including the design of all components of the main canal system including design of drop and check structures, turnouts, earthwork, seepage and erosion control works. Material specifications and the shape of canal cross sections were also generally outlined in the design criteria. This manual was then used by all the consultants on the Project.

Meetings were held with the Alberta Road Builders Association to obtain the input of contractors on earthwork specifications, tendering and contract administration. This input was largely responsible for dividing the work into manageable size contracts which would give most Alberta road builders the opportunity to bid on the work. The timing of calls for tender was also important due to the critical scheduling requirements of construction work.

Contractors soon discovered that it was important to complete the earth work portions of the work as soon as possible prior to the onset of cold temperatures and frozen ground conditions. Other work such as the placement of gravel armour protection could be later.

Similarly, some cold weather construction techniques had to be developed for large reinforced concrete drop and check structures which were installed under freezing conditions. Generally the entire structure site was enclosed in a protective cover and heated with propane until the concrete was properly cured. The first time that this technique was used was on the replacement of drop structures on the north east lateral in the western division.

Where lining of main canals was required for seepage control a whole new set of design criteria and construction methods were developed. After extensive review and testing the material selected for use as a buried membrane liner was 20 mil. PVC. The PVC liner was covered by a protective layer of granular material carefully graded and placed in a thickness of 300 mm.

Foundation preparation prior to placement of the PVC liner involved normal earth moving, placement and compaction techniques. All of the soft and organic materials within the existing canal bottom and side slopes were removed and replaced with suitable backfill material. 3H:1V side slopes were used for membrane lined sections to ensure slope stability under draw down conditions.

One of the contractors developed new equipment which used a central conveyor system with two cantilevered conveyor arms to spread the granular cover material. This conveyor equipment spanned the entire canal cross section, spreading the gravel uniformly to the required depth while slowly but continuously moving ahead along the canal. One of the advantages of using this type of equipment was that the granular material was placed very gently on top of the membrane. As a result, the possibility of the membrane becoming damaged during placement was minimized. This placement method also greatly increased the rate of production to about 260 metres length of canal per day.

OTHER INNOVATIONS

During the first 10 years of the program the emphasis was on canal enlargement and upgrading as well as modernizing of water control structures. The use of concrete replaced wood in the construction of drops and checks. The overshot gate was developed for the St. Mary main canal. The first one was introduced and was subsequently used on other large canals. After most of the main canal system was rehabilitated attention shifted to automated and remotely operated gates. One of the major innovations was the development of the overshot gate to replace the radial gate which had been in use since the St. Mary Project was began in 1898. The first one was installed on the entry structure to Sauder Reservoir .The next step has been the application of computers to the task of automating water management.

CONCRETE LINING TO BURIED PIPELINE

The use of concrete as a lining for laterals which had predominated in the 1970s was revaluated beginning in 1979 when severe frost conditions during the winter of 1978-79caused cracking in concrete lined sections of canal. The implications of this development were evaluated by UMA Engineering with the assistance of the Irrigation Division Alberta Agriculture at the request of five major irrigation districts including the SMRID. UMA undertook an inventory of all concrete lining in Southern Alberta and a statistical analysis of its conditions. This study indicated that 20% to 25% of lined ditch sections were severely damaged. The result was that maintenance cost to the districts were therefore going to be higher than anticipated. In an effort to solve this problem the construction modified to incorporate wire mesh into the concrete to provide some reinforcement. A machine to undertake this task was invented by Peter Langemann. During the summer of 1980 a slip form owned by Dyck Bros. Construction was modified to permit the unrolling of a wire mesh into the slip form concrete as a test methods. In 1981, three miles of reinforced concrete lining was built for the purposes of determining is durability.

Machine used to lay re-inforced concrete lining (SMRID)

Despite the fact that reinforced concrete was a significant improvement, the economics of pipelines because of new material made them a better alternative. The use of pipelines in the SMRID had first been evaluated in 1966. The cost at that time and the expectation that concrete laterals would be satisfactory ensured that they would not be built. After 1966 concrete lined laterals predominated and only limited use was made of pipelines. In 1972 a 18 inch buried concrete pipe line was built. The SMRID made the switch to pipelines in 1982 with the construction of southern Alberta's first pressure pipeline which replace Bow Island Lateral 12.

Transition structure going from earth canal to inlet of 18" concrete pipe on Pavka farm in the Chin area, 1971.

ADMINISTRATION OF THE REHABILITATED SMRID

Staff changes during the rehabilitation era included the departure of Jake Thiessen. He left the position of SMRID manager on 1 July 1978 to become Director of Alberta Environment's Design and Construction Division. He also served as the chairman of the Irrigation Implementation Committee which was responsible for the development of major water conveyance and storage systems. From 1988 to1991 he was seconded to Alberta Public Works, Supply and Services during which time he was the project manager of the Oldman River Dam. His promotion to senior positions ensured that he would continue to play a role in the St. Mary Project. He was succeeded by Jim Brown who had begun his career in Saskatchewan where he had worked for the Conservation and Development Branch of the Saskatchewan Department of Agriculture from 1956 to 1974. In 1974 he became the first manager of the Bow River Irrigation district after it was transferred from the PFRA to the water users. He became manager of the SMRID on 1 June 1978.

The rehabilitation program brought about the first major changes in the administration since the creation of the SMRID in 1968. In 1978 an Engineering Department was created with the hiring of Larry Martin who served as District Engineer. His task was to provide engineering services for district rehabilitation projects and co-ordinate the services of consulting engineers engaged by the District on larger projects.

The increased pace of activity was reflected in the need for more office space. The office on Mayor Magrath Drive which the SMRID inherited from the SMRD had become too small by 1981. Increased cost sharing expenditures, the addition of an engineering staff to handle the cost sharing program and the main canal and Forty Mile projects all caused a need for additional office space. In 1981 the Board of Directors purchased a lot at the corner of 36th Street and 12th Avenue North. The tender was awarded for the construction of the new office in September 1982 and the new building occupied on 18 May 1983.

The rehabilitation of the main canal was also reflected in the gradual lifting of the moratorium which began in 1982. In that year it was lifted on a pro-rata basis on three blocks of the District below Stafford. An additional 1,000 acres was provided between Stafford Reservoir and Horsefly Check

St. Mary Irrigation District (SMRID) main office building situated in Lethbridge, Alberta and headquarters for Irrican Power.

while 4,300 acres was provided between Horsefly Check and Forty Mile Coulee and 4,700 acres was made available between Forty Mile Coulee and Sauder Reservoir. Lifting the moratorium below Sauder and Murray Reservoir was not possible because of main canal capacity. In 1984 it was lifted along the main canal up to Horsefly Check and a further 700 acres were allowed to be developed below Sauder Reservoir and above Murray. It was anticipated in 1983 that the construction program would allow expansion to the 360,000 acre level in the near future and 400,000 acres in the far future.

LICENSE

The Canadian North-West Irrigation Company had been issued a permit to divert water in 1899. This permit was amended in 1902 when the right to divert water from the Belly River was added. This permit was renewed on a regular basis because of the ongoing expansion of the St. Mary Project which meant that it was never completed. The CPR made no effort to convert the permit into a permanent license. The conversion of the permit into a license was initiated in 1975 when the SMRID made its first application for a license and paid the required fee. No action was taken, however, until 1987 when Alberta Environment began a very serious drive to update the license of all irrigation districts. By late 1990 agreement had been reached on the final wording of a license. Final acreage has been set at 372,000 acres with a review to be made in 2000.

THE VERDIGRIS EXTENSION

The most controversial expansion came in 1977 when a group of water users along Verdigris Coulee petitioned to join the SMRID. Water was supplied to them through Middle Coulee from the south outlet on Ridge Reservoir. The incorporation of Verdigris Lake into the St. Mary Project had been under consideration since 1902 but had been dropped from the project in 1950. By 1981 the project was completed and had a design capacity of 9,000 acres. Since operation began in 1983 there has been a water quality problem at the lower end of Verdigris Reservoir and in Verdigris Slough. The number of acres was reduced and the flow into Verdigris Coulee increased beyond the design capacity of Middle Coulee to improve the water quality. The increased flow through Middle Coulee created erosion problems.

HYDRO POWER

Using the St. Mary Project to generate hydro electric power was first suggested in 1894 when the federal government act creating the Alberta Irrigation Company authorised the company to furnish water power to any municipality, corporation or individual. Power development was also suggested in the federal government development plans for the St. Mary Project in the 1940s. The federal government identified three possible sites for power development. The idea was suggested again by Jake Thiessen during his term as manager. The idea was finally carried through to implementation by Jim Brown when he became Manager of the SMRID in 1978. In 1988 the Board of Directors budgeted $50,000 for a preliminary look at hydro electric generation at three possible sites which included the Raymond Chute and Drops 1 and 15 on the main canal. A financial analysis was also initiated to determine costs and projected income. The development of a hydro project was given further incentive in 1989 with passage of the Small Power and Development Act which guaranteed the rate for the sale of power from small facilities at 5.2 cents per kilowatt hour for 20 years. In 1989 the Raymond and Chin Chutes were selected for development and preliminary engineering reports commissioned. With the amendment of the Irrigation Act to permit an irrigation district to engage in activities other than the conveyance of water, application was made for approval of the district's power proposal. Construction began in 1992 at the Raymond site and the first power generated in 1994. In 1993 the Irrigation Canal Power Co-operative Ltd (IRRICAN) was established to administer the two projects. IRRICAN is jointly owned by the SMRID which holds 75% of its shares, the TID which holds 15% of its shares and the RID which holds 10%.

HEADWORKS IMPROVEMENTS AND BTAP

The major changes to the St. Mary Project after 1968 had taken place between Ridge Reservoir and Medicine Hat. These included the new distribution system and the improvements to the distribution system. The headworks system west of Ridge Reservoir to Waterton Dam had remained unchanged through to the late 1980s. The first change came in 1990 when the wood stave syphon across Pinepound Coulee was replaced with a buried concrete structure. In 1996 construction began on a new spillway and additional sealing of the dam's abutments. It was completed in 2000. To meet current dam safety requirements, the new spillway has a discharge capacity of 2,500 cubic metres per second compared to 1,500 cubic metres per second for the old one. The total cost of the new spillway is higher than the amount spent on the entire St. Mary Project when it was expanded in the 1950s

The development of irrigation on the Blood Reserve is another recent development. The irrigation potential of the Blood Reserve was first noted by Arthur Wheeler during his 1895 irrigation survey. It was not until the signing of the Blood Tribe Irrigation Project Agreement (BTAP) on 24 February 1989 that actual development began. The BTAP Agreement committed Alberta and Canada to provide major funding for the development of the necessary diversion and conveyance capital works to support an irrigation operation of 25,000 acres of Blood Tribe land. In addition to providing capital funding, Alberta and Canada also provided technical expertise and guidance to the project. The Blood Tribe is responsible for development of the on-farm operations, electrical grid, farm drainage and road network infrastructure.

NINETEEN
THE TABER IRRIGATION DISTRICT
1946 -2000

The post-World War Two era was one of continuity and change for the TID. Because the TID had always operated with a full-time staff, there was no need to make the shift from lateral associations to a centralized system of water management which occurred in the rest of the St. Mary Project after 1968. The availability of stored water was also not new for the TID since it had enjoyed that advantage from 1921 when it went into full operation. The post-war expansion of industrial development in Taber was also anticipated because of the 1939 agreement to build the third plant in Taber between the Taber Beet Growers Association and the Canadian Sugar Refiners Ltd. The expansion of the TID into the Big Bend area was also the culmination of plans made before the Second World War.

The post- Second War changes in the TID were in the size of its operations and the scale of investment made in the industries based on irrigation agriculture. The new plants established after Second World War represented a significant increase in the level of investment as compared to those plants established in the 1930s. The first of these new large-scale plants was begun in 1949 by the Rogers Sugar Company. The construction of this plant, which had been anticipated since 1939, was constructed at a cost of over two million dollars. The canning factory constructed in 1935 and leased to the Broder Canning Company was built at a cost of $30,000. The construction of the new plant to process sugar beets has been followed in more recent times by the construction of potato processing facilities by Lamb-Weston and McCains. The size of the staff was increased and reorganized to accommodate the new demands made by increased expenditures on rehabilitation and the larger size of the District. The increased expenditures on rehabilitation had a dramatic change on the rates which began to increase in 1969 and eventually stabilized in 1988 at $12.00.

In addition to the increase in size was the reconstruction of the system and the new construction using new materials and methods in the conveyance of water. The system built by the CPR in 1919-20 and maintained by the district virtually unchanged for the next 29 years was systematically dismantled and replaced with new canals, structures and construction materials which increased the efficacy of water delivery. The most important symbol of this transformation was the replacement of the Barnwell Flume with a canal in enbankment. The expansion fulfilled the expectations of Sundal who suggested in 1920 that the east limit of the TID should be in the general area of Purple Springs.

Taber Irrigation District, inlet structure (PFRA-13858)

The physical transformation of the TID took place in two phases beginning with the construction of a new main canal from Chin to Medicine Hat and the development of the Bid Bend area in the late 1940s and early 1950s. The new canal was built parallel to the existing Taber main canal. The expansion of the St. Mary Project at this time implemented plans which dated back to the 1920s.

The second phase in the transformation came with the implementation of the rehabilitation funding program in 1969. Funds were made avialable on the basis of a matching formula with the district's share being 14%. The work undertaken though this program resulted in the construction of a new main canal and the abandonment of the original TID main canal which had been built in 1919-20. This program brought about the construction of concrete laterals and eventually pipelines. The use of pipe lines also replaced many miles of concrete lined open ditches. The willingness to adopt new methods of construction also had its disadvantages with respect to the use of concrete as a lining for laterals.

The third feature of the post-Second World War transformation of the TID was the introduction of sprinklers. In 1949 Ted Francis, chairman of the TID acquired one of the first hand move sprinkler systems to be used in Alberta. From 1949 to 1961, the number of sprinkler systems in operation in the TID went from 1 to 182. By 1979, 90% of the irrigable acreage was being irrigated by means of sprinkler systems.

A fourth element of change was the TID's new role in the overall administration of the St. Mary Project. After 1960, the TID actively participated in the overall advisory committee created as a result of the rate dispute. Beginning in 1958, the TID also became involved in the overall review of Alberta irrigation through its submission to the Hanson Committee. Participation in the review process continued into the 1960s both through individual submissions to government and through its participation in the Alberta Irrigation Projects Association (AIPA).

ADMINISTRATION

Ted Sundal continued as the TID's Secretary-Treasurer and was the most enthusiastic supporter of the TID's role as a specialty crop producer until his retirement in 1966. Every Annual Report from

Barnwell flume (PFRA-13849)

1947 to 1967, contained at least one article on his own research, research by others or new commercial uses of specialty crops which were grown in the TID. Corn was the subject of most of these articles. His interest in this crop went back to 1913 when he introduced its cultivation while Secretary of the Taber Agricultural Society. In addition to simply canning corn and using it to manufacture brooms, Sundal suggested that it could be used to manufacture fuel, plastic, nylon cloth, insulation board, syrup, corn flakes and as a canal lining. Corn may be processed into power alcohol which will give 60% efficiency in present tractors ad trucks. By redesigning the manifold Sundal indicated that a much higher efficiency could be secured. He also had a number of suggestions on the way the St. Mary Project should be expanded. These included the construction of a dam on the Oldman River west of Lethbridge.

He was succeeded by Ken Anderson who had joined the staff as Sundal's assistant. Anderson brought a different style to the job and different interests which was evident in the changed format of the Annual Report. The articles on the many uses of corn disappeared. Other staff changes included the hiring of Ron Lloyd as construction foreman in may 1973 in order to coordinate the construction work in the district being done under the government cost-sharing program. The size of the Board was increase from three to five people in 1955 to accommodate the addition of land to the district in the Cranford, Big Bend, East Horsefly and North Fincastle areas.

In 1965 the TID moved its shop and office from a location in town to a site on the southern outskirts of town purchased from John Fleming. In 1974 a new shop was constructed on this site. It was attached to the old shop that was on the property when it was purchased from Fleming. The addition was a Butler steel building with a concrete floor. It was fully insulated heated with 4 large overhead gas furnaces, completely wired for flourescent lighting, power outlets and electrical welder outlets. It was used not only for shop work and machinery repair but to manufacture pre-cast concreted structures for the TID system. The shop was also used to manufacture some metal structures use in rehabilitation work. The ditch rider staff were kept on staff year around to help in the manufacturing and production program. In 1982 the TID relocated the general office to a building shared with the Town of Taber.

The TID began the post-war period in excellent financial condition with the bonded debt reduced to $81,600 which was fully paid off in 1955. The rates of the TID began to increase in 1969 as a

New turnout to Taber Irrigation District (PFRA-13859)

result of the rehabilitation program. The TID choose to take full advantage of its share of these funds which were provided on a cost-share basis. The issue was first discussed at the 1973 annual meeting. In preparation for this meeting, the Board of Directors sent the water users an information letter on the cost of rehabilitation on the project and how this might affect the annual rate structure depending on the pace that the work was done. The responses as compiled in the Annual Report ranged from caution because it would place a financial hardship on the water users to a willingness to accept a substantial increase in order to get the system rebuilt and modernized as soon as possible. The result was a vote of confidence in the elected board which were directed to deliver water as economically as possible while the same time maintaining the district works in a good state of repair.

EXPANSION

The expansion of the TID before the Second World War had been undertaken at its own initiative. The western extension was made up of land purchased by the district from the Schools Land Branch and then re-sold by the TID. The construction and financing of the system plus its settlement was managed entirely by the TID. The role of the federal government, in this situation, was providing the land and a plan for the construction of the irrigation system.

The TID began the development of the Big Bend addition just prior to the start World War Two. In 1938 the TID approached the CPR with the suggestion that it increase the capacity of Chin Reservoir and then provide the additional water to the TID. The CPR was not interested at that time in spending extra funds on the Chin Coulee. In January 1939 the Board met with Sam Porter in Calgary to discuss the expansion issue with the CPR. In anticipation of this meeting, Fred Cross prepared a review of the operations of the contract between the TID and the CPR. These comments reflect his view that the TID had been the main beneficiary of this contract and his suspicions that it was not entirely living up to its terms. The CPR indicated that this time that they were interested in proving more water as well as selling additional land.

In 1938 the TID also considered expansion east of the Town of Taber in the area of Reliance Lake. Ted Sundal and John Barton examined Reliance Lake for possible use as a storage reservoir. This investigation indicated that it had possibilities as a tail-out reservoir and that, by means of pumps, this water could be used to irrigate additional land to the northeast. Sundal was immediately

Ted Sundal and others plantng tomatoes (Taber Irrigation Impact Museum)

authorised to make a full survey of the lake in order to determine its possibilities for development as a commercial reservoir using TID waste water. This decision was influenced by the signing of an agreement in the same year which committed the Canadian Sugar Factories Ltd. to building a sugar beet processing plant in the Taber area. Four copies of the plan of the reservoir prepared by the Water Resources Branch were sent to the Company as part of the TID's efforts to attract the company to Taber.

Development work on the lake included the construction of dams in 1940. Development work on the reservoir and plans to build a distribution system were delayed by the Second World War and the transfer of the St. Mary Project from the CPR to Alberta The Second World War prevented and further construction.

Expansion, after World War Two, was courtesy of the PFRA and the Alberta government which had decided to undertake a joint program to expand the St. Mary Project based on the plan developed during World War One. On 29th July 1948 Ted Sundal met with PFRA officials Wally Foss and W.B. Thomson as well as N.M. Sutherland who was from the office of the provincial director of water resource. to discuss the expansion of the TID. The meeting was called at the inititiative of Foss who was anxious to make a start on the development of the distribution system since water was going to be available from the Jensen Reservoir in 1949. The construction of the new sugar factory at Taber was anther reason for the choice of the Big Bend area as the location for this expansion. The development plans for the area, as already outlined in chapter 16, included a new main canal with a capacity of 1,700 cfs from Chin Reservoir to Horsefly Lake which was to be converted into a reservoir with a capacity of 10,000 acre-feet. Taber Lake was to be enlarged from 3,500 acre-feet to 7,000 acre-feet. A main canal and a distribution system to irrigate 12,000 acres of irrigable land was then to be built.

The PFRA prepared the plans and supervised the work while the actual construction was handled by the Province of Alberta. Construction of the system began in September 1949 by Shannon Brothers of Didsbury. After Alberta and Canada had negotiated a master agreement to govern the overall expansion of the St. Mary Project, expansion of the distribution system in the Big Bend area was continued and new distribution systems begun in the vicinity of Fincastle and Horsefly Lake Reservoir.

The TID considered the inclusion of the Big Bend area in the TID at a special meeting held on 18 March 1952. At that time, the Board of Directors proposed an ambitious plan of rapid expansion beginning with the Big Bend area which was to be added in 1952. Other additions proposed for 1952 included ten quarter sections in the Cranford district area. This land already a had a partially developed ditch system which conveyed surplus water to the area when it was not needed by the TID. The board of directors wanted to provide full water right service as soon as water was available from the expanded St.Mary Project. Other miscellaneous parcels of land in and bordering the present Taber Project which have been served with surplus water were also proposed for inclusion. The final phased in this expansion program was to occur 1953 and 1954 when the north and east Fincastle units were to be added top the TID. These

Flood caused by canal break, 1974 (Taber Irrigation District)

extension would expand the TID to 60,000 acres which would extend from Chin Coulee in the west to a point three miles east of Purple Springs. The advantage of the extension was that it would lower the per acre operating cost.

The water users, however, were not as anxious to expand as the Board of Directors. The decision was made to only include the Big Bend area along with the miscellaneous parcels. The inclusion of the north and east Fincastle units were tabled pending further consideration. In 1965 these areas were transferred from the SMRD to the TID. This added a total of 11,627.1 acres to the TID. This step was taken at the initiative of the water users in these areas. These water users felt that they were more a part of the TID than of the SMRD. By 1965 the TID had expanded to the size predicted by Sundal in 1921.

In 1990 the Alberta government set a cap on the number of acres that could be irrigated in each of the irrigation districts in southern Alberta. The irrigated acreage cap for the TID was set at 82,200 areas. In 1996 the TID had reached within 3,000 acres of that total. The district has decided that it will probably want to retain some of these acres for existing water right parcels that may want to increase the number of acres that they irrigate in the future.

REHABILITATION

The rehabilitation of the existing system prior to 1969 began with the replacement of the Barnwell flume. The topic of replacing this structure first came up for general discussion at the 1952 annual general meeting. It was reported to the meeting that the PFRA had estimated the cost of a earth filled dam to replace the flume at about $13,000. The decision was made at that time to try and maintain the flume and use it for a few more years. The flume was operated until 1958 when excessive maintenance costs finally made its replacement by an earth fill as suggested in 1952 a necessity. It was completed in the spring of 1959 with the cost shared between the provincial government and the TID. This was the first time in the history of the TID that it had received a grant from any government for infrastructure construction.

The Big Bend area was also in need of immediate rehabilitation because of seepage problems. The high rate of seepage was the result of a change to more sandy soils. The Department of Agriculture subsequently rebuilt the system and installed canal lining to reduce seepage.

Taber Irrigation District Shop
(Taber Irrigation District)

The second phase in the transformation of the system began in 1969 when the Province of Alberta began to fund an irrigation rehabilitation program. The main canal rehabilitation program, which has already been discussed in detail in chapter 16, resulted in the construction of one canal to replace the old TID main canal and the main canal built by the SMRD between 1950 and 1955.

The Province of Alberta and the TID signed their first agreement to begin the rehabilition of the TID distribution system on 15 July 1969. This agreement made $55,165.55 available for capital improvements on the project. These funds were used for lateral relocation and the backfilling of the old main canal. In 1972 the installation of concrete lining began in earnest as 3 ½ miles of open concrete lateral and buried concrete pipe was completed on lateral C. Concrete lining was a significant part of the 10 year plan prepared in 1975.

WATER LICENSE AND WATER SUPPLY

The TID was issued a separate water license for the first time in its history in 1982. Prior to 1982 only permits to divert water had been issued first to the AR&ICo. and then to the PFRA under the terms of the 1950 agreement to expand the St. Mary Project. The TID license was based on the original request to divert water from the St. Mary River made by the CNWICo. on 7 February 1899. It also included the 12 July 1919 agreement for the delivery of water to the TID by the AR&ICo. and the 1950 and 1973 agreements between Alberta and Canada for the construction and management of the headworks. It permitted an annual diversion of 150,000 acre-feet from the St.Mary River. The license was revised in 1991 to permit the diversion of an additional 8,000 acres- feet.

TWENTY
THE MAGRATH IRRIGATION DISTRICT 1946 - 2000

The post-Second World War development of the MID was in sharp contrast to its earlier years during which time it struggled to survive because of the burden of debt and an unreliable water supply. After 1946 it was secure financially and had a stable water supply thanks to a system of dams which it had helped to bring about through the persistence of Chris Jensen. The MID was thus able to expand in size and deal with other issues related to its development. The most important issue was drainage which was first investigated in 1948 by staff from the Dominion Experimental Farm, in 1962 by the PFRA as part of the ARDA irrigation studies and in 1982 by the Irrigation Branch as part of the preparation of a new master plan.

Like the Lethbridge-Coaldale area and Taber, the MID's development revolved around the changes in the way the St. Mary Project was administered given the involvement of the province of Alberta and the federal government in its expansion, rehabilitation and on going operation after 1946. It also moved steadily away from being administered on a part-time basis by the Board of Directors from their own homes to having its own shop and office facility, owning its own equipment and having full-time staff to deliver water directly to the land and carry out other administrative tasks required under the 1968 Irrigation Act.

The SMRD brought the first changes to the MID when it terminated the two tier system of water contracts and gave the MID greater flexibility in its administration of the water rights under these contracts. It was also encouraged by the Manager of the SMRD, P.M. Sauder, to adopt a system of full water rights in order to make maximum use of the increased water supply. The MID also benefited from the assistance provided by the SMRD, along with other government agencies such as the PFRA and Dominion Experimental Farm to help operate the district and investigate problems such as drainage. The MID was fortunate in the way it undertook its post-1968 rehabilitation program. The problems with concrete laterals became evident before it undertook this type of construction which had been included in the 10 year plan adopted in 1975. The MID was thus able to use its resources for pipeline construction which dealt more effectively with its particular needs since seepage was eliminated in the transport of water.

Unlike the other three irrigation districts, the MID became less involved in the overall administration of the St. Mary Project after 1968 when it did not participate in the main canal rehabilitation program. Since the MID was supplied directly from the headworks, there was no necessity for it to do so.

Despite its small size, the MID played an active role in the irrigation industry review during the early 1960s through the activities of George Dudley who was an early advocate of the installation of drainage systems in irrigation districts. The operations of the MID were reviewed as part of this review.

WORKING WITH THE CROWN CORPORATION

The post-Second World War history of the MID began on 6 May 1946 when the MID was officially informed that the SMRD was now the administrator of water agreements 1374 and 1375. These contracts had been negotiated in 1926 to provide water to the holders of old and new water contracts in the MID. Thus there was different rates for the water supplied under each contract and different obligations with respect to the collection of those rates by the MID. In the case of the old water contracts, the MID acted as agent for the CPR. The SMRD reduced the charge to 50 cents per irrigable acre for both contracts which immediately reduced the cost of the water supply by $1,344.50. When the land owned by Charles Buras was removed from the assessment roll in 1947, the charge was further reduced to $3,515.00. The SMRD also turned over the total administration of contract 1374 to the MID thus it was not required to remit any of the funds collected from these water users to the SMRD. The SMRD thus gave the MID a free hand to terminate any of the old water contracts because of nonpayment of rates and sell the water to other farmers at the same price as the new water users were paying. Adjustments were also made by the SMRD to accommodate circumstances which made it difficult for the MID to pay this amount. In 1952, owing to the delay in harvesting of the 1951 crop and the consequent delay in collections, no interest was charged on arrears This procedure was first carried out in 1949. The SMRD was willing to make these changes because P.M. Sauder, the Manager of the SMRD, intended to terminate their use in the operation of the project and introduce a more flexible approach to establishing the water rates which would reflect the actual cost of operation. Setting a uniform price of 50 cents an acre-foot for each contract and letting the MID administer both contracts as they were the same eliminated the distinction between "old" and "new" water. With the passage of the St. Mary and Milk Rivers Water Agreements (Termination) Act in 1960, the old water contracts were officially cancelled and the land incorporated into the MID.

Grand openingof Magrath Irrigation office, March 30, 1988 (Magrath Irrrigation District)

In 1957 the SMRD increased the water charge to the MID from 50 cents to 75 cents per acre. In 1958 they were increased to 85 cents per acre. In 1959 charges were levied at 25 cents per acre-foot for water delivered plus a canal conveyance charge of 19 cents per irrigable acre. After many meetings of protest and negotiations the rate structure was revised. A charge of 25 cents per acre-foot was made for water delivered and the canal conveyance charge was reduced to 10 cents per irrigable acre was agreed to. After the transfer of the SMRD to the SMRID, the MID obtained its water without charge from the Department of the Environment which operated the headworks after 1973.

The SMRD also provided technical assistance. The first time this occurred was in 1948 when Sauder and Brown advised the Board on how to deal with an obsolete flume. In 1950 they advised the Board again on how to deal with a section of a canal which was in danger of sliding into the St. Mary River.

POST-WAR ADMINISTRATION

The first significant change in the administration of the MID came in 1947 with the retirement of Christian Jensen as Chairman of the Board of Directors. He had been actively involved in the organization of the district between 1922 and 1926 and had been Chairman since 1929. His retirement came with the fulfilment of his desire to deal with the water supply problem. The new Pothole Dam was appropriately renamed Jensen Reservoir in recognition of his years of service and dedication to the issue of water storage on the St. Mary Project. Other board members who retired at this time included Ralph Thompson who had served as a trustee since 1925 and David Whitt who had been a trustee since 1938. The new group of board members who directed the MID's affairs into the 1950s included Mike Schneyder, Harry Bly, Paul Taylor, David Keeler and C.C. Spencer.

The 1960's also brought a change in the membership of the board. This new group included George Dudley, Cyrus Card, and Lalovee Jensen. George Dudley served as trustee for one year before becoming chairman 1962. He remained chairman until 1987 during which time the MID initiated its rehabilitation program. This rehabilitation program dealt with the issue of drainage in which he had taken considerable interest. George Dudley was also an important link with the origins of the Town of Magrath since his family had been the first settlers.

The Board of Directors continued to be involved in the day to day operations of the district up until 1999 when Low Hansen was appointed as the first full time manager of the MID. Prior to this decision, the Chairman carried out this task. This role was formalised when John Weing became a member of the Board in 1977. During his term as chairman he was also the director in charge of operations. In that capacity, he prepared detailed reports for the annual general meeting which outlined the work for the year.

On 4 November 1946 the St. Mary and Milk Rivers Development and the MID signed an agreement which permitted the MID to retain use of the DNR excavator based on the conditions that the MID would defray charges of labour and repairs to place the machine in good working condition for the coming year of 1947. The MID was not charged a rental fee on the understanding that when it was required by the SMRD that it was to be returned in as good condition as when received. It was used until July 1954 at which time the Sugamato Brothers from Raymond were hired to clean the canal. The machine was eventually given to the Town of Magrath and is currently on display in the park.

Between 1954 and 1961 the MID rented spraying equipment from the municipality for weed control or contracted out the task of ditch cleaning to the Prairie Ditching Company. The idea of the district purchasing its own back hoe was first suggested in 1959 but it was not until 1962 that such a piece of equipment was acquired. In 1962 it also purchased a new truck. This equipment was used strictly for district work since the rate payers were opposed to its use for custom work. In 1964 the MID purchased an old DNR excavator from the CPR. In 1965 the MID purchased a cat and a dozer from Nieboer Farm Supplies for $4,637. With the introduction of the rehabilitation funding program the MID took the opportunity to use its own equipment and to be reimbursed for its use by the government. This practice was one of the reasons for the financial health of the district which was out of debt by 1949.

In 1963 it purchased its first shop to house this equipment. In 1986 the MID acquired 8 acres of land for $12,000.00 to build a permanent building, storage and maintenance facility. The MID also manufactured pre-cast structures at the site for use in the system.

ADDITIONAL WATER AND DISTRICT EXPANSION

The first actual expansion of the system was initiated on 6 November 1951 when Horace Ririe, Secretary-Treasurer of the Magrath Irrigation District was informed by P. M. Sauder that it was now possible to deliver the legal duty of water to all water users in the MID. Under the regulations in force at that time, the duty of water set by the Director of Water Resources, each water user was entitled to one and one half acre feet per acre for

George Dudley (Magrath Irrrigation District)

the area covered by his agreement and assessment. Sauder noted further that "as the assessment on most of the quarter sections in your irrigation district are less than the irrigable area, water for an additional 5,000 acres had been allotted to your district. Sauder also indicated that "your water users will not in future be permitted to irrigate a larger area than they pay for but as stated above, they will be entitled to one and one half acre feet per acre for the area actually paid for. In due course the land will all be reclassified. Sauder also noted that "we are also asking for an appropriation to pay for the reconstruction of your west lateral, but the appropriation of this money is contingent on your district completing the sign-up for the additional area at an early date. I would, therefore, like to arrange to have the sign-up completed as soon as possible. We can loan Mr. Wyman to assist in getting the sign-up started.

The MID originally distributed water on the basis of shares. This approach permitted the water users to irrigate as much land as they wanted. In 1960 the MID changed this system and introduced new rules for water distribution which were similar to the ones introduced into the Lethbridge-Coaldale area by Sauder. Under the new system, no farmer was to irrigate more than his acreage according to records of the MID. The water user was charged $1 per acre for any land irrigated beyond this limit. The second rule was that a farmer could not buy his neighbour's water. The third rule was that each farmer should take out what water he wants or expects to use now. The final regulation was that everyone should ask the watermaster at least 24 hrs. in advance for water and that the water master would turn the water on or off.

DRAINAGE

The main concern for the farmers, Board of Directors and the government agencies involved in the administration of the MID after World War Two was drainage. The first people to take action were the farmers themselves who were very interested in the problem due to the increased alkali areas on their farms. Many of these farmers were also in a financial position to take limited steps to remedy these conditions. One farmer, who lived 4 miles west of Magrath, had actually begun the construction of a drainage system for his land which had become non-productive because of alkali. He installed approximately 600 feet of tile drains and reclaimed about 8 to 10 acres. He and other farmers in the area contacted the PFRA in the fall of 1948 and the spring of 1949 requesting technical assistance in their efforts to deal with the problem.

In response to these requests, an investigation of the drainage problem was initiated by staff from the Dominion Experimental station and the PFRA in the spring of 1949. A rectangular area 2 miles by 4 miles, which included the townsite of Magrath and its immediate vicinity was selected for study. The investigation was started at this location because the water table appeared to be the highest and because the land had been irrigated for the greatest number of years. Piezometers of 3/8 black pipe, 11 feet in length were installed throughout the selected area in May 1949. The Piezometers were installed be jetting using water pressure. A piezometer was installed every half mile along road allowances as field conditions did not permit installations in the centre of the sections. A few piezometers were plugged or damaged during installation and were replaced using the driving method. Readings were taken from 13 June to 28 December 1949.

Subsoil borings were made using a Buda drill. This drill was able to take samples down to a depth of 24 feet. Samples were taken at almost every piezometer location and samples of each change of material were sent to the Soil Mechanics Laboratory at Saskatoon.

The data obtained from these tests showed that the water table was generally 4 feet or less throughout the area which was advantageous to the accumulation of salts on the surface of the soil. The maximum elevation of the water table occurred in June and stayed fairly constant during the 7 month period. The minimum elevation occurred in December. The average difference between the maximum and minimum water table elevations was approximately 3 feet. The conclusions drawn was that the water table in the Magrath area was too high for continued high crop production. The report briefly outlined a drainage plan which could reduce the water table.

The study of the problem was continued in the summer of 1953 by R.A.. Milne and T.W. Peters who organized their investigation as part of what they hoped would be the first steps in a study to be made of presently irrigated areas in Alberta. The need for this type of survey, as noted by Milne and Peters, was emphasised at a meeting of the technical committee on irrigation and drainage held in Lethbridge early in 1953.

IRRIGATION REVIEW

The MID was part of the irrigation industry review undertaken in the early 1960s through the ARDA program. It was the subject of a study by the PFRA entitled "The Annual Cost of Magrath Irrigation District" which was completed in August 1967. This document summarized the history of the district, described the projects's works and summarized costs of operation. The MID in 1967 as described by this report, contained approximately 23,000 acres of which approximately 8,000 acres were irrigable. The district had approximately 12 miles of main canal with a capacity of 50 cfs or over. The canals were maintained in good condition with periodic cleaning and trimming. There were no structures in the MID that were classed as major structures. Some road crossing were timber bridges of 8, 10 or 12 feet spans. Corrugated pipe was being used when replacement was necessary. The distribution system contained 271 structures including checks, turnouts, culverts, small bridges, flumes and paved (grouted rip rap) ditch sections. Most of the original untreated timber drop structures had been replaced with rip rap or field stone dumped into the ditch to prevent scour. In many cases the rip rap was cemented with grout. This method, the report noted, of replacing drops and maintaining canal sections on steep grades appeared to be well suited to the needs of smaller districts and provided as adequate and economical replacement for small drops. This technique was developed by the TID in the early 1920s.

Of the 271 structures inspected 25 or 9% were in poor condition and required replacement within two years, 113 or 42% were in fair condition and were expected to last an additional 5 to 10 years, while the remaining 133 structures were in good condition and were expected to last 10 years or more. The MID did not have a inventory of structures nor a record of repair and replacement. Since the estimated life of a timber structure was 10 years, an annual replacement of 10% of the structures was necessary to maintain the system in good operating condition. Since the MID had only 9% of its structures requiring replacement within the next two years it was concluded that their renewal program was adequate. The District was also using creosote treated timber in its replacement program which, doubled the life of the structures over that formerly obtained by untreated timber construction.

The main deficiency was the district's lack of a adequate network of drainage ditches. The result was that by 1954, 48% of the irrigable acreage in the District and approximately 27% for the total acreage within the District was affected to some extent by salt accumulation and water-logging. By 1965 further deterioration had taken place with the result that about 23% of the irrigable area was now completely non-productive while 12% of the dryland was in the same category.

The main distributaries for the western part of the District were located on high ground and the canal in many cases were excavated to bedrock. Seepage from these canals ran down the slopes to the flatter land below and also surfaced on the slopes at various locations. Much of the irrigated area is in the flat land which has heavy soils with poor surface and internal drainage. These conditions, the report concluded, was one of the causes of the serious deterioration of land within the district. The PFRA report repeated the recommendation of the 1965 report of the Alberta Department of Agriculture that a complete surface drainage system and extensive canal lining should be constructed, the estimated cost of which was $610,695.00.

The report then made an estimate of the average annual cost of operating the system. This figure took into consideration the cost of water and rehabilitation. Based on these various input costs, it arrived at a figure of $2.92 cents per acre. This figure included 68 cents for replacement of major structures and bridges.

The report also included proposals for improvements to the district. Its main recommendation was the elimination of the use of Pothole Creek to deliver water. The total requirements it suggested should be made via the high line canal.

George Dudley also gave a paper at the Taber Irrigation Policy Conference held on 12 April 1965. In this paper he reviewed the history of the MID from 1924 to 1965 noting the problems of debt and water supply. In 1965 the biggest problem for the MID was:

> that much of our land is deteriorating due to the accumulation of salts. In 1954, 48% of the irrigated portion of this area was affected to some degree, and 27% of all the land within the irrigated district was effected. This condition is progressively getting worse. A large percentage of the land effected has been under the ditch since 1901. This was part of the oldest project other than private enterprise in the province. An undertaking to remedy this is beyond the financial scope of our district. Are

> we going to spend fortunes on developing projects to reap what benefits we can from the land until it becomes unproductive and then move on or as history in the past has usually revealed, or should we aim for permanent developments? In many areas drainage is considered part of the initial capital undertaking. With the world population explosion, and the tremendous amount of good productive land being taken out of production for industrial and urban development, can we justify a short term policy?[281]

REHABILITATION

The MID signed its first cost-sharing agreement with the provincial government on 15 July 1969. Under the terms of this agreement, the provincial contributed $6,825.00 or 86% of the cost of rehabilitation work undertaken by the district for that year. The MID contributed a further $1,111.20 or 14% of the cost of rehabilitation. The first project was the Magrath Spur Drain which involved deepening of a natural drain to make it more efficient in the disposal of water. The existing natural course was very shallow, meandered and flowed all year. Wherever the natural drain was not well defined or was lacking in depth a high ground water table existed and farming was no longer possible. The MID deepened and straightened the natural channel. Additional rehabilitation on work undertaken to 1975 included bridge repair and structure replacement.

At the annual general meeting held on 16 March 1974 the MID considered the construction of concreted lined lateral. Ron Francis of the Irrigation Branch attended the meeting In October 1975 the MID submitted a 10 year plan to the Irrigation Council. This plan indicated that $3,227,000 would be required for rehabilitation and development of the district up to 1985. It consisted of 24 projects, eleven of which were the installation of concrete lining on Lateral 31 at a total cost of 2.3 million. The remaining projects were for the expansion of the district to the north and south. The feasibility of concrete lining was questioned after the plan was submitted to irrigation council. The problem of concrete lining cracking became a serious problem in other districts. The plans for expansion were also modified because of the tight water supply. The MID Board of Directors, therefore, defer the concreted lining projects. The projects that were built included the MID first pressure pipeline to serve 1040 acres south of Magrath. It was completed in 1982 at a cost of $547,953.00.

In March 1980 the MID submitted a revised plan which indicated that the district had tried to follow the 10 year plan as outlined in 1975, however, inflation, problems with concrete construction and fixed funding provisions had resulted in delay. This report proposed a revised schedule of projects for completion by 1985.

In the summer of 1981 the Department of Agriculture requested through the Irrigation Council that a General Rehabilitation Plan be prepared. The purpose of this report was to provide a comprehensive planning report which included proposal for systems improvement and costs estimates to upgrade the distribution and drainage systems to present day engineering standards. These proposals represented the district's requirements in the way of rehabilitation and enlargement through to 2000. This report was completed in March 1983. It indicated that the MID had a number of problems which included seepage and salinity. It was estimated that there are approximately 5100 acres affected by seepage, salinity and water logging adjacent to the irrigation works in the area. This amounted to 18% of the total irrigable land area within the district boundaries. These problems occurred in the district largely because of the lack of adequate natural or artificial drainage outlets, shallow bedrock, over-irrigation and spillage of excess water. A second problem noted was excessive vegetative growth in canals particularly on small laterals having a flat gradient and shallow depth of water.

TWENTY-ONE
RAYMOND IRRIGATION DISTRICT 1946 - 2000

The construction of the Jensen and St. Mary Dams and the elimination of its debt to the CPR, immediately after World War Two, removed the two major barriers to the progress of the RID. The post-Second World War period thus provided the RID with the opportunity to build a progressive and efficient irrigation system, given the advantage of an assured water supply and freedom from debt.

The post-Second World War progress of the RID came about in two stages with the 1968 Irrigation Act being the dividing line between these two periods. The first stage involved the changes to the RID as a result of the expansion of the St. Mary Project. Construction of Ridge and Raymond Reservoirs and the new main canal north to the Lethbridge- Coaldale area replaced Pothole creek, the 1889 main canal from Magrath and Nine Mile Coulee as the means by which water was delivered to the RID and transported further north to the Lethbridge-Coaldale area. The additional water provided to the RID by the new headworks necessitated the enlargement of the RID canal system, the construction of a new main lateral to serve the west Raymond and Welling areas and the modification of some structures such as the Stirling flume to carry the increased amount of water. After 1968, the emphasis shifted to the rehabilitation of the distribution system through the introduction of concrete structures and pipelines.

The expansion of the St. Mary Project and the rehabilitation of the RID brought with it new ideas about its administration which P.M. Sauder was also trying to introduce into the Lethbridge-Coaldale area. The most important of these ideas was the elimination of partial water rights which had been a distinctive part of the RID since its creation in 1925. The RID was thus involved in ongoing negotiations with the SMRD over how and where the new canals were to be built, the administration of the new water, the disposition of the surplus revenue account and the price for water. These negotiations would ultimately serve to give the RID a role in the management of the St. Mary Project which had never been the case when it was administered by the CPR. Its role in the main canal administration committee provided a way for it to contribute to the overall operation of the St. Mary Project while preserving its independence.

The administration of the RID affairs after World War Two went through a fundamental change. It went from a system in which the Board of directors managed the main canals and the water users managed their own laterals to a centralized system of water management with a full- time manager and other staff to supervise maintenance and the delivery of water. While this shift was made over a number of years, the Irrigation Act 1968 was the critical step because it repealed the 1933 legislation which had justified the retention of the previous system. The shift was reflected in number of important staff changes and in Board membership which began in 1945 with the retirement of John W. Evans and the assumption of his duties as Secretary-Manager by Rulon Dahl. The extra work and increased responsibilities of this position after 1968 culminated in the hiring of Gordon Zobell in 1980 as the first full time Manager. The establishment of a permanent office complete with a shop and the adoption of a system of ditch riders completed the transformation of the RID into the modern institution it is today.

In addition to accommodating the expansion of the St. Mary Project while at the same time preserving its own independence, the Board also dealt with other issues such as drainage. The drainage problem was particularly severe in the north portions of the RID, in the area of Stirling Lake.

This post-Second World War period is also distinctive for the changes made in the way the system was maintained. This change involved the use of machines to replace water users and horses in the maintenance of the main canal.

An important element of continuity during the post-Second World War era was the RID's continued role in providing people who took leadership roles in the industry. Doug Paxman served on the 1957 Irrigation Study Committee while Rulon Dahl was active in the development of the Alberta Irrigation Projects Association which was the successor to the South Alberta Water Conservation Council.

POST-WAR DEVELOPMENT

The first post-Second World war changes in the administration of the RID involved the election of new people to the board of directors and new staff to replace individuals who retired, many of whom had served since the creation of the district in 1925. Jesse Wilde retired in 1952 after serving on the Board since the creation of the district in 1925. The other retiring member was Wilford Brandley. In recognition of the increased workload for members of the Board, they were paid for their supervisory activities.

Staff changes included the retirement of William J. Evans as the Secretary-Treasurer in 1945. His association with the development of irrigation and the sugar beet industry went back to the time of the Knight Sugar Company when he was their field supervisor. Rulon Dahl assumed his new duties as the Acting Secretary on 7 April 1945. He combined the duties of Secretary-Treasurer and continued as a board member until the election of Doug Paxman in 1948. Dahl not only served as Secretary-Treasurer but made his house available for use an office. His was paid $500 a year for his services. His pay was increased in 1953 after he pointed out to the Board that the Secretary-Treasurer's salary had remained the same since the organization of the district in 1925. It was not sufficient for the work required especially in view of the fact that his home was used as the office. The only time that this approach to managing the districts affairs was questioned prior to 1968 occurred in 1955. At a meeting of the Board of Directors, the Chairman of the Board Doug Paxman reported receiving complaints over the present arrangements of a part-time secretary and the office hours at the home of the Secretary. Some water users felt that it was sometimes difficult to contact the Secretary to get information or to make a water rates payment. It was explained that the secretary arranged to have the office open for 3 or 4 days after beet payments were made in the fall to take care of beet cheques which might have the districts name on them. Alternatives to paying rates at the office were also available since they could be paid by mail or by depositing the amount to the credit of the RID at the Bank of Montreal. Following the discussion of the issue, a motion was passed approving the present arrangement of a part-time secretary and the use of his home as an office since it would be more expensive to rent and maintain an office and hire a full time secretary. Rulon Dahl was the son of Arthur Dahl who had been very active in the effort to save the Knight Sugar before World War One. In the 1920s he had also played an important role in the re-establishment of a new sugar beet plant in Raymond.

John F. Salmon, who had served as ditch rider since 1926 retired in 1947. In his letter of resignation, he indicated that his work with the Board had always been most pleasant but due to the fact that he was 75, he felt that he should make way for someone younger who might need the work. The resignation apparently came as a surprise to the Board which requested him to stay on long enough for a replacement to be hired and trained.

Record keeping was also improved with the purchase of new ledgers. Information as to expense items and cost of operation would be much more easier to obtain especially where comparison with previous years operations was desired.

The RID never returned to the situation it was

in before World War Two with respect to its financial affairs. It was able to build up a capital reserve with which it purchased bonds. The problem of arrears returned periodically but the extent of the problem never threatened the survival of the RID or required it to take extreme measures to solve the problem. The solution adopted was to have the beet cheques for those farmers in arrears to be made out to the RID rather the use of the enforcement procedures as outlined in the Irrigation Districts Act.

The financial health of the district was reflected in the water rates. Following the large increase in the rates in 1946 to cover the cost of paying off their debt to the CPR, the Board was able to substantially reduce rates in 1947 and maintain them at that level through to 1968.

One reason why rates could be maintained at this level was the continued role of the water users in maintaining the own laterals. The use of Rulon Dahl's house was a second reason. A third reason was the reduction in the cost of water by the SMRD. The increase in the charge for water after 1955 did not require a rate change. The cost-sharing program brought about the first significant changes in rates.

The cost of operating the district gradually increased because of increases in wages. The rate paid to members of the board for their supervision went from $5.00 per day in 1948 to $8 per day in 1950. In 1950 a mileage allowance of 10 cents per mile was provided.

The operation and maintenance of the RID, like the SMRD, was divided between the water users who maintained their own laterals and the Board of Directors who maintained the main canal. Up to 1946, the Board hired their own water users to carry out the maintenance of the main canal. After 1946, however, water users and teams of horses were replaced by machines rented by the Board to carry out specific tasks. The first recorded use of machines for maintenance purposes occurred in the fall of 1946 when L. L. Christensen, who operated a construction Company, was hired to use his dragline for canal cleaning. He worked a total of 153 hrs. at $6.00 per hour and completed about 3 miles. In April 1949 the Board rented the municipal grader and in 1950 Mr. Dunsmore of Cardston used his equipment to work 9 mile of canal at a rate of $7.50 per hour. Between 1951 and 1957 the RID rented a Gregerson ditcher from the Eastern Irrigation District for ditch cleaning purposes.

The idea that the RID acquire its own equipment was first suggested by L.L. Christensen in October 1947. He proposed that the District purchase a levelling machine which he would pay for and operate. The board decided at that time that it did not need such as machine because its responsibility was only to distribute water to farmers headgates and not provide equipment to them for such purposes as land levelling. The Board was prepared to recommend L.L. Christensen's application for the machine with the understanding that he would make his own arrangements with farmers for land levelling. In May 1949 the Board discussed the possible purchase of equipment from the Union Tractor Company. The Board had previously made a trip to the St. Mary dam site to evaluate the type of equipment which was being sold by this company.

The first machine purchased was a pile driver which was obtained in November 1950 from the SMRD. This purchase was followed in September 1953 with the purchase of a machine manufactured by the Kirchner Machine Co. In January 1958 the Board decided to purchase the Gregerson ditcher rather than renting it from the Eastern Irrigation District. By 1968 two back hoes had been acquired along with other machinery. This machinery was rented to the water users to clean their laterals .

DEALING WITH THE CROWN CORPORATION

The immediate impact on the RID of the transfer of the St. Mary Project to Alberta came on 6 May 1946 when the Board of Trustees was officially notified by the SMRD that under the provisions of the St. Mary and Milk River Development Act, the interests of the ARICo. in water agreements 1340 and 1341 had been transferred to the government of Alberta and that monies due under these agreements were collectable by the Manager of the St. Mary and Milk Rivers Development. The RID had fully paid both contracts with the exception of a small balance of $269.59.

The SMRD's administration of these contracts was very different from that of the CPR. The reason for this different approach by the SMRD was that P.M. Sauder, the Manager of the SMRD intended to abolish their use in the administration of the St. Mary Project. He wanted to introduce a more flexible system in which the rate would be based on the actual cost of the service provided. This approach had a number of immediate benefits for the RID, the first of which was a reduction in the rate which the RID had to pay for water. The rate was set at 50 cents for both agreements which required that the RID pay $4,284.00 under agreement 1340 and $3,221.00 under agreement 1341. The SMRD, in effect, did away with the distinction between old and new water which had been a part of its history since 1926. The SMRD gave the RID a free hand in the administration of the old water contracts by not interfering with their effort to convert them into new water contracts.

THE CONSTRUCTION BOOM AND THE RID

The next phase in the evolving relationship between the RID and SMRD began in the summer of 1950 after the provincial and federal governments had successfully negotiated a comprehensive agreement for the expansion of the St. Mary Project. The construction program undertaken between 1945 and 1950 had not directly affect the main canal and distribution system of the RID since the additional water available from the Jensen reservoir was delivered using the existing system. The first delivery of stored water to the RID was made in March 1948. The construction program launched by the SMRD and the PFRA in 1950, however, had an immediate impact on the RID since it meant the abandonment of the canal from Magrath which had originally been built in 1899-1899. The RID was now to be supplied with stored water from a canal running north from Ridge Reservoir. The issues raised by this change included the location of the new main canal south of Ridge Reservoir, the role of the Raymond Reservoir in the expanded St. Mary Project and the location of the canal to serve the Welling and west Raymond areas. The expansion of the RID based on the availability of new water also raised a number of issues which included the administration of the new water rights and the role of the SMRD in the enlargement of the main canal and laterals of the RID itself.

The expansion of the RID after World War Two was first considered in November 1951 when P.M. Sauder informed Rulon Dahl that it would be possible to deliver the legal duty of water to all water users in the RID. Each water user was thus eligible to receive one and one half acre-feet per acre covered by their agreement and assessment. Sauder further pointed out that in the future water would only be supplied to the area actually covered by water right. While the Board was happy to receive the additional water, it objected to the water right regulation. As a result, Sauder indicated the distribution of water would be left up to the Raymond Irrigation District but the SMRD would expect the Board to see that water users did not irrigate more land than they had water right to cover. Following the resolution of this issue District purchased 5,000 additional shares of water from the SMRD with the basis of settlement being $1.00 per share paid to the SMRD and $9.00 of the $10.00 water rights share being retained by the District and the water users. The Board was to retain $5.00 to cover the expenditures on canal improvements while the water users got $4.00 per share to cover the improvements to the laterals.

DRAINAGE

Drainage was also a concern of the water users immediately after the Second World War. The first of a long list of drainage issues dealt with by the RID was the result of a dam constructed north of Raymond across a natural water course to capture run-off. Permission to undertake the development had been requested in August 1946 but it was withheld pending consultation with Ben Russell, Director of Water Resources and George Brown, assistant manager of the SMRD. Russell's response was that the Water Resources Branch had no objections to the development, provided it would not adversely affect the district or land owners in the area.

The first indication that the adjacent land owners were being affected came in October 1946 when the legal firm of Rice and Paterson who were acting on behalf of various owners sent a letter dated 16 October 1946 to George Brown. Rice & Paterson were advised to communicate with Rulon Dahl, who Brown said would probably be able to give them information on this matter.

Before this advise could be followed, the RID board dealt with the matter at their 2 November meeting which was prompted by the presence of two other area farmers objecting to the flooding of their land because of the dam. Their presence prompted the Board to obtain information on the drainage responsibilities of the District. The immediate response of the Board was that the Raymond Irrigation District Act of 1933 relieved them of any responsibility after the water left the canal right of way and the responsibility for waste water rested with the individual water user. It was also noted by Jesse Wilde that when property in this area had been flooded on a previous occasion that the Board had surveyed a drain ditch on the understanding that the farmers in the area would build and maintain it but that this was never done. A copy of the 1933 act was to be procured in order to see if the text of the act would clarify the matter.

When it became evident that there was no legal basis for action against the RID Board or the SMRD, a less confrontational approach to solving the problem of drainage north of Raymond was taken. On 11 July 1950 a special meeting was held regarding the cleaning and maintenance of the proposed ditch from Brandley Lake to Stirling reservoir after it joined the factory waste water. The people at the meeting included the Board of Directors and F. George Wood and F.R. Taylor of the Sugar Factory. George Wood presented the agreement between the A R & I and the Raymond Irrigation District and the Sugar Company giving them the right to spill their waste water into Raymond (also referred to as middle) coulee. Wood indicated that in view of these agreements, he was not prepared to undertake any agreement where they would in any way be responsible for artificial ditches or diversions from the natural water course. He felt that the landowners in the Lake Bottom area had diverted this water for their own benefit and protection and would therefore have to be responsible for these ditches. It was decided upon checking the area that it might be advisable to have the water from the Brandley reservoir area run into Stirling Lake without joining up with the factory waste as the drainage water running in the summer would not present any problem of freezing and overflowing the banks. Board agreed that they would accept the responsibility of cleaning and maintaining the proposed drainage ditch from the Brandley Lake to a point where it would join with factory waste water or to its entrance into Stirling Lake if the ditch did not join up with the factory waste water.

In 1950 Ducks Unlimited became involved when they under took a survey of the area. Following this survey Ducks Unlimited declined to become involved because the project was too big handle considering the benefits they might expect but that they may assist financially and their surveys would be made available to the district.

Given the fact that Ducks Unlimited would not undertake full responsibility for the Project and given the way the SMRD was handling the issue of drainage in the Coaldale area it was decided to call special meeting to get the approval of the water users in the drainage area for construction of a ditch which was to be paid for by those using the ditch. The special meeting, chaired by Doug Paxman, was held on 4 July 1951. The meeting feature a presentation by George Brown, Assistant Manager of the SMRD, who outlined the drainage project being carried out in the Coaldale area where the farmers paid the cost of the construction and the SMRD assumed the responsibility for maintenance. E. Burkell, of Ducks Unlimited, discussed the survey work undertaken to date by his organization and recommended that it be continued east of Stirling, thus permitting the water to flow through to Etzikom Coulee. He indicated that Ducks Unlimited could complete this work next year. Brown's presentation dealt with the issue of responsibility and Ducks Unlimited provided the engineering data required to solve the problem. The meeting decided to create a committee which would have the responsibility to get an organization set up to put through a drainage ditch and that they were empowered to make an equitable assessment of the landowners to cover construction.

The committee created at the meeting had their first meeting on 9 July 1951 at which time Doug Paxman was elected as permanent chairman and Tucker Hironaki was selected as the secretary treasurer. The action taken involved a request to Paxman to secure information on water rights in the proposed drainage area which could be used to arrive at an assessment figure. Paxman and Bennet were chosen to meet with the municipal council to discuss the municipal drainage problem in the area under consideration.

The future management of the Stirling Lake was eventually solved by a 1987 agreement which preserved Stirling Lake for duck habitat while compensating the adjacent farmers for crops losses and the RID for the water supplied to maintain the water level.

DRAINAGE AND A NEW ROLE FOR THE BOARD

Discussion about specific drainage problems very quickly turned into a debate about the role of the Board of Directors in the administration of the RID. This shift took place at the 6 January 1947 meeting the Board of directors when M.H. Holmes and J. B Roberts expressed the view that the board should accept the responsibility for drainage. Holmes and Roberts both owned land which was being flooded by the dam across the coulee. The immediate problem with the implementation of this idea, as explained by the board, was that most of the holders of water rights in this area had old AR&ICo. contracts where there was no provision made for drainage nor additional assessment against these water agreements to cover the additional cost that drainage would require. The Board indicted that the only reasonable solution was the formation of a drainage district or drainage areas within the RID. In order to ensure that some action would be taken on this suggestion Holmes and Roberts succeeded in passing a motion which requested the board to bring the matter of drainage before the annual meeting with the idea of organizing into drainage districts.

The 31 March 1947 annual meeting the discussion of the drainage issue was lead by M.H. Holmes who gave a "constructive talk on the care and use of water as well as the necessity of proper drainage, citing many instances of improper use of water ponding, over irrigation and neglect." The result of the discussion which followed his presention was the creation of a five member committee to make a survey of the drainage problem within the district and bring recommendations to the Board with a view to helping to solve the problem of waste water and damage to the land thru seepage and flooding thus creating alkali area. The committee appointed included Holmes as chairman while Ralph Meldrum, D.A. Bennet, Ike Roberts and Wilfred Brandley served as the other members of the committee.

The Drainage Committee held three meeting which not only included the committee members but other water users interested in the issue. It recommendations were discussed at annual meeting held on 27 March 1948. The Board indicated that some survey work had been completed in the fifteen and eighteen mile lakes area by Mr. Winters who had made recommendations for necessary work to deal with drainage and flood problems in the area. The recommendation that the board make a special assessment to raise sufficient money to properly put in drainage works throughout the district was more difficult to implement. As secretary Rulon Dahl explained, the Board could not assess the old water users above the $1.00 service charge because their rate was set by the contracts these water users had signed with the AR&ICo. The only solution was to exchange old water agreements for new water rights so that there would be one classification of water within the district.

In the discussion that followed some water users severely criticised the Board for not accepting responsibility of making and cleaning ditches through out the district and putting in proper drainage. Jesse Wilde countered this criticism by pointing out that some years ago the district had applied for special legislation to make sure that the districts was not responsible except to deliver water into farm laterals. He support this approach because in his opinion the "individual water users organized into water user associations could clean and maintain the ditches much cheaper and with more satisfaction to the water users in general than if the District tried to do all this work and assess for it". He was supported in this view by fellow board member Wilford Brandley who felt that "the matter of management and supervision would be expensive and it might be difficult for the District to get the work done to the satisfaction of the water users or in time when all the laterals in the district had to be taken care of." I B. Roberts and M. H. Holmes continued, however, to argue for the board accepting the responsibility of ditch cleaning, maintenance and assessment to cover the costs. They pointed out that the ditches were not being properly cleaned because the work came at a time when the farmers were busy all would not come out on an appointed day. They also warned that unless some adequate measures are taken to properly drain the district agriculture will be jeopardized and that we might cease to exist as a profitable producing area. The debate at this meeting was resolved when Doug Paxman suggested the appointment of a committee to approach the holders of the old water agreements to have them change to the new water agreement as it now operates and will apply under the new SMRD extension.

The committee established to interview old water users on their attitude to exchange old AR&ICo. contracts so that the RID could operate as one unit so that any work of construction, maintenance or drainage could be done if the water users so desired and an equal assessment could be levied on all water users to cover such costs reported the results of their interviews at the annual meeting on 12 March 1949. Jesse Wilde reported that most water users that he had approached did not wish to be make any transfer at present and would rather maintain their own laterals rather than have the work done by the District and assess them the costs. Doug Paxman reported that most farmers in his district favoured transfer of old AR&ICo. agreements for new agreements so that what ever work was necessary in the district could be carried on by the Board and the costs assessed to all water users under one class of water right. Wilford Brandley stated that about half of the old water users he had approached were prepared to transfer old contracts to new while others were not familiar enough with the proposed change and its benefits to declare themselves on the matter.

Louis Brandley expressed the opinion that the matter should not be dropped and moved the following resolution. That a committee of three be appointed by the board to further investigate the possibility of bringing old water users into Raymond Irrigation District by having them transfer. The committee was given the responsibility of contacting all old contract holders. The issue of the cancellation of the old water agreement was eventually solved by the passage of the St. Mary and Milk Rivers Water Agreements (Termination) Act in 1960

DRAINAGE SURVEYS

Another initiative taken by the Board to deal with the issue of drainage was to make use of the services of the provincial Department of Agriculture, the PFRA and the Experimental farm to undertake surveys to determine the extent of the problem. The idea of undertaking these types of surveys was first suggested by William Jacobson who was the guest speaker at the 31 March 1951. Jacobson was the former secretary of the Water Resources Committee who was now an irrigation specialist at the experimental farm. Jacobson discussed the work that was being done in the Magrath area. He suggested that it might be in order to request the Dominion Government through the PFRA and Experimental Farm Service to make a survey of the area for drainage and a soil survey.

Further interest in drainage was generated in August 1952 upon Farrel Nelson's return from a trip to the Fairfield area of Montana where he had gone as part of a tour organized by the Experimental Station to look over their drainage program. This trip introduced the RID representatives to current methods of drainage. The return of the delegation was followed by the creation of the Central Drainage Committee which was made up of staff from the PFRA and the Experimental Farm and representatives from the various irrigation districts.

The decision to request a formal survey was initiated at the annual meeting held on 18 March 1953. The resolution was drafted by the Drainage Committee which had been established to deal with the drainage north of the District. The special resolution noted the problems being caused by poor drainage, the necessity of subsoil surveys to determine the steps that had to be taken to successfully cope with the situation which had been observed in Montana and the availability of PFRA staff to carry out the work. After this preamble, it requested that the RID ask the Water Resources Branch of the provincial government to contact the PFRA with the view of getting them to make such a soil survey on this area. A second motion instructed the Board to apply for the services of the Provincial Drainage engineer to give them advice on the most practical work that could be done in our district in the matter of surface drainage.

IRRIGATION INDUSTRY REVIEW

The RID also played an active role in the review of the irrigation industry which was initiated in 1958 with the creation of the Irrigation Review This committee included Doug Paxman as one of its members. Paxman was good choice since he had considerable experience in dealing with such issues as drainage in his capacity as a member of the RID Board. The ongoing review of the irrigation industry in the 1960s through the ARDA program. The recommendations that came out of the ARDA studies had both positive and negative implications for the RID. The most serious negative recommendation from their point of view was the idea that each system or geographical area in a system should be encouraged to amalgamate into a single irrigation district was the recommendation which had the most serious implications for the RID. The potential impact of this idea was the disappearance of the RID as an independent district as a result of its amalgamation with Magrath or with the those part of the St. Mary Project west of Chin Coulee or with the entire St. Mary Project from Magrath to Medicine Hat. The positive recommendation was its suggestion for the funding of capital works. The RID made an immediate response to both of these recommendation. The motion passed at the 1967 annual meeting held on 8 April indicated that the RID was separated from every other district by natural topography and geographical boundaries and that it had operated its system successfully since 1925 following its purchase from the CPR . It was, therefore, resolved by the water users to:

> recommend to our Board of Trustees, to the Government Irrigation Policy Committee, to the Minister of Agriculture, the Hon. H.E. Strom, and to the Irrigation Council of Alberta that the presently organized and operating Raymond Irrigation District remain as a separate operating unit, with the right to effect some minor extensions or additions within or immediately adjacent to its present boundaries and that . . . the Board of Trustees take any steps necessary to qualify for Capital Work Cost Sharing, and negotiation such agreement with the Minister as may be needed or required under the recommendation of the Irrigation Policy Committee.[282]

Welling Main Canal before and after Rehabilitataion (RID)

1968 AND THE SHIFT TO FULL-TIME MANAGEMENT

The provincial government responded to the recommendations of the Irrigations Policy Committee with the passage of the 1968 Irrigation Act and the rehabilitation funding program This legislation did not terminate the existence of the RID which had been recommended by this Committee but it did repeal the Raymond Irrigation District Act of 1933. The RID board of directors now had a number of new responsibilities which could not be handled in the long run on a part-time basis. These new responsibilities included the delivery of water to the individual parcels of land and not simply the head gate of the lateral. Pipelines and sprinklers were also factors in the process of centralizing water management in the hands of the board of directors and a full-time staff.

The RID responded to the new legislation by first reaffirming in 1968 the commitment made in

RID office & shop (RID)

Raymond Hydroelectric Station (SMRID)

Pipeline installation (RID)

1967 to its existence as a separate district. The task of handling its new responsibilities began at the 19 March 1968 meeting of the Board of Directors when they discussed the feasibility of sharing ditch cleaning and ditch rebuilding costs with the farmers. . Further discussion took place at the 1968 annual meeting held on 23 March when it was decided that the board should pay ½ the cost of ditch cleaning and ditch rebuilding and replacement of structures approved by the ditch organizations. The assessment notices for 1968 included farmers share of ditching costs for the year. Implementation of the parcel system of assessment was begun in 1969. In addition to determining the exact amount of land for which the farmer held water rights, the RID also decided to charge people $2.00 per acre for any acreage over and above the specified acreage under water agreement that was irrigated on any farm in 1968.

The RID also took advantage of the cost sharing program. The most dramatic impact of this program was the removal of the original canal from the Town In 1982 the original canal constructed in 1899 which went through the townsite was removed. After 1982 the water supply for the area east of Raymond was through the Six Mile Coulee.

Canal construction (RID)

The Six Mile Coulee was stabilized and a new turnout from the SMRID main canal was constructed to provide the water. . The construction of a pipeline system was first considered in 1980 with the installation of the first system in 1982. This was the first pipeline installed in Alberta as part of the Irrigation Capital Works Program.

The staff changes associated with this new era began with the retirement in 1970 of Rulon Dahl as Secretary-Manager. The RID continued to be administered on a part-time basis by his successor Howard Card until 1980. On 1 January 1981 Gordon Zobell became the first full-time manager of the RID. He had originally joined the RID in 1977 and had served as the ditch rider in the Welling district. In 1981 the RID also acquired a permanent home with the construction of a shop and office on property located west of the Town. The last phase in the post-Second World War transformation of he RID was the implementation of a full ditch rider system.

STIRLING WATER SUPPLY

The water supply for the Village of Stirling had been supplied since 1902 on the basis of a contract between the RID and Village. The Village distributed the water to the residents by using a ditch system in the town. At the 13 April 1968 meeting the Board was informed that the Village of Stirling were planning to take the irrigation water and delivery ditches out of the village on all land east of Kipp Coulee. The RID decided that the landowners could have the water right transferred to them from the Village and they could deal directly with the RID for the delivery and payment of water. The Village representatives were assured that the Board would negotiate an agreement with the Village for domestic water to be delivered to the Stirling water supply reservoir. The terms of the agreement and cost of water would be negotiated when the Village had made its final decision regarding domestic water supply and the cancellation of flood irrigation rights to village lots.

TWENTY-TWO
REVOLUTIONIZING IRRIGATION AGRICULTURE AND INDUSTRIAL EXPANSION 1946 - 2000

The previous four chapters have made reference to a technological revolution which changed the way water was delivered and applied to the land and the way some crops were seeded, cultivated and harvested. The water users and managers of the St. Mary Project were an important part of this revolution. Sustained research and application of the results did not begin until after World War Two. The PFRA, the Lethbridge Experimental Farm and the Water Resources Branch of the provincial Department of Agriculture launched this technological revolution in Alberta irrigation. The federal government took the lead on this issue because:

> experience has shown that the majority of the salinity problems on irrigated land have resulted from not having adequate drainage. In irrigating western soils which are naturally high in salt content and with water that may contain varying amounts of salt, it is inevitable that salt concentrations will build up in the root zone of the soil. Some excess irrigation is a prerequisite to leach away the salts, an drainage is necessary (both surface and sub-surface) to remove the excess irrigation water. In fact, there area practically no irrigation soils which have sufficient natural drainage to remain permanently irrigable under normal irrigation practice, and it is essential that all proposed developments be investigated to determine the amount and kind of drainage required and reclamation costs will be economically feasible in relation to the value of the crops to be produced.[283]

The PFRA response to this situation was the creation of a Drainage and Reclamation Division in 1949. The headquarters for the new division was established at Vauxhall which was the administrative centre for the Bow River Project which was purchased from the Canada Land and Irrigation Company by the federal government in 1950. The rehabilitation of the Bow River Project was the major task undertaken by this new division which established a laboratory in Vauxhall as part of its research on this issue.

On the St. Mary Project canal seepage problems were investigated in the Taber region in 1953. These investigations indicated the need for a fairly impervious lining material or the installation of a drainage system to cope with seepage. In the same

Lapping new sheet to previous length using sealing tape (PFRA-17450)

Land leveller in operation at Barnwell (PFRA-1842)

year studies were carried out in Medicine Hat area to determine the saline conditions at three locations prior to irrigation. These studies showed that only moderate amounts of toxic salts were present at depths where they would not affect crops. In 1955 preliminary investigations were made in connection with seepage areas along the St. Mary Project main canal in the Meek-Hutterite, Koppenstein and Grassy Lake Districts. Some additional information was collected in 1956 and a final report prepared with recommendations.

As part of their program to study this problem, the PFRA also brought in a number of American consultants to examine the situation in Canada and advise them on the action required. The first consultants to be brought up were O. W. and O. Allen Israelsen in July and August 1949 and Charles Maierhofer, the director of the Office of Drainage and Groundwater Engineering of the United States Bureau of Reclamation in Denver. In 1954 he made an inspection of the irrigated and potentially irrigable areas of Alberta and Saskatchewan at the request of L. B Thomson, the Director of the PFRA. His report covered a wider range of topics including the susceptibility of some existing irrigation projects.

Be Ge leveller in operation at Taber (PFRA-1853)

Based on this advice and their own research, the PFRA initiated the first systematic program in Canada to test various types of canal lining materials. The PFRA tested asphalt, shotcrete (pneumatically applied mortar), and concrete for lining purposes. The engineering challenge was the development of a cheap material which would make it economically feasible to lining the vast network of canals.

The first material which appeared to meet this criteria was polyethylene plastic. In 1956 polyethylene plastic was installed in a test section of a lateral on the Bow River Project and in two dugouts. In 1958, based on these results, the PFRA expanded their use of polyethylene liners of .006 to .008 inch thickness in an attempt to solve the problem of water seepage. By 1972 over 120 miles of various sized canals were lined using this material.

By the mid-1960s, problems in the use of this material were becoming evident. The most important drawback, being the difficulties it created in cleaning silt from canals. Polyethylene could also be damaged by stock if they were permitted to use the canal bank for watering purposes. The replacement of polyethylene lined canals was concrete lined canals. The use of concrete as a canal lining material had been used by the CPR in the Strathmore area as early as 1914 and had been used

Irrigating sugar beets by flooding near Lethbridge (PFRA-4207)

Sprinkler irrigation, 1949 (Taber Irrigation Impact Museum)

for several years in the USA and in British Columbia but it had only been used to a limited degree in Alberta prior to 1970. By 1970 concrete lining was a major part of the rehabilitation of in the SMRID and TID. By 1978, concrete lining has also proved unsatisfactory because of cracking caused by winter. This problem was solved by the introduction of wire mesh. The problems encountered in the use of concrete lead to the use of plastic pipelines as an alternative. Plastic pipelines were cheaper than concrete pipeline which had been used to a limited degree since the 1960. Pipelines also had the added advantage of eliminating seepage and the making more land available for crop production. Pipelines were also more appropriate to the use of sprinklers.

While concrete failed as a lining material for canals, it proved to be invaluable as a replacement for wood in the construction of diversion structures, drops and checks. The value of concrete for these uses had always been recognized. The problem had been the cost of initial construction and installation. The availability of prefabricate concrete structure has also promoted its increased use .

Furrow irrigating using syphons (PFRA-23437)

Improvements in the construction of canals and the introduction of pipelines were complemented by the introduction of new ways to control the water which flowed through these canals. The major innovations were the overshot, Langemann and LOPAC gates. The overshot gate was developed during the rehabilitation of the main canal as a replacement for the radial gate. The overshot gate principal was applied by Peter Langemann to develop a gate which automatically adjusted the water level in the canal to the actual needs of the farmers. The LOPAC gate was also developed by Langemann for use on smaller laterals. Peter Langemann transferred the patent rights to gate over to the SMRID.

SPRINKLER REVOLUTION

The application of water to the land had traditionally been done by flooding using various techniques. After World War Two this approach was given a new license on life because of the availability of large earth moving equipment to undertake land levelling. Land levelling was also encouraged by the provincial government which made funds available. The use of plastic syphon tubes were also used after 1946 as a way of improving flood irrigation.

Hand move sprinkler system (Sir Alexander Galt Museum & Archives, Lethbridge, Alberta)

The second option was to use sprinkler system. The sprinkler revolution began in 1948 with the demonstration of the first hand move systems by such companies as Major Aluminium. The company worked with farmers such as E.N. Francis, chairman of the TID, who purchased one of the first systems to be put into operation in Alberta. In 1949 at least 38 hand move systems were in operation on farms, with thirteen of them being located in the Taber area, while eight were in the Coaldale

Wheelmove sprinkler system (Sir Alexander Galt Museum & Archives, Lethbridge, Alberta)

First pivot in Alberta owned by Don Campbell (Sir Alexander Galt Museum & Archives, Lethbridge, Alberta)

area and six were in the Raymond area. These systems irrigated a total of 1,545 acres or 5% of the total sugar beet crop in 1949.

The Rogers Sugar Company and the Experimental Farm took particular interest in their introduction. According to the fall 49 edition of *Silver Sunshine* one of the chief reasons for the increased use of sprinklers in 1949 was that "growers were quick to observe the possibility of successfully watering dry hills and rough land, of exercising a more rigid control over the rate of irrigation, of avoiding excessive irrigation and ponding and of the advantage of sprinklers in irrigating crops already up." The Experimental Station purchased four in order to carry out evaluations of their effectiveness. Their initial evaluation indicted that:

> this method of applying water may be most useful in this region as a supplement to surface irrigation for watering higher-priced crops and for irrigating land not suited to surface irrigation where economic conditions permit . . . Sprinkler irrigation affords a high degree of moisture control over a wide range of soil types and topography. This can be accomplished without the aid of skilled irrigators if wind velocities are not excessive or too changeable. The main disadvantages of sprinkler irrigation for prairie conditions are high initial cost, high operating costs, wind interference and the aversion of skilled irrigators to carrying pipe through muddy fields and wet foliage. The economic feasibility of establishing irrigation farming wholly on the basis of sprinkler irrigation under prairie conditions has yet to be determined, except where crops of sufficiently high value per acre are grown.[284]

The introduction of hand moved sprinklers was, however, only the first stage in the development

Langemann Gate (Aqua Systems 2000 Inc.)

of sprinkler technology. By the mid-1950s sprinkler system had become mechanized with the introduction of wheel move and central pivot systems.

The introduction of pivot systems to Canada was the result of the initiative of Burdett area farmer Don Campbell. He first found out about self propelled sprinkler systems in 1955 when he read an article on these machines in a U.S. farm magazine called the *Country Gentlemen*. He wrote to the Valley Manufacturing Co. for more details and in 1961 travelled to Rexburg, Idaho to see the Valley pivots in action. In November 1961 he set up a meeting in Burdett at which time a representative from the Valley Manufacturing Company gave a slide presentation to several water users from the district who remained skeptical about the use of this machine. Don and his brother Wyman, however, retained their interest and each purchased a pivot which were in operation by 21 May, 1962.

The use of sprinkler irrigation spread much faster than was predicted. By 1959, even before pivots were introduced, approximately two- thirds of the water right area in the Taber Irrigation district of 31,600 acres was irrigated using sprinklers. The reason for the continued expansion of sprinkler irrigation, according to the 1959 annual report of the Taber Irrigation District, was the:

> difficulty in securing casual labor to do a proper irrigation job with a shovel. The shovel job is usually a 16 to 18 hour daily effort, in other words two shifts, and if the stream is left to run wild the remaining six hours of the night, it can flood out your neighboring fields and into road borrow pits.[285]

SUGAR BEET INDUSTRY

The expansion, mechanization and use of improved seed have been the main features of the sugar beet industry since 1946. The expansion of the sugar beet industry began on 13 August 1946 when Rogers Sugar announced that construction of a new plant would begin in Taber. The company announcement indicated that the final decision was reached after careful studies of beet areas, water supply and drainage problems had been made. Initially the company intended to locate the plant in Barnwell and had purchases land for this purpose. Problems with water supply and disposition of effluent resulted in the move to Taber. The incentives for moving to Taber was the offer of a site by the town which was close to the Taber Reservoir of the TID. Contracts for factory engineering have been let to Stearns-Roger Mfg. Co. of Denver Colorado and for building construction to the Dominion Construction Company of Vancouver B.C. Designs and placement of all machinery will be supervised by the sugar company's own engineering staff headed by Frederick H. Ballou, chief engineer. The investment by the sugar company will exceed $2,000,000. The Taber sugar factory will be the most modern on the continent combining many original and unique process recently developed in the beet sugar industry. It will include not only the usual refining processes but also those for Steffenizing, molasses for sugar production, and for utilization of molasses by-products. There will

Harvesting sugar beets by hand (Taber Irrigation Impact Museum)

Harvesting sugar beets in Taber area (PFRA-1397)

Cultivating potatoes on Mr. Sakamoto's farm (PFRA-11120)

be a pulp dryer, instead of the conventional pulp silo , and it is anticipated that the dried pulp will prove a convenience to all farmer feeders as in concentrated form it can be transported to much greater distances and fed to livestock over all the irrigated and stock feeding area of Alberta. Since 1969 the industry has been centralised in Taber with the closure of the plants in Raymond and Picture Butte.

Mechanization and improved seed was drive by the need to reduce the need for hand labour. Growing sugar beets was very labour intensive. The Knight Sugar Company dealt with this issue by bring in Japanese and Belgium workers. When the industry was re-established in 1925, labour was one of the first issues to be dealt by the new Alberta Sugar Beet Growers Association. In the late 1920s and early 1930s, the irrigations districts were also involved with the CPR in a program to bring people from eastern Europe to work in the beets fields. The Mennonite community in the Coaldale area was established as a part of this program. German prisoners of war and Japanese, who had been removed from the west coast, met the labour needs of the industry during the Second World War. After the Second World War European immigrants were recruited along with native people.

The solution to the need to recruit labour was mechanization and the introduction of monogerm seed which did not require thinning. The first commercial harvesters were introduced into southern Alberta in 1945. In the 1953 season about 23 % of the beet acreage was harvested mechanically while in 1955 the percentage had increased to 44%. This transition period created considerable hardship for those beet workers who remained. As the mechanization process continued, the situation became worse for the workers which prompted the comment in the 1955 annual report that "for the present, however, it is almost as much of an embarrassment as of advantage since it has the general effect of reducing the earning power of hand labourers but not completely replacing them."

Harvesting sugar beets, October, 1954

Harvesting squash on Broder farm in Coaldale area (PFRA-16718

Mechanization was strongly supported the Rodgers Sugar Company as early as 1940 when it converted to this method of harvesting the beet seed grown in the Ladner area of B.C. at the insistence of the growers. It carried the mechanization message to Alberta by using its publication, *Silver Sunshine*. The first article dealing with this topic appeared in fall 1944 issue and dealt with a mechanical loader developed by H.M. Haney and his son Leonard who farmed 320 acres of irrigated land in the Picture Butte area. By 1947 articles on mechanization had become a part of every issue.

Sugar factory in Taber, 1956 (PFRA-12283

The effort to reduce labour also involved the development of a type of seed which reduced and then eliminated the need for thinning. The seed development program of the Rogers Sugar Company was based in the Ladner district of British Columbia which was located in the lower main, near Vancouver. The mild winters permitted the over-wintering of the stecklings. This program was established in 1940 as a result of the Second World War. The development of seed which required no thinning was a three stage process which began with the introduction of segmented seed in 1946 and decorticated seed in 1949. The introduction of monogerm seed was introduced gradually into Alberta beginning in 1959 when 160 growers were each supplied with sufficient seed to grow 5 to 6 acres of beets. The introduction was accompanied by an ongoing research program designed to ensure maximum results. This program included numerous drill test, improvement in seed processing machinery to provide more uniform seed size and shape and the development of high yielding monogram varieties well adopted to Alberta conditions.

Pea vining near Magrath (PFRA-13671)

Sugar beets, Coaldale, Alberta (PFRA-16784)

Pea thrashing at Taber (PFRA-7226)

harvesting canning peas near Barnwell (PFRA-17404)

Hay baling near Fincastle (PFRA-3528)

Truck loads of sugar beets being delivered (PFRA-7228)

Sugar beets, Coaldale, Alberta (PFRA-16783)

Pumpkins at the cannery at Magrath (PFRA-16772)

CONCLUSION

The St. Mary Project has been an integral part of the history of southern Alberta since the 1890s. It is the thread which links the history of the Cardston area to that of Medicine Hat. Its history has two main periods which are divided by the construction of the St. Mary Dam. Before 1946, the St. Mary Project was a vision of a better life which the farmers struggled to implement. The Irrigation Development Association and the South Alberta Water Conservation Council have been part of this struggle. The period after 1946 is the realization of that vision through the construction of the St. Mary Dam. The challenges faced to implement this vision have been the search for a formula which would provide the funds for its complete construction.

SOURCES

The sources used in the preparation of this history include a number of original documents which were obtained from archives in Britain, United States and Canada. The National Archives of Canada and the Provincial Archives of Alberta have the surviving records of the Department of the Interior which relate to water resource development. The PFRA has retained the records on the expansion of the St Mary Project from the late 1940s to1968. These records will eventually be transferred to the National Archives of Canada. The four irrigation district which make up the system all have their own archives which include minutes, correspondence files and engineering documents. The information on the individual districts included in their respective chapter reflected the types of records which have survived.

A limited of theses, articles and books have also been written which relate directly or indirectly to the history of the St Mary Project. This literature includes local histories of all communities which provides background information on the community and on the people who brought about its development. Biographies have also been written of some of the key individuals in the history, such as William Pearce, Alexander and Elliott Galt and Charles Ora Card.

THESES

den Otter, A. A. "Sir Alexander T. Galt and the Northwest: A Case Study of Entrepreneurialism on the frontier" Ph.D, University of Alberta, 1975

Hicken, John R. "Events Leading to the Settlement of the Communities of Cardston, Magrath, Stirling and Raymond, Alberta." M.A. diss., Utah State University, 1968.

Mitchner, E. Alyn. "William Pearce and Federal Government Activity in Western Canada, 1882-1904." Ph.d. University of Alberta, 1971.

Petker, Don "The irrigation frontier in southern Alberta, 1885 -1915" University of Calgary, 1996.

Tagg, Melvin S. "A History of the Church of Jesus Christ of Latter-Day Saints in Canada, 1 8 3 0 - 1963." Ph.D. diss., Brigham Young University, 1963.

Wilcox, Archie. "Founding of the Mormon Community in Alberta." M.A. diss., University of Alberta, 1950.

ARTICLES AND BOOKS

Cardston and District Historical Society. *Chief Mountain Country*. Cardston: Cardston and District Historical Society, 1987.

Hartley, William G. "Mormon Sugar in Alberta: E.P Ellison and the Knight Sugar Factory, 1901 - 1917." *Journal of Mormon History* (1997): 1- 29.

Hicken, John, Kay Redd and John Evans. *Roundup*. Raymond: Raymond Chamber of Commerce, Centennial Committee, 1967.

Magrath and District Historical Society. *Irrigation Builders*. Magrath: Magrath and District Historical Society,1974.

Otter, Andy den. Civilizing the West: The Galts and the Development of western Canada. Edmonton: The University of Alberta Press, 1982.

– – – – *Irrigation in Southern Alberta: 1882 - 1901*. Lethbridge: Whoop-up Country Chapter, Historical Society of Alberta, 1975.

Palmer, Byron C. and Craig J. Palmer. "Mormons in Western Canadian Agriculture: From Irrigation to Agribusiness." in *The Mormon Presence in Canada*. ed. Brigham Y. Card, Herbert C. Northcott, John E. Foster, Howard Palmer and George K. Jarvis. Edmon ton: University of Alberta Press, 1990.

Stirling Sunset Society. *Stirling: its Story and People*. Stirling: Stirling Sunset Society, 1981.

Town of Raymond History Book Committee. *Settlers, Sugar and Stampedes: Raymond Remembered*. Raymond: Town of Raymond history Book Committee, 1993.

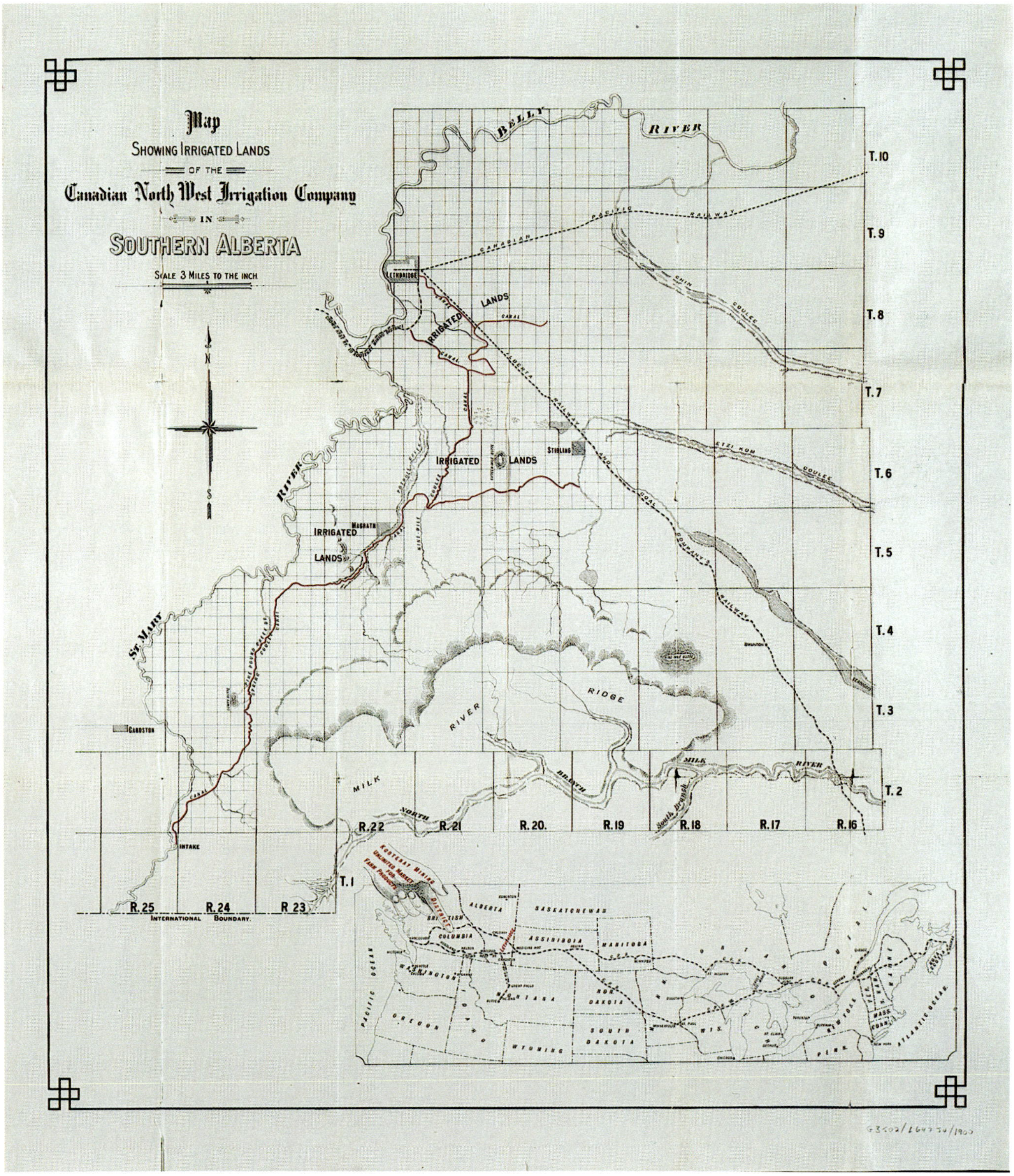

Map of North-West Irrigation Company, 1902 (Glenbow Archives, Calgary)

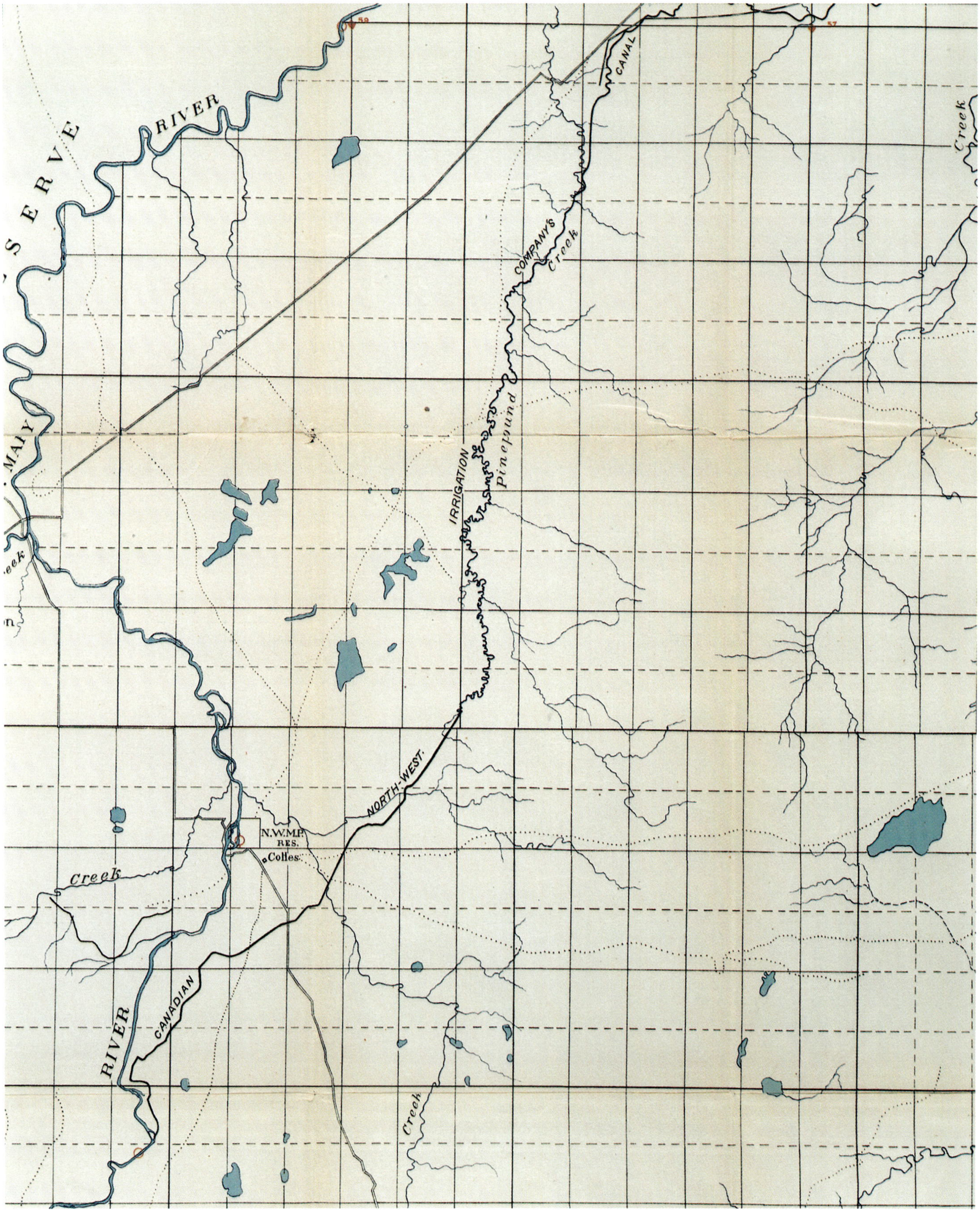

Kimball to Spring Coulee, 1902 (Glenbow Archives, Calgary)

Magrath area showing Magrath Lateral and main canal to Nine Mile Coulee (Glenbow Archives, Calgary)

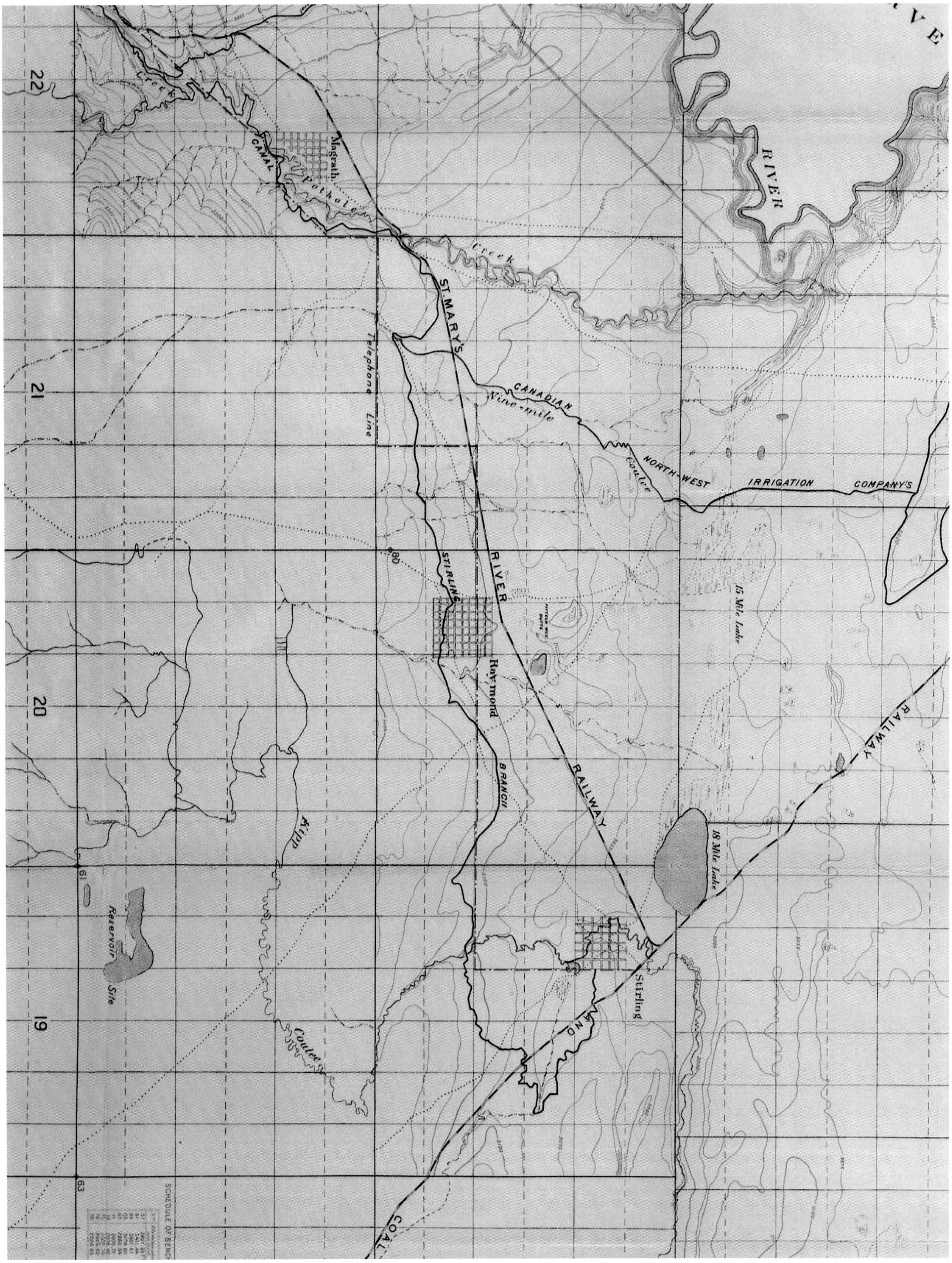

Raymond Lateral, 1902 (Glenbow Archives, Calgary)

Main canal to Lethbridge, 1902 (Glenbow Archives, Calgary)

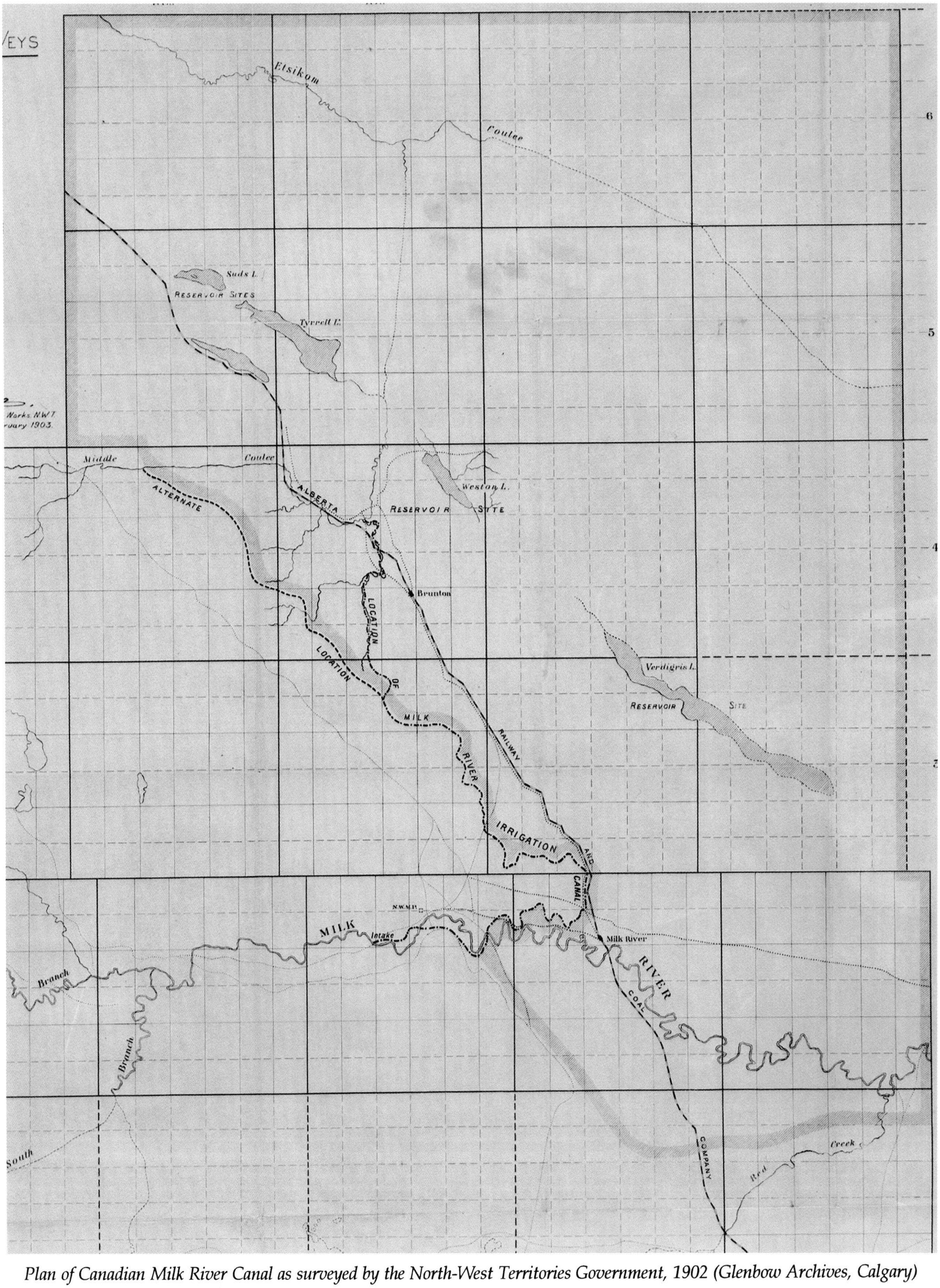

Plan of Canadian Milk River Canal as surveyed by the North-West Territories Government, 1902 (Glenbow Archives, Calgary)

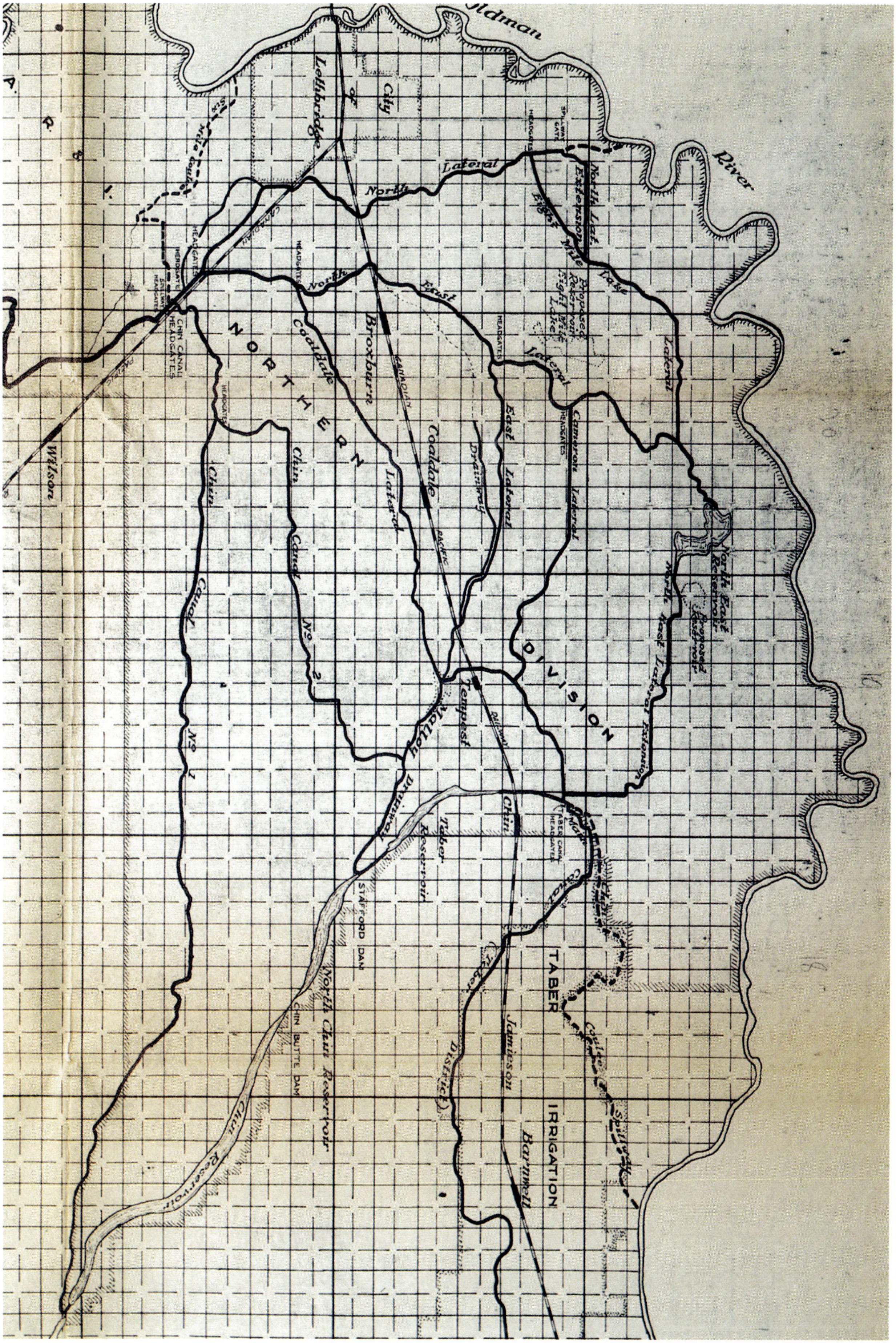

Lethbridge Coaldale distribution system, circa 1940 (Glenbow Archives, Calgary)

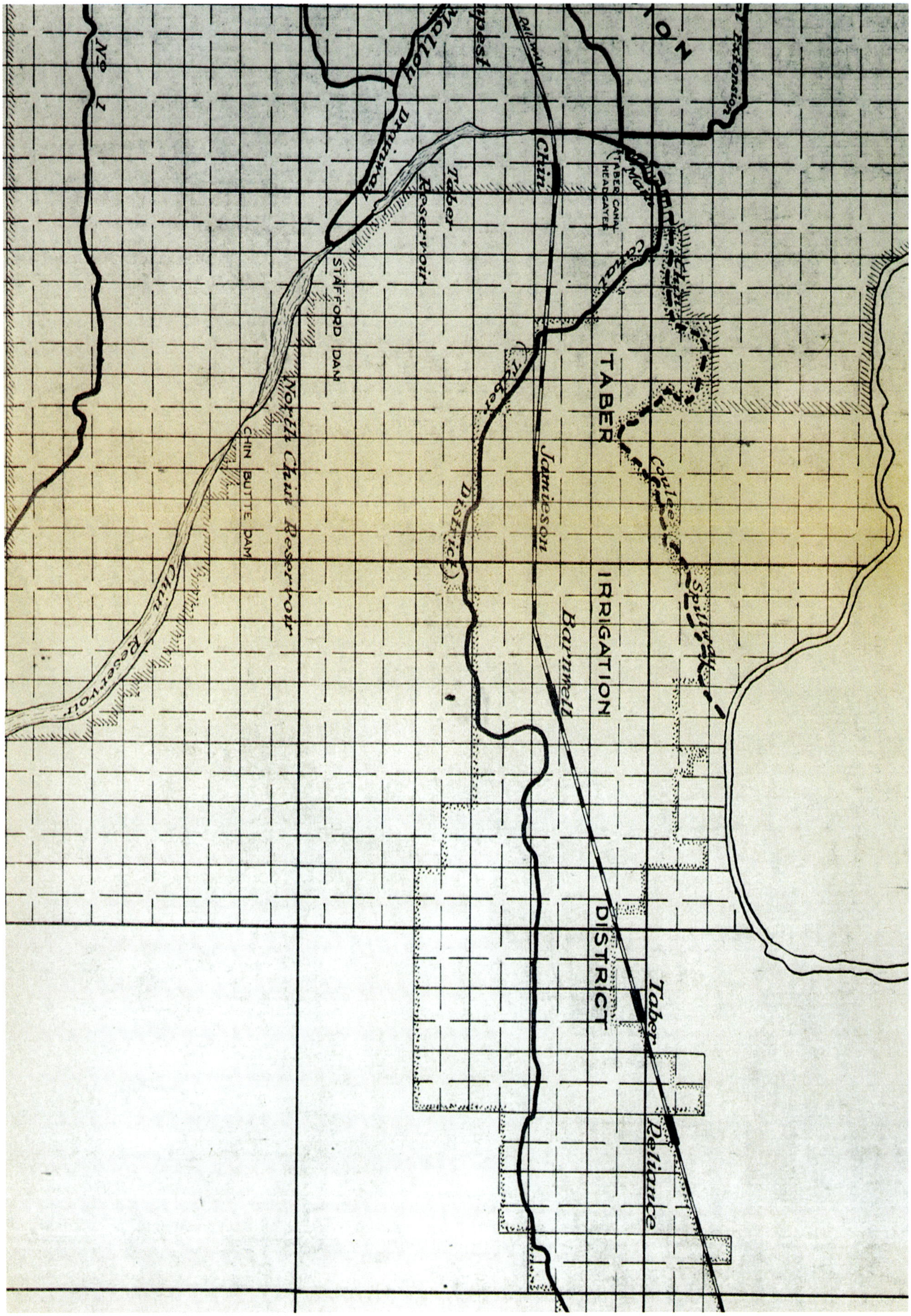

Taber Irrigation District, 1943 (Glenbow Archives, Calgary)

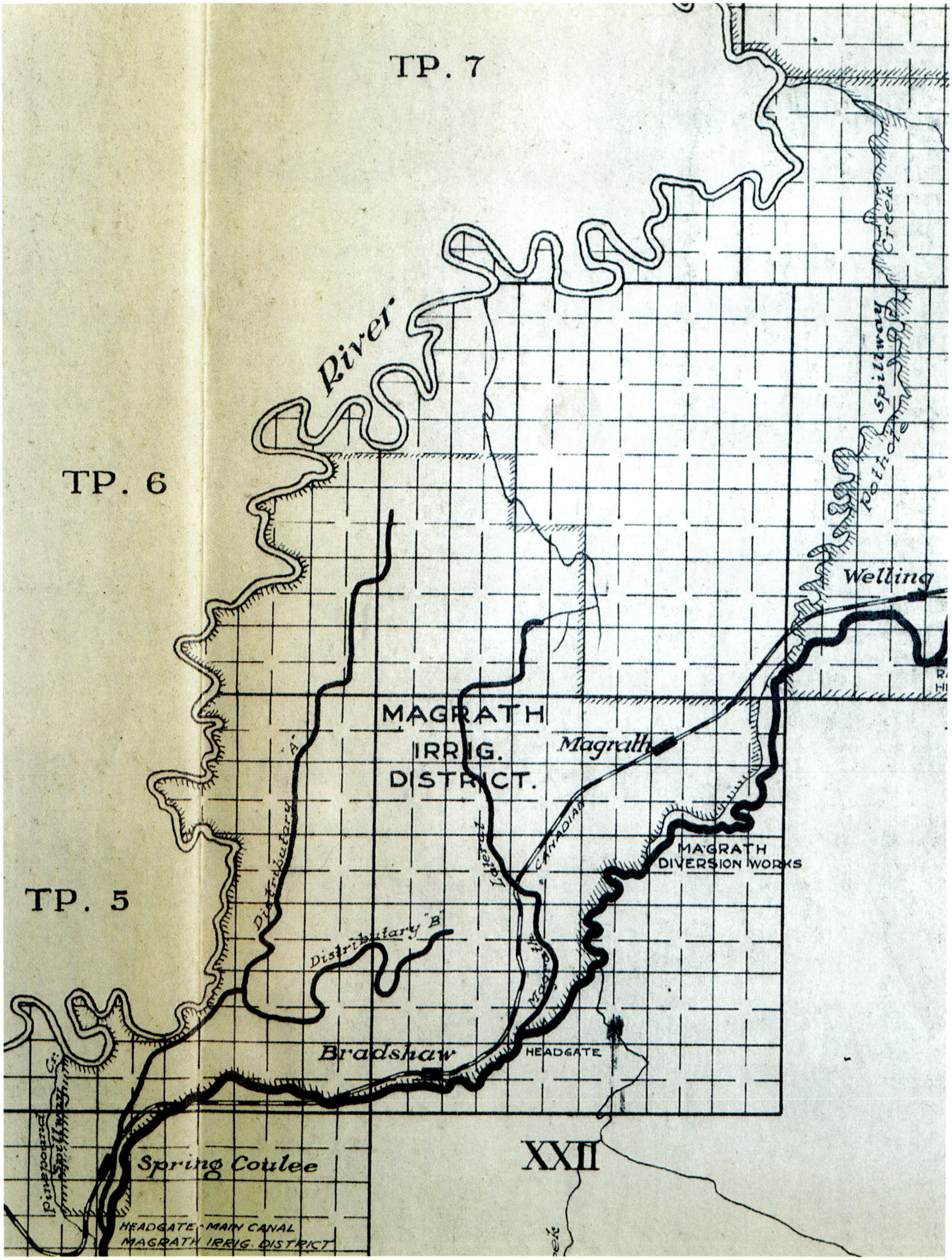

Magrath Irrigation District showing High Line Canal, circa 1943 (Glenbow Archives, Calgary)

OWN
YOUR OWN
HOME IN CANADA
and apply for a
READY-MADE FARM
to the nearest
CANADIAN PACIFIC AGENT
CANADIAN PACIFIC
EYRE & SPOTTISWOODE LTD H.M. PRINTERS, LONDON.

Above and facing page: CPR promotional poster circa 1912 (Glenbow Archives, Calgary)

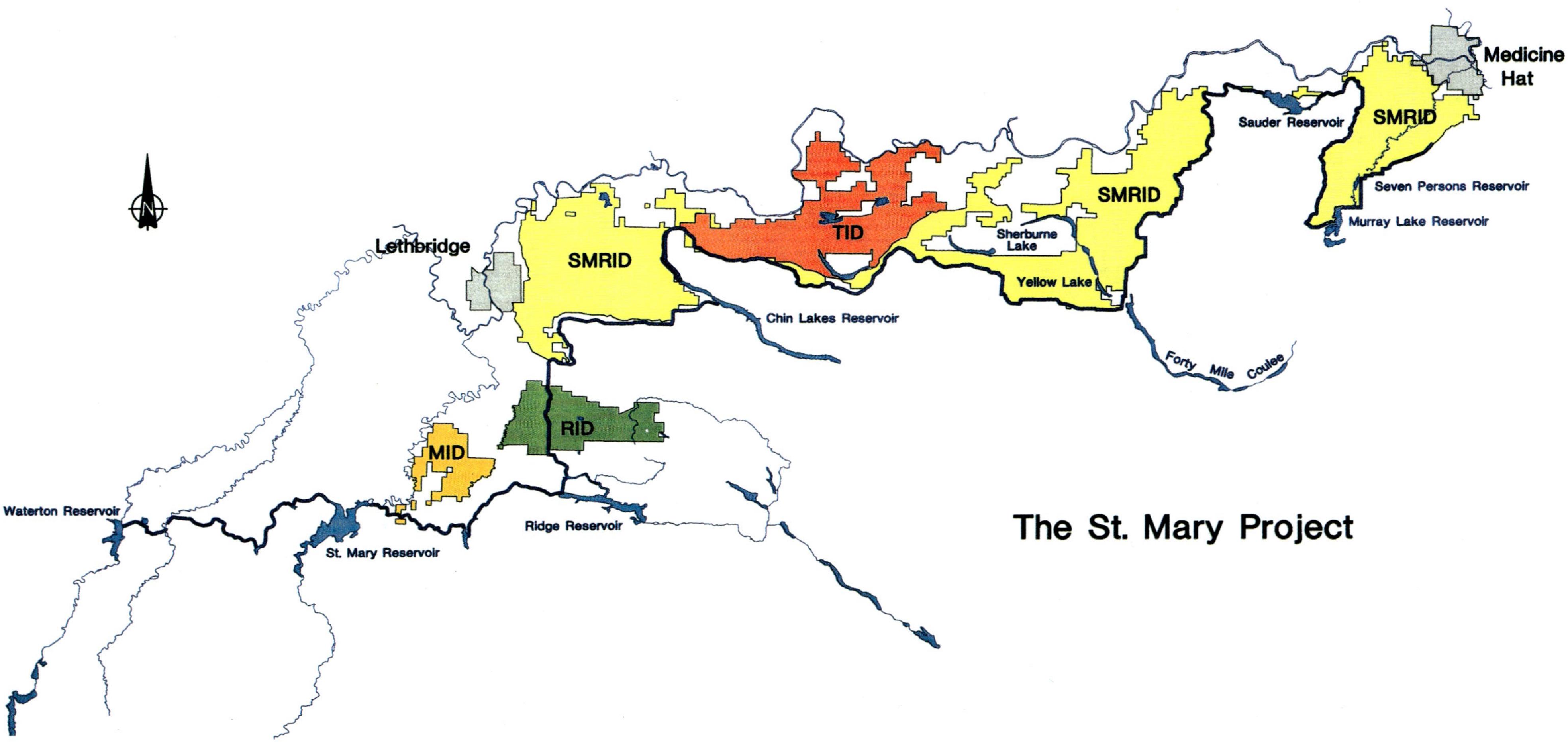

(Irrigation Branch, Agriculture, Food and Rural Development)

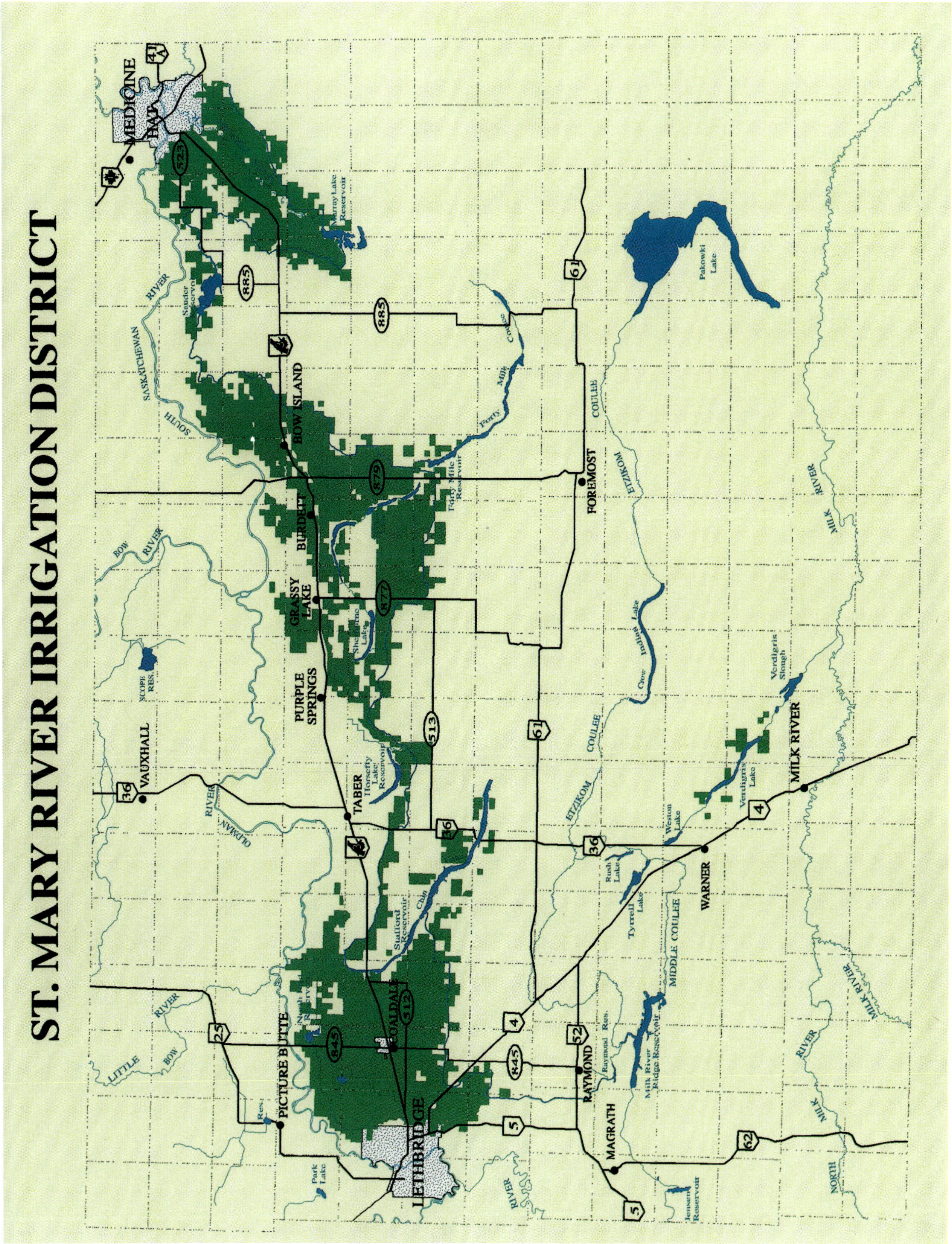

(St. Mary River Irrigation District)

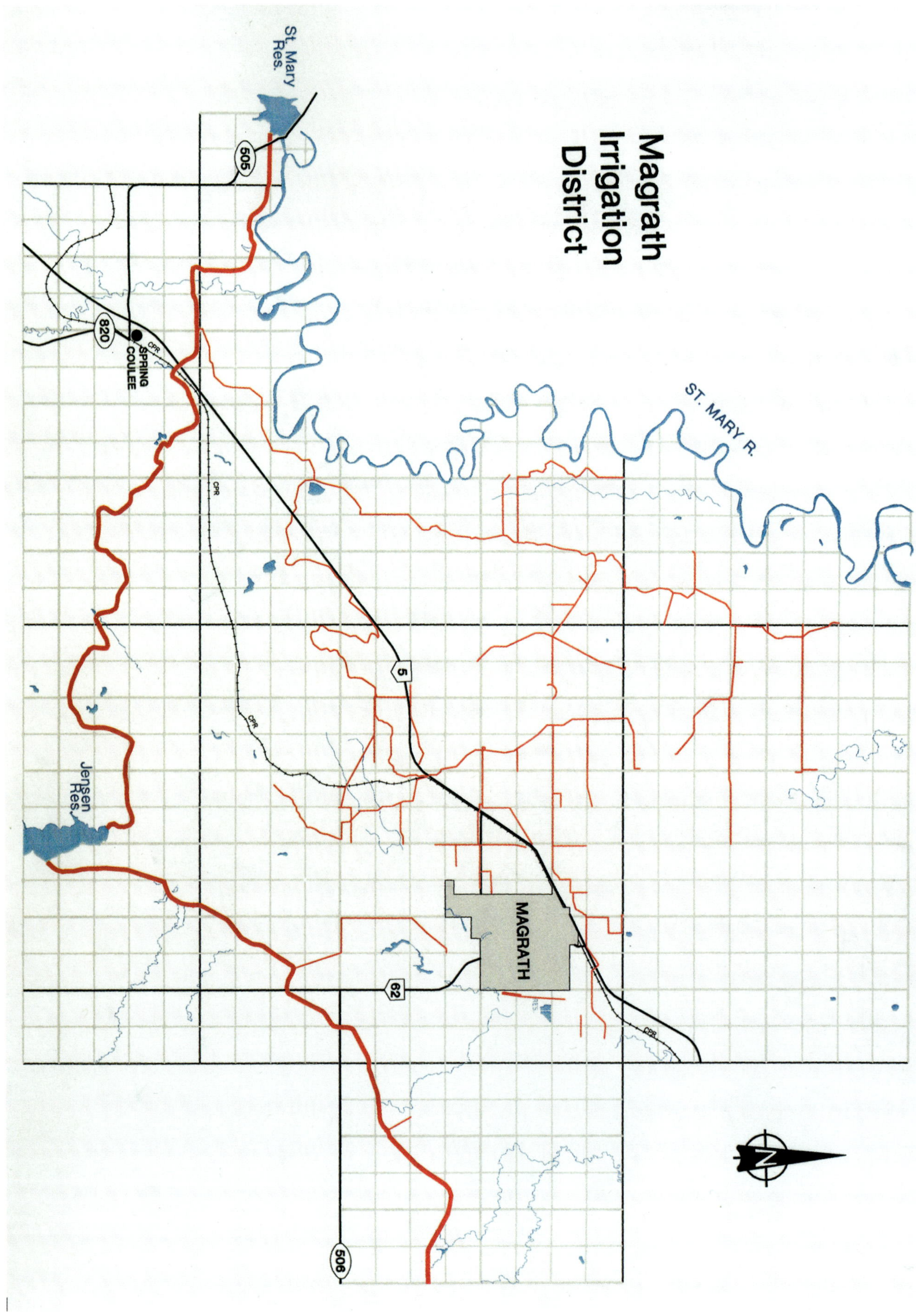

(Agriculture, Food and Rural Development)

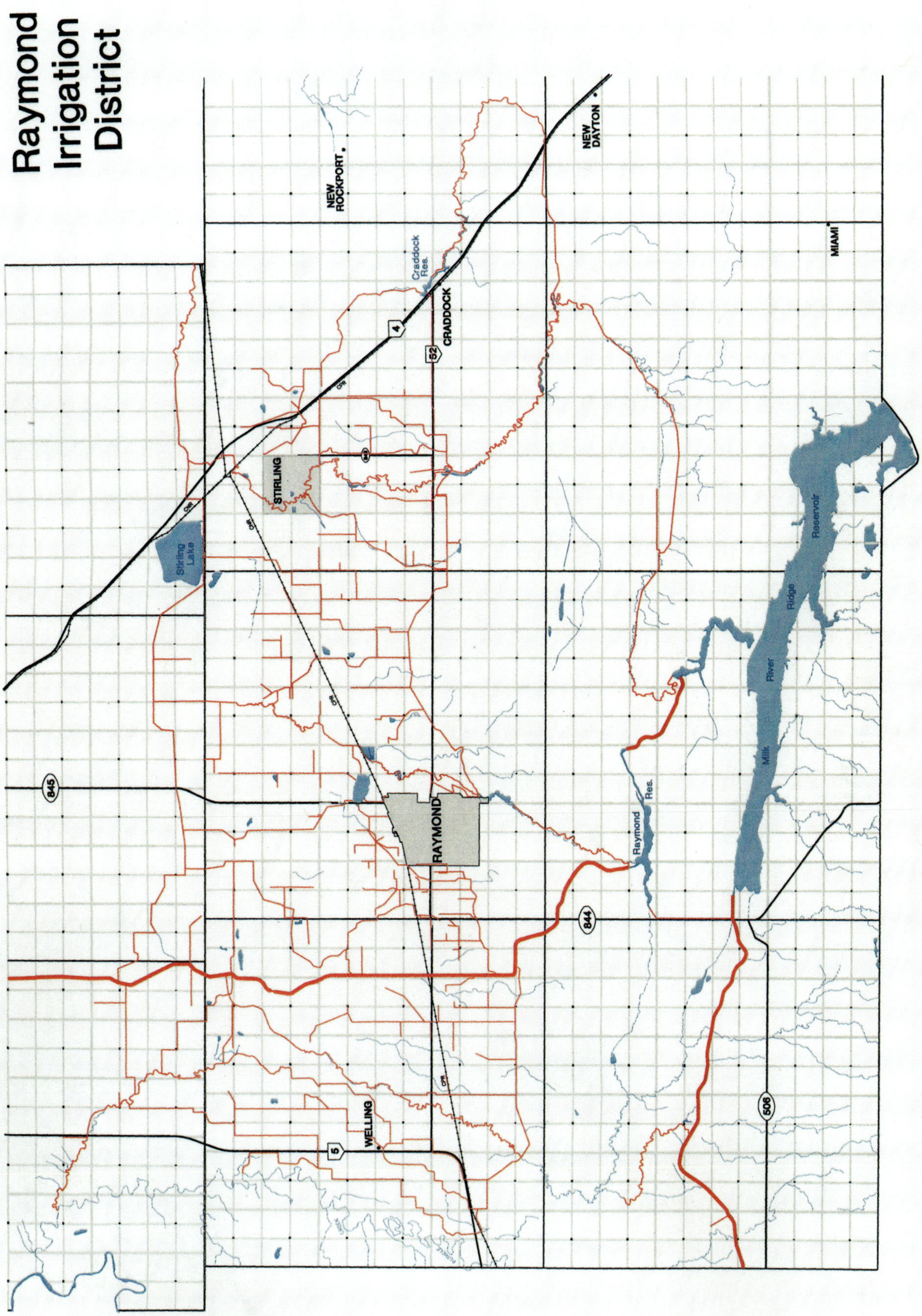

(Agriculture, Food and Rural Development)

Taber Irrigation District

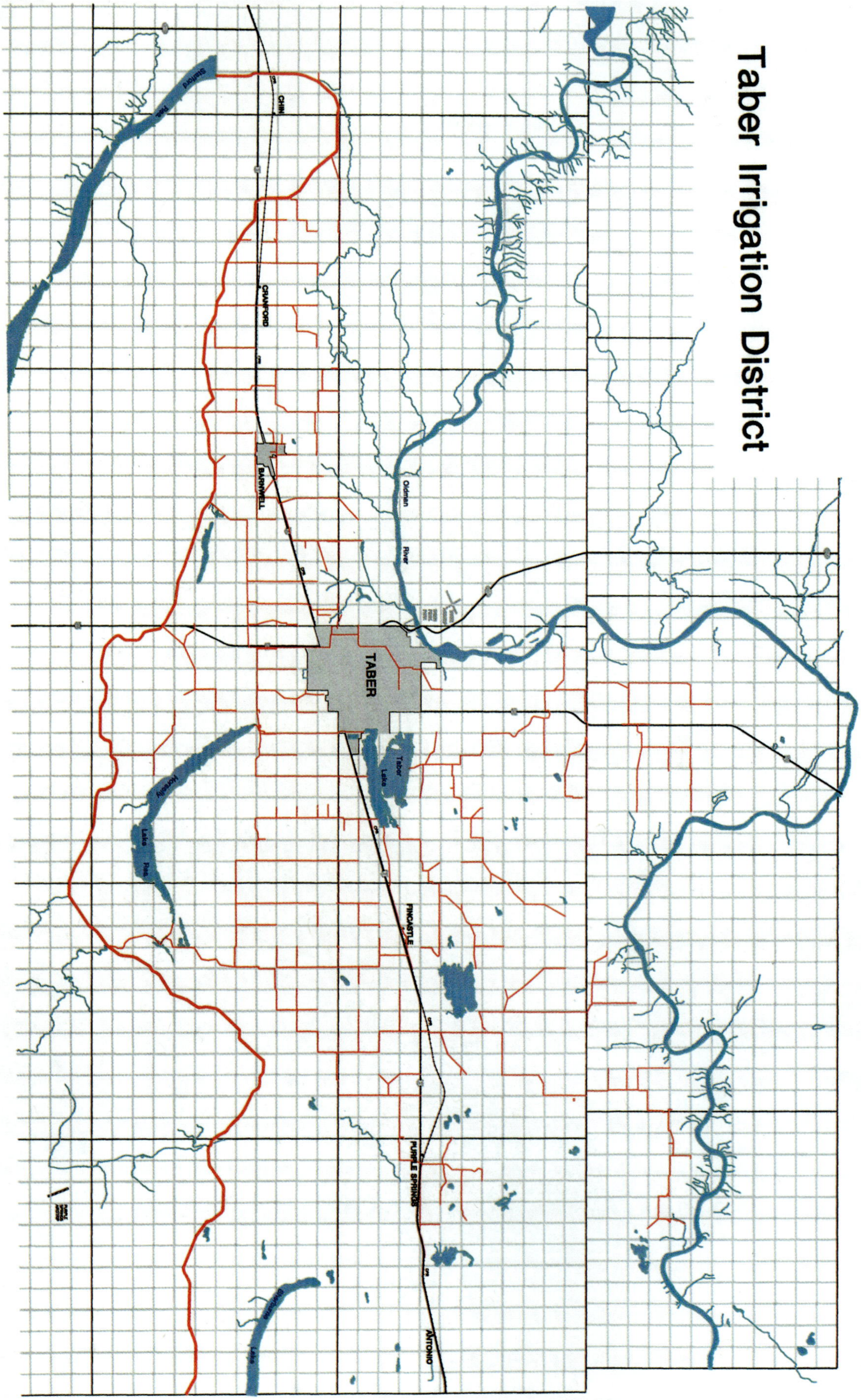

ST. MARY PROJECT WATER RATES

	SMRID	TID	MID		RID	
			OWA	NWA	OWA	NWA
1902	$1.00**					
1903	$1.00**					
1904	$1.00**					
1905	$1.00**					
1906	$1.00**					
1907	$1.00**					
1908	$1.00**					
1909	$1.00**					
1910	$1.00**					
1911	$1.00**					
1912	$1.00**					
1913	$1.00**					
1914	$1.00**					
1915	$1.00**					
1916	$1.00**					
1917	$1.00**					
1918	$1.00**	$0.10				
1919	$1.00**	NA				
1920	$1.00**	$1.50				
1921	$1.00**	$2				
1922	$1.00**	$2				
1923	$1.00**	$2				
1924	$1.00**	$1.50				
1925	$1.00**	$1.50				
1926	$1.00**	$1.80	$1	$1	$1.00**	
1927	$1.00**	$2	$1	$3.50	$1.00**	
1928	$1.00**	$2	$1	$3.50	$1.00**	
1929	$1.00**	$2.60	$1	$5	$1.00**	
1930	$1.00**	$2.45	$1	$5	$1.00**	
1931	$1.00**	$2.29	$1	$5	$1.00**	
1932	$1.00**	$2.29	$1	$5	$1.00**	
1933	$1.00**	$2.20	$1	$5	$1.00**	
1934	$1.00**	$2.20	$1	$5	$1.00**	
1935	$1.00**	$2.10	$1	$5	$1.00**	
1936	$1.00**	$2.10	$1	$5	$1.00**	
1937	$1.00**	$2	$1	$2.50	$1.00**	
1938	$1.00**	$2	$1	$2.50	$1.00**	
1939	$1.00**	$2	$1	$2.50	$1.00**	
1940	$1.00**	$2	$1	$2.50	$1.00**	
1941	$1.00**	$2	$1	$2.50	$1.00**	
1942	$1.00**	$2	$1	$2.50	$1.00**	
1943	$1.00**	$2	$1	$2.75	$1.00**	
1944	$1.00**	$2	$1	$2.50	$1.00**	
1945	$1.00**	$2	$1	$2.75	$1.00**	
1946	$1.00**	$2.50	$1	$2.75	$1.00**	
1947	$1.00**	$2.50	$1	$5	$1.00**	
1948	$1.00**	$2.50	$1	$6	$1.00**	
1949	$1.00**	$2.50	$1	$2.50	$1.00**	
1950	$1.00**	$2.50	$1	$2	$1.00**	

	SMRID		TID	MID		RID	
				OWA	NWA	OWA	NWA
1951	$1.00**		$2.50	$1	$2	$1.00**	$0.50
1952	$1.00**		$2.50	$1	$2.50	$1.00**	$1.00**
1953	$1.00**		$2.50	$1	$2.50	$1.00**	$1.00**
1954	$1.00**		$2.50	$1	$2.50	$1.00**	$1.00**
1955	$1.00**		$2.50	$1	$2.50	$1.00**	$1.00**
1956	$1.00**		$2.50	$1	$2.50	$1.00**	$1.00**
1957	$1.00**		$2.50	$1	$2.50	$1.00**	$1.00**
1958	$1.00**		$2.50	$1	$2.50	$1.00**	$1.00**
1959	$1.00**		$2.50	$1	$2.50	$1.00**	$1.00**
1960	$1.00**		$2.50	$2.50			
	WD	ED					
1961	$1.75	$2.50	$2.50	$2.50		$1.00**	
1962	$2	$2.50	$2.50	$2.50		$1.00**	
1963	$2.25	$2.50	$2.50	$2.50		$1.00**	
1964	$2.25	$3	$2.50	$2.50		$1.00**	
1965	$2.25	$3	$2.50	$2.50		$1.00**	
1966	$2.29	$3	$2.50	$2.50		$1.00**	
1967	$2.29	$3	$2.50	$2.50		$1.00**	
1968	$2.29	$3	$2.50	$2.50		$1	
1969	$2.29	$3.30	$3	$2.50		$1	
1970	$2.29	$3.30	$3	$2.50		$1	
1971	$2.29	$3.50	$3	$2.50		$1.50	
1972	$2.50	$3.50	$3.25	$2.50		$2.50	
1973	$2.50	$3.50	$3.25	$2.50		$2.50	
1974	$4		$3.75	$2.50		$2.75	
1975	$4		$4.50	$3		$3	
1976	$4.50		$4.80	$3.50		$3.25	
1977	$5.90		$5	$3.75		$4.25	
1978	$6		$6	$4		$5	
1979	$6.50		$6.75	$4.50		$5.25	
1980	$8.19		$7.50	$5		$5.75	
1981	$10		$9.75	$5.50		$6.50	
1982	$12		$10.75	$6		$6.25	
1983	$12		$11	$6		$6.50	
1984	$12		$11	$6		$6.50	
1985	$12		$11	$6		$6.50	
1986	$13		$11	$6		$6.50	
1987	$13		$11	$6		$6.50	
1988	$13		$12	$6		$6.50	
1989	$13.25		$12	$6		$6.50	
1990	$14		$12	$6		$6.50	
1991	$14		$12	$6.00 + $.10 PSI*		$6.50	
1992	$14		$12	$6.00 + $.10 PSI*		$6.50	
1993	$14.25		$12	$6.00 + $.10 PSI*		$6.50	
1994	$15.25		$12	$6.50 + $.10 PSI*		$7	
1995	$16.14		$12	$7.00 + $.10 PSI*		$7	
1996	$16.14		$12	$7.00 + $.10 PSI*		$7.50	
1997	$16.14		$12	$7.00 + $.10 PSI*		$7.50	
1998	$16.64		$12	$7.50 + $.10 PSI*		$8.50	
1999	$17		$12	$7.50 + $.10PSI*		$8.50	
2000	$17.50		$12	$7.50 + $.10PSI*		$8.50	

OWA = OLD WATER AGREEMENT NWA = NEW WATER AGREEMENT
* ADDITIONAL CHARGE FOR ACRES SERVICED BY A PRESSURE PIPELINE
** ADDITIONAL CHARGES WOULD HAVE BEEN LEVIED BY THE LATERAL ASSOCIATION

ST. MARY PROJECT
ASSESSED ACRES

	SMRID	TID	MID	RID	TOTAL
1965		49,321			
1966		50,641			
1967		53,300			
1968		56,039			
1969		56,461			
1970	197,540	57,484	8,506	20,847	284,377
1971	199,572	59,172	8,105	20,349	287,198
1972	204,051	59,833	9,042	20,269	293,195
1973	216,136	61,692	9,182	20,647	307,657
1974	237,244	63,823	9,077	23,257	333,401
1975	252,019	62,692	8,871	29,920	353,502
1976	262,231	66,423	9,069	28,350	366,073
1977	274,301	68,178	9,480	31,626	383,585
1978	284,151	68,815	10,737	31,570	395,273
1979	287,329	69,828	10,797	31,524	399,478
1980	293,126	70,368	10,797	33,681	407,972
1981	299,548	70,819	10,963	35,385	416,715
1982	301,446	71,529	11,647	39,130	423,752
1983	313,728	72,623	12,357	39,148	437,856
1984	319,712	72,971	13,047	41,729	447,459
1985	328,063	73,063	14,218	44,990	460,334
1986	331,493	73,314	14,579	44,950	464,336
1987	334,285	73,654	14,885	44,407	467,231
1988	339,091	73,602	15,030	44,196	471,919
1989	342,451	74,898	15,569	44,144	477,062
1990	349,849	74,568	15,099	44,044	483,560
1991	350,108	77,740	16,665	44,305	488,818
1992	351,393	78,177	16,391	44,279	490,240
1993	353,039	78,412	16,775	44,279	492,505
1994	353,466	78,629	16,785	44,219	493,099
1995	356,618	78,676	17,908	43,678	496,880
1996	358,399	79,069	18,169	44,315	499,952
1997	360,659	79,788	18,300	44,810	503,557
1998	360,780	80,455	18,300	45,533	505,068
1999	367,431	81,984	18,300	45,751	513,466
2000	367,950		18,300		

ST. MARY PROJECT
ACRES ACTUALLY IRRIGATED

	SMRID	TID	MID	RID	TOTAL
1911	47,000				47,000
1912	76,200				76,200
1913	76,200				76,200
1914	76,200				76,200
1915	76,200				76,200
1916	71,281				71,281
1917	91,780				91,780
1918	102,900				102,900
1919	112,569				112,569
1920	60,893	2,000			62,893
1921	75,550	10,352			85,902
1922	75,558	13,122			88,680
1923	72,345	3,625			75,970
1924	71,700	13,051			84,751
1925	81,110	13,472			94,582
1926	78,399	14,385			92,784
1927	20,365	576	1,235		22,176
1928	21,153	3,171	800	6,800	31,924
1929	25,713	10,094	4,000	6,634	46,441
1930	67,004	19,322	5,161	7,000	98,487
1931	68,089	19,914	3,400	9,000	100,403
1932	68,000	18,998	4,000	10,000	100,998
1933	70,000	18,854	4,000	12,000	104,854
1934	70,000	18,490	4,000	13,000	105,490
1935	70,000	19,471	4,000	13,000	106,471
1936	75,274	19,123	4,000	13,000	111,397
1937	72,250	21,296	4,000	13,000	110,546
1938	75,749	20,880	4,000	12,000	112,629
1939	76,492	18,485	3,700	12,000	110,677
1940	76,639	21,391	3,700	13,000	114,730
1941	76,579	15,103	3,700	14,000	109,382
1942	57,575	14,108	3,700	10,000	85,383
1943	75,707	20,935	3,500	12,000	112,142
1944	75,725	21,325	3,500	12,000	112,550
1945	75,766	21,218	3,500	12,000	112,484
1946	76,013	21,222	3,500	12,000	112,735
1947	110,848	19,141	3,500	12,000	145,489
1948	76,351	17,720	3,500	12,000	109,571
1949	70.117	21,533	3,500	12,000	37,103
1950	77,674	21,864	3,500	12,000	115,038
1951	83,682	8,739	3,500	12,000	107,921
1952	86,297	21,251	3,500	12,000	123,048
1953	76,624	22,503	5,000	13,000	117,127
1954	102,266	27,224	7,000	16,000	152,490
1955	81,100	20,530	7,000	16,000	124,630
1956	98,647	21,009	7,000	15,500	142,156

1957	116,713	21,579	7,000	16,000	161,292
1958	123,440	22,854	7,000	15,500	168,794
1959	155,155	25,426	7,000	15,800	203,381
1960	179,477	29,448	5,000	15,200	229,125
1961	157,300	30,747	6,000	18,300	212,347
1962	155,500	36,029	7,000	18,500	217,029
1963	170,000	33,902	7,250	18,500	229,652
1964	170,000	39,015	7,500	18,000	234,515
1965	165,000	26,005	3,000	15,000	209,005
1966	165,000	25,904	4,000	16,000	210,904
1967	160,000	38,235	5,000	17,000	220,235
1968	154,336	45,420	4,400	15,200	219,356
1969	124,879	45,980	4,500	15,000	190,359
1970	134,982	50,094	5,000	15,000	205,076
1971	149,444	52,185	5,500	15,000	222,129
1972	147,884	52,470	5,750	15,000	221,104
1973	176,120	58,826	5,600	15,000	255,546
1974	190,847	60,981	6,000	18,000	275,828
1975	171,883	44,606	6,000	21,500	243,989
1976	213,085	59,116	9,070	22,680	303,951
1977	237,660	64,115	7,000	30,000	338,775
1978	198,755	56,940	6,000	31,060	292,755
1979	249,232	62,635	8,000	31,264	351,131
1980	251,914	63,202	8,000	19,137	342,253
1981	259,564	66,206	8,650	19,462	353,882
1982	268,916	67,305	8,500	25,169	369,890
1983	288,969	68,474	9,000	28,055	394,498
1984	300,071	69,847	10,200	34,728	414,846
1985	305,560	70,133	13,000	36,286	424,979
1986	307,875	69,928	13,000	36,008	426,811
1987	305,964	69,413	12,400	36,008	423,785
1988	316,223	69,581	13,436	36,496	435,736
1989	323,400	70,278	13,555	35,766	442,999
1990	338,274	73,329	14,000	36,911	462,514
1991	319,745	73,169	15,000	32,847	440,761
1992	324,477	74,229	15,240	36,788	450,734
1993	262,718	66,158	0	0	328,876
1994	330,949	73,949	11,425	36,291	452,614
1995	289,173	72,108	3,250	16,473	381,004
1996	339,098	74,766	14,341	39,662	467,867
1997	342,234	76,083	13,479	39,484	471,280
1998	342,758	76,872	11,189	32,259	463,078
1999	355,988	79,166	14,722	38,966	486,482

ST. MARY RIVER IRRIGATION DISTRICT
BOARD OF DIRECTORS

* Served as Chairman

*Frederick (Jack) Brewin	1968-86, 1990-Present
Stanley B. Tiffin	1968-78
Robert Skiba	1968-69
Peter Shologan	1968-69
Paul Joss	1968
Gifford Woodcock	1968-77
Theodore Eiserman	1968-73
Paul Edlund	1968-79
Leroy Niedermier	1969-76
Frank Pavka	1969-73
William Doram	1973-76
Wilfred Mayer	1973-78
Norman Ober	1976-87
*Peter Langemann	1976-97
Peter J. Biemans	1977-85
Casey Aasman	1978-87
John Ontkean	1978-83
David Geldreich	1979-84
Steve Tokariuk	1983-86
Donald J. Campbell	1984-99
Jean Lehr	1985-91
Garth Bekkering	1986-90
Frank Malec	1986-Present
*R. James Csabay	1987-Present
Rockford Aman	1987-90
Wayne Schlenker	1990-95
Otto Gross	1991-Present
John Mann	1995-96
Robert Nemeth	1996-Present
Blaine Takeda	1997-Present
David Wallwork	1999-Present

BOARD OF TRUSTEES/DIRECTORS OF THE TABER IRRIGATION DISTRICT *

1915-1919	Carl C. Cook (Chairman 1915-1918)
1920-1922	John T. Willard (Chairman 1922)
1923-1925	Samuel Ayers
1926-1944	Carson B. Hill
1944-1951	Wilbern Johnson
1951-1965	Frank P. Stevens
1915-1920	John W. Anderson(B.A. Stringam filled in temporarily in 1917)
1921-1946	William Valgardson (Chairman 1923-1946)
1946-1965	Edwin N. Francis (Chairman 1946-1965)
1915-1921	Lawrence Peterson (Chairman 1919-1921)
1922-1923	Thomas W. Harris
1923-1925	Benjamin A. Stringam
1925-1937	Henry N. Peterson
1937-1944	Bryant R. McMullin
1944-1965	Nephi Jensen

1965 Increased from 3 trustees to 5 trustees with 5 nominating zones designated as follows:

i. Cranford and Barnwell to include Lateral 5
ii. Taber west - from Lateral 6 to 13
iii. Taber east - Laterals 13 to 21
iv. Big Bend unit
v. North Fincastle and Horsefly Lake unit

Zone 1	1965-1969	Frank P. Stevens
	1969-1989	P. William Kirkvold
	1989-present	Dennis Fenske
Zone 2	1965-1969	Edwin N. Francis (Chairman 1965-1969)
	1969-present	Keith Francis (Chairman 1975-present)
Zone 3	1965-1971	Nephi Jensen (Chairman 1969-1971)
	1971-1974	Al Machacek
	1974-present	Rudy Gregus
Zone 4	1965-1985	Mike D. Reti
	1985-1991	Roy Reti
	1991-1995	Richard Ferguson
	1995-present	Brian Valgardson

Zone 5	1965-1983	Norman Anderson (Chairman 1971-1975)
	1983-1989	Jake Stolk
	1989-present	Richard Visser

Under the new Irrigation Act of 1968, Board of Trustees was changed to Board of Directors.

EMPLOYEES OF THE TABER IRRIGATION DISTRICT

Secretary-Treasurer/Manager

1915-1966	Ted Sundal
1966-1987	Kenneth Anderson
1987-1989	Allan Herbig
1989-present	M. Kent Bullock

Assistant Secretary-Treasurer

1951-1966	Kenneth Anderson

Operations/Maintenance Superintendent

1937-1964	John G. Barton
1964-1981	Reed Peterson
1980-1988	John Sparks
1989-present	Ronald Lloyd

Construction Superintendent

1973-1988	Ronald Lloyd

Assistant Superintendent

1982-present	Barry Jensen

Engineer

1981-1989	M. Kent Bullock
1989-present	Ronald Hadden

Engineering Technicians

1981	Baljinder Boora
1981-present	Fred Williams
1982-1985	Loren Francis
1985-present	Dan Podstawka

Office Manager

1974-1989	Audrey Mereski
1989-present	Joyce Peterson

Land Administrator/Clerk

1986-1992	Judy Lamond
1992-present	Pat Goruk (part-time)

Ditchriders

1920	A. Peterson	1920	H. C. Tufteland
1921-1937	Parley Palmer	1921-1922	Chester Asplund
1923-1927	George Allison	1925-1936	John G. Barton
1927	Ingwall Sundal	1929	Jim Fuller
1930-1947	Edwin Palmer	1930-1961	Joe Haynes
1937-1941	Howard Wood	1938-1947, 1951-63	Reed Peterson
1942-1943,1946-56	Walter Valgardson Jr.	1944-1945,1948-64	Harold Haynes
1948-1949	Orland Johnson	1950-1952	Lewis A. Martin
1952-1978	Loren Easthope	1953-1980	Boyd Anderson
1962-1975	John Chudik	1964-1979,1989-93	John Sparks
1965-1984	Henry Wiebe	1965-1975	Frank Burbank
1965-1987	Herb LeRoy	1976	Evan Evanson
1976-1980	Martin Kolibar	1979-1988	Tony Kudilek
1980-present	Nick DeBona	1981-1993	Delbert Johnson
1981-present	Randy Sparks	1985-1996	Wes Dunz
1992-present	Alan Anderson	1994-present	Charlie Kanomata
1997	Glen Wada	1998-present	Gerald Layton

Dragline Operator

1940-1946	Clifford Mably (Harry Bland-helper)
1947-1951	Roy Valgardson (Wilfred Valgardson-helper)
1952-1956	Wilfred Valgardson
1956-1984	E. Arvid Nordquist

Equipment Operators

1967-1979	Joe Raca	(Service Man 1980-1988)
1976-1985	Angelo Skariot	(Utility Man 1985-1990)
1978-1993	Charlie Hoffart	
1980	Ray Van Schouwen	
1981-1992	Lynn Turnbull	
1985-present	Fred Gross	

Welder

1976-1992	Taras Nawolski

Utility Man - 1976-1979 Tony Kudilek

MAGRATH IRRIGATION DISTRICT

BOARD OF DIRECTORS

Ernest Benion*	1924 - 1927
Leslie E. Pharis	1924 - 1925
Isaac Sabey	1924 - 1931
Ralph S. Thompson	1925 - 1947
Christian Jensen*	1928 - 1947
John H. Bridge	1932 - 1937
David L. Whitt	1938 - 1946
Mike Schneyder	1947 - 1955
Harry Bly*	1948 - 1950
Paul B. Taylor*	1948 - 1960
David H. Keeler	1951 - 1954
C. C. Spencer *	1954 - 1962
John Austin	1956 - 1967
George Dudley *	1961 - 1987
Cyrus Card	1963 - 1974
Lalovee R. Jensen	1968 - 1976
John Bevers	1975 - 1988
John Weing*	1977 - 1988
Dean Anderson*	1988 - present
Neal Miller	1989 - present
Robert Wocknitz	1989 - present

* chairman

RAYMOND IRRIGATION DISTRICT

DIRECTORS

Heber S. Allen*	**1925 - 1928**
Arthur E Fawns	**1925 - 1939**
Jesse H. Wilde *	**1925 - 1952**
Francis B Rolfson*	**1928 - 1941**
Wilford W. Brandley	**1939 - 1950**
Douglas Paxman*	**1949 - 1963**
Leslie Chipman	**1952 - 1963**
Fay H Walker	**1963 - 1969**
Farrell Nelson*	**1950 - 1980**
Willard Paxman	**1969 - 1981**
Robert H. Wilde*	**1963 - 1996**
Robert L. Fraser	**1981 - 1988**
Paul F. Schneider	**1980 - present**
Alan Heggie*	**1988 - present**
Byron Wilde	**1996 - present**

*** chairman**

NOTES

Chapter 1

1. The literature on the Great Plains includes *The Great Plains* by Walter Prescott Webb and *Whoop -Up Country* by Paul Sharp.

2. *The Great Plains* by Walter Prescott Webb p.152-160.

3. Ibid.

4. Ibid.

5. *The Canadian Prairies* by Gerald Friesen p.107.

6. Journal of Palliser Expedition p.538.

7. Quoted in interpretive sign text at Stirling National Historic Site.

8. *Report of the Geology and Resources of the Region in the vicinity of the Forty-Ninth Parallel* by George Mercer Dawson p. 320.

9. BT 31 Public Record Office Kew Gardens, England.

10. An Act to empower The North-Western Coal and Navigation Company (Limited), to construct and work a line of railway between Medicine Hat and the Company's mines on the Belly River, and for other purposes.

11. An Act to incorporate the Alberta Railway and Coal Company

12. An Act to incorporate the Alberta Railway and Coal Company

13. *Civilizing the West* by A.A. den Otter p. 151

14. File 9 /2/7/1/9 MG 9 William Pearce Papers, University of Alberta Archives.

15. MacLeod Gazette 5 December 1889

16. Ibid.

17. Report of W. M. Pearce Superintendent of Mines, 31 October 1887 in the Department of the Interior, *Annual Report*, 1887, Part 1, Dominion Lands, 13. See Mormon Paper

18. *Medicine Hat Times*, 12 December 1889.

19. *Journals of the Second Legislative Assembly of the North-West Territories* 12 January 1892, 79-83.

20. Ibid.

21. William Pearce "The Proposed Northwest Saskatchewan Irrigation Project and Some Reminiscences of Irrigation in Western Canada" William Pearce Papers, University of Alberta Archives, MG 9, File 2/7/4/6 Pearce recollections in item from U of A

22. House of Commons, *Debates*, 14 July 1894, 6032-6033.

Chapter 2

23. Report of William Pearce, Superintendent of Mines 1 December 1891, Department of the Interior, *Annual Report*, 1891, Part I Dominion Lands, 11.

24. Ibid.

25. Originally quoted John R. Hicken, "Events Leading to the Settlement of the Communities of Cardston, Magrath, Stirling and Raymond, Alberta" (M. A. diss. Utah State University, 1968), 62.

26. Ibid.

27. National Archives of Canada RG 89, Vol. 38, File 17 Part 1 .

28. Ibid.

29. National Archives of Canada. RG 89, Vol. 38, File 17 Part 1

30. Annual Report of Department of the Interior, 1895.

31. National Archives of Canada. RG 89, Vol. 38, File 17 Part 1

32. National Archives of Canada. RG 89, Vol. 38, File 17 Part 1

33. Charles Magrath. *The Galts Father and Son: Pioneers in the Development: Pioneers in the Development of Southern Alberta and How Alberta Grew Up: Brief outline of Development in the Lethbridge District*, 1936.

34. Ibid.

35. University of Alberta Archives, Pearce Papers

36. George G. Anderson, "Report on the St. Mary's Irrigation Canal, Alberta, N.W.T. 1897, Canadian Pacific Railway Archives RG 2/45236. Anderson Report

37. Ibid.

38. Ibid.

39. Ibid.

40. Ibid.

41. Ibid.

42. Ibid.

43. Ibid.

44. Glenbow Archives, M 2431 File 22.

45. Ibid.

46. Ibid.

47. Ibid.

Chapter 3

48. Originally quoted by John R. Hicken in his thesis entitled "Events Leading to the Settlement of the Communities of Cardston, Magrath, Stirling and Raymond, Alberta" (M. A. diss. Utah State University, 1968)

49. Glenbow Archives, M 2431 File 22.

50. *Lethbridge News* 1 December 1898

51. Ibid.

52. Ibid.

53. Ibid.

54. *Lethbridge News* 16 February 1899.

55. Ibid.

56. *Lethbridge News* 23 March 1899.

57. *Dessert News* 5 August 1901.

58. Ibid.

59. *Lethbridge News* 15 March 1900.

60. *Lethbridge News* 19 October 1900

61. *Lethbridge News* 20 September 1900

Chapter Four

62. Provincial Archives of Alberta. Accession 80.168 file

63. *Lethbridge News* 29 November 1900.

64. Glenbow M3748

65. Ibid.

66. Ibid.

67. *Lethbridge News* 29 March 1900

68. *Annual Report of the Dominion Experimental Farm* 1900.

69. *Lethbridge News* 5 June 1899

70. Ibid.

71. Glenbow Archives M 3761, file 143.

72. *Lethbridge News* 1 October 1903

73. *Lethbridge News* 19 October 1900.

Chapter Five

75. RG 89 Vol. 38 File 17 Part 1 National Archives of Canada or Document P19731815001-43 Sir Alexander Galt Museum and Archives

76. Ibid.

77. Department of the Interior file dealing with the St. Mary and Milk River Diversions. Sir Alexander Galt Museum and Archives

78. Ibid.

79. National Archives of Canada RG 89 Vol. 38 File 17 Part 1

80. National Archives of Canada RG 89 Vol. 38 File 17 Part 1

81. National Archives of Canada RG 89 Vol. 38 File 17 Part 1

82. *Lethbridge Herald* 16 September 1921

Chapter Six

83. RG 89 Vol. 53 File 29 Part 2

84. Canadian Pacific Railway, Department of Natural Resources Annual Report of Operation and Maintenance, Lethbridge Section, 1918

85. Ibid.

86. Ibid.

87. Ibid.

88. Ibid.

89. *Lethbridge Herald* 1 December 1916.

90. As quoted in *Prairie Promises* by John Gilpin

91. Canadian Pacific Railway, Department of Natural Resources Annual Report of Operation and Maintenance, Lethbridge Section, 1918

Chapter Seven

92. *Lethbridge Herald* 23 February 1906.

93. *Lethbridge Herald* 16 February 1914

94. Ibid.

95. As quoted in "Mormon Sugar in Alberta: E.P Ellison and the Knight Sugar Factory, 1901 - 1917." by William G. Hartley *Journal of Mormon History* (1997): 1- 29

96. *Lethbridge Daily Herald* 16 October 1913

97. *Lethbridge Daily Herald.* 5 March 1914

98. *Lethbridge Daily Herald.* 31 July 1913

99. Ibid.

100. Ibid.

101. *Lethbridge Daily Herald.* 6 August 1913.

102. *Lethbridge Daily Herald.* 22 August 1913

103. *Lethbridge Daily Herald.* 4 September 1913.

104. *Lethbridge Daily Herald.* 10 February 1914

105. *Lethbridge Daily Herald.* 1 May 1914

106. Ibid.

107. *Lethbridge Daily Herald.* 22 January 1915.

Chapter Eight

108. *Annual Report Department of the Interior*

109. *Annual Report Department of the Interior*

Chapter Nine

110. *Taber Times* 14 January 1915

111. Ibid.

112. Ibid.

113. *Taber Times* 11 February 1915

114. Ibid

115. Provincial Archives of Alberta Accession 80.163.

116. Provincial Archives of Alberta Accession 80.163

117. *Taber Times* 1 July 1915

Chapter Ten

118. Canadian Pacific Railway, Department of Natural Resources Annual Report of Operation and Maintenance, Lethbridge Section, 1918

119. Provincial Archives of Alberta Accession 80.163

120. Ibid.

121. Ibid.

122. Ibid.

123. Ibid.

124. Ibid.

125. Ibid.

126. Ibid.

127. Ibid

128. Ibid.

129. Ibid

130. Ibid

131. Ibid

132. Ibid

133. Ibid

134. Ibid

135. Ibid

136. Ibid

137. Ibid

138. Ibid

139. Ibid

140. Ibid

141. Ibid

142. Ibid

143. Ibid

144. Ibid

145. Ibid

Chapter Eleven

146. *Lethbridge Herald* 22 September 1939

147. Provincial Archives of Alberta Accession 80.163.

148. Provincial Archives of Alberta Accession 80.163

149. Provincial Archives of Alberta Accession 80.163

150. Provincial Archives of Alberta Accession 80.163

151. TID Annual Report 1921

152. Provincial Archives of Alberta Accession 80.163.

153. TID Annual Report 1921

154. Provincial Archives of Alberta Accession 80.163.

155. TID Annual Report 1933

160. Contract file at SMRID

161. Minutes TID

162. Minutes. TID 27 January 1921.

163. Minutes. TID

164. Minutes. TID 1936

165. TID Annual report 1921

166. TID Annual report 1921

167. Provincial Archives of Alberta Accession 80.163.

168. Ibid

169. Ibid.

170. Ibid.

171. Ibid.

172. TID Annual report 1933

173. Provincial Archives of Alberta Accession 80.163.

174. Ibid.

175. Ibid.

176. Ibid.

177. Ibid.

178. Ibid.

179. Ibid.

180. Ibid.

181. Ibid

182. Ibid

183. Ibid.

184. TID Annual Report 1921

185. Ibid

186. Ibid.

187. Ibid

188. TID Annual Report

189. TID Annual Report

190. Annual Report

191. Annual Report

192. Annual Report

Chapter Twelve

193. MID 1922 Correspondence file

194. Ibid

195. Provincial Archives of Alberta Accession 80.168.

196. MID 1922 Correspondence file

197. Provincial Archives of Alberta Accession 80.168.

198. Ibid

199. *Lethbridge Herald* 12 June 1927.

200. Ibid

201. Ibid

202. Ibid

203. Ibid

204. Ibid.

205. Ibid

206. Correspondence files of MID

207. MID Water Contract file

208. Ibid.

209. Ibid

210. Ibid

211. Ibid

212. Ibid

213. Ibid

214. MID Minutes

215. PFRA Records

216. PFRA Records

Chapter Thirteen

217. RID File at SMRID

218. Raymond Irrigation Act

219. RID correspondence

220. RID correspondence

221. Ibid.

222. Ibid.

223. Ibid.

224 RID correspondence

225. Ibid.

226 Ibid.

227. Ibid.

228. Ibid.

229. RID correspondence

Chapter Fourteen
230. Glenbow M 2237

231. CPR Annual Report

232. Ibid

233. CPR Annual Report

234. CPR Annual Report

235. CPR Annual Report

236. CPR Annual Report

237. CPR Annual Report

238. CPR Annual Report

239. National Archives of Canada RG 89 Vol. 256 File 3280

240. Ibid.

241. Ibid

242. Ibid.

243. SMRID records

244. SMRID records

245. SMRID records

246. SMRID records

247. SMRID records

248. Annual Report

249. Annual Report

250. SMRID records

251. SMRID records

252. SMRID records

253. Annual Report

Chapter Fifteen
254. A Handbook of Information about Irrigated Farm Lands Improved and Unimproved in Southern Alberta p. 2.

255. *Lethbridge Herald* 10 January 1923.

256. Ibid

257. Ibid

258. *Lethbridge Herald* 16 January 1923

259. Ibid

260. *Lethbridge Herald* 12 February 1923

261. Glenbow Archives M 2263

262. Ibid.

Chapter Sixteen
263. Glenbow Archives M 3760

264. Ewing Commission Report

265. Ibid.

266. Glenbow Archives M 3760

267. SMRID records

268. Lethbridge Herald 2 July 1937

Chapter Seventeen
269. PFRA records

270. Ibid

271. PFRA records

272. Annual Report of the Department of Agriculture

273. SMRID records

274. Ibid.

275. Ibid.

276. Ibid

277. Ibid

278. SMRID records

279. Hanson Committee Report

Chapter Eighteen
280. Rehabilitation File

Chapter Twenty
281. MID Correspondence file
Chapter Twenty-One

282. Minutes of the RID 8 April 1967

Chapter Twenty-Two
283. PFRA Annual Report

284. Annual Report Dominion Experimental Farm, 1949.

285. TID Annual Report 1959

INDEX